ADVANCES IN
X-RAY ANALYSIS

Volume 24

ADVANCES IN
X-RAY ANALYSIS

Volume 24

Edited by

Deane K. Smith

The Pennsylvania State University
University Park, Pennsylvania

and

Charles S. Barrett, Donald E. Leyden,
and Paul K. Predecki

University of Denver
Denver, Colorado

Sponsored by
University of Denver Research Institute
and
Department of Chemistry
University of Denver
and
JCPDS-International Centre for Diffraction Data

PLENUM PRESS • NEW YORK AND LONDON

The Library of Congress cataloged the first volume of this title as follows:

Conference on Application of X-ray Analysis.
Proceedings 6th-   1957-   [Denver]

    v. illus. 24-28 cm. annual.
No proceedings published for the first 5 conferences.
Vols. for 1958-   called also: Advances in X-ray analysis, v. 2-
Proceedings for 1957 issued by the conference under an earlier name: Conference
on Industrial Applications of X-ray Analysis. Other slight variations in name of con-
ference.
    Vol. for 1957 published by the University of Denver, Denver Research Institute,
Metallurgy Division.
Vols. for 1958-   distributed by Plenum Press, New York.
Conferences sponsored by University of Denver, Denver Research Institute.
    1. X-rays–Industrial applications–Congresses.   I. Denver. University.
Denver Research Institute   II. Title: Advances in X-ray analysis.
TA406.5.C6   58-35928

Library of Congress Catalog Card Number 58-35928
ISBN 978-1-4613-9992-6   ISBN 978-1-4613-9990-2 (eBook)
DOI 10.1007/978-1-4613-9990-2

Softcover reprint of the hardcover 1st edition 1980
Proceedings of the Twenty-Ninth Annual Conference on Applications of
X-Ray Analysis held in Denver, Colorado, August 4–8, 1980
© 1981 University of Denver
Denver, Colorado

Plenum Press is a division of Plenum Publishing Corporation
233 Spring Street, New York, N.Y. 10013

FOREWORD

Deane K. Smith
Department of Geosciences
The Pennsylvania State University

Computer automation of x-ray powder diffraction has been one of the dominant topics of this conference for many years. In fact, the first description of such instrumentation dates back to 1967, Rex (1). The modern instruments are considerably more sophisticated than this early unit, but the goals of automation are essentially unchanged. They are to obtain better data at a faster rate with less effort than is possible with manual instrumentation. Indeed "laziness is the mother of invention."

The emphasis of most of the papers on automation has been toward hardware-controlling systems and achieving accurate d values and good intensities for effective phase identification and phase characterization. Tests of good data include successful pattern searching and matching. Indexing by computer methods or accurate lattice parameters through least-squares fitting procedures with resulting small residuals is a good test of d value accuracy. Intensity accuracy is much harder to test unless a theoretical data set is available.

In most of the reported studies, the emphasis has been more on the data acquisition than on the specific problems to which the data is to be applied. In organizing this plenary session, an attempt was made to focus on applications, especially those which are at the forefront of materials studies. In addition many other applications papers were encouraged with the result that a good variety of such topics are included in the program this year.

To bring the role of automation into perspective, it may be useful to review the information contained in diffraction data and how this information can be used. The information can be divided into three areas: the line positions, their intensities, and the detailed shapes of their profiles. Each of these areas is expanded in the accompanying tables. One should always remember the parameters which control each of the bits of information. They indicate how the data must be interpreted.

The line positions, Table I, are the easiest quantities to measure accurately especially when using well aligned instruments and reference standards. Consequently, most diffraction studies are centered around this information. Phase characterization and identification and other applications of accurate d-values and lattice parameters are probably the most important and certainly have received much attention in the design of automated systems. Lattice changes have received less attention because the control systems should also control the peripheral equipment involved, such as thermal and pressure stages. However, here are several areas which could benefit from automation. In particular the field of pressure and thermal effects could utilize the new rapid scanning diffractometer or position sensitive systems.

Table I Line Position

Controlled by:

1. Geometry of unit cell (lattice)
2. Wave length of x-ray beam
3. Instrumental and sample aberrations

Applications:

Accurate lattice parameters
   Least-squares refinements
   Phase characterization
   Identification (Qualitative anal.)
      Recognition by familiarity
      d-value searching (PDF Fink)
      Cell data (Crystal Data)

Lattice changes
   Lattice shifts
      Composition changes
      Thermal expansion
      Compressibility
      Homogeneous strain
   Phase changes (polymorphism)

Line intensity measurement, Table II, has much to gain from automation because it now becomes routine to measure integrated intensity values. Peak intensity values are difficult to relate to sample properties except under very limited conditions. Quantitative analysis in particular now becomes a standard procedure in the analytical laboratory. Standardized methods are

used in many industries, and quantities such as respirable silica
and crystallinity in polymers are commonly measured. Details of
the crystal structures of materials which occur only in the polycrystalline state such as tooth and bone materials and catalysts
are now possible. Chemical kinetics is also feasible with the new
fast scanning systems.

Table II   Line Intensity

Controlled by:

1. Types of atoms and atomic arrangement
2. Amount of sample which can diffract
3. Intensity correction factors

Applications:

Identification (Qualitative anal.)
    d-I search/match (PDF Hanawalt)
    Intensity changes with element substitution

Crystal structure analysis
    Differences in fine grained state from
        single crystal state
    Structures of materials which only occur
        in fine grained state

Quantitative analysis
    Phase composition
    Order-disorder ratios
    Percent crystallinity
        polymers
        calcine reactions
    Chemical kinetics

State of polycrystalline aggregate
    Preferred orientation
    Texture

The measurement and utilization of detailed line profiles,
Table III, are the areas where automation shows much potential.
Considerable theory already exists on the analysis of the diffraction profile, and the present limitation of its use lies in the
difficulty of obtaining data of sufficient accuracy especially in
the tail region of the profile. There has long been a controversy
concerning the analytic shape of the profile but numerical methods
are available for unfolding the profile into its components. The
main use of profile analysis in the present automated systems is to
obtain the best position of the line, and fitting of analytic expressions or digitized tables of data are employed with considerable
success. The stripping of the α_2 component is now commonly done
before interpreting the profile. However, the goal of these latter
methods is the more accurate determination of position and intensity
information rather than the interpretation of the profile shape

itself. An extreme example of line profile usage is the simultaneous determination of crystallite size and strain which will certainly be facilitated by more accurate measurements of the profiles.

Table III   Line Profile

Controlled by:

1. State of crystallite perfection and
     sample homogeneity
2. Instrumental aberrations
3. Spectral distribution

Applications:

Crystallite size
Size distributions

Lattice distortions
   Crystal defects
Inhomogeneous strain

Sample inhomogeneity

Crystallinity
   Amorphous state

Automation has certainly come of age. Developments in hardware are still evolving, but the major development in the next several years will be in software. The advances in software will not only include better algorithms to accomplish the present goals, such as obtaining accurate d values, but also to utilize the details of the line profiles. It is for these reasons that we need to examine how to apply the theories which exist to real problems.

REFERENCES

1. Rex, Robert W., "Numerical Control X-ray Powder Diffractometry," Advances in X-ray Analysis, 10, 366-373, Plenum Press (1967).

PREFACE

This volume constitutes the proceedings of the 1980 Denver Conference on the Applications of X-Ray Analysis, 29th in the series. The conference was held August 4-8, 1980 at the University of Denver and was cosponsored by the JCPDS-International Centre for Diffraction Data and the University of Denver. The local conference chairmen were D. E. Leyden and P. K. Predecki, with C. S. Barrett and J. B. Newkirk as honorary chairmen.

The invited conference chairman, D. K. Smith of The Pennsylvania State University, organized and chaired the plenary session of the conference entitled "Practical Applications of Automated Analysis of Diffraction Data."

In addition to the plenary and regular contributed sessions, a special session on "XRD Applications of Position Sensitive Detectors" was organized and chaired by C. O. Ruud of The Pennsylvania State University.

This year we were pleased to honor two investigators for their outstanding contributions to the X-ray applications field. These men are L. S. Birks of the U.S. Naval Research Lab and J. L. de Vries of the Philips Company, Eindhoven, The Netherlands, and we take this opportunity to thank them for participating in the conference.

The names of invited speakers on the program and the titles of their papers are listed below.

M. T. Abell, D. D. Dollberg, and J. V. Crable, "Quantitative Analysis of Dust Samples from Occupational Environments Using Computer-Automated X-Ray Diffraction."

L. S. Birks, "Historical Development of X-Ray Spectrochemical Analysis."

J. L. de Vries, "X-Ray Powder Diffraction in Europe."

H. E. Göbel, "The Use and Accuracy of Continuously Scanning Position-Sensitive Detector Data in X-Ray Powder Diffraction."

D. J. Johnson, "Crystallinity, Crystallite Size and Lattice Perfection in Fibrous Polymers."

G. T. Kokotailo and John L. Schlenker, "Porotectosilicate Structure Determination from Model Building."

J. H. Konnert, P. D'Antonio and J. Karle, "Analysis and Interpretation of Diffraction Data from Amorphous Materials."

R. A. Young and D. B. Wiles, "Application of the Rietveld Method for Structure Refinement with Powder Diffraction Data."

Tutorial workshops on various topics in diffraction and fluorescence were held during the first two days of the conference. These are listed below together with the names of the workshop organizers and instructors.

(1)   "Search/Match Techniques for Qualitative Powder Diffractometry: (a) Manual Search; (b) Computer Search." Sponsored by JCPDS-International Centre for Diffraction Data, Swarthmore, PA.

        H. D. Bennett, NASA, C. M. Foris (chair), E. I. du Pont de Nemours Co.; R. P. Goehner, General Electric Company; M. Holomany, JCPDS-International Centre for Diffraction Data; C. R. Hubbard (chair), National Bureau of Standards; G. G. Johnson (chair), The Pennsylvania State University; G. J. McCarthy, North Dakota State University; N. Panagiotopoulos, JCPDS-International Centre for Diffraction Data; C. O. Ruud, The Pennsylvania State University; and P. A. Stoll, U.S. Steel Corp.

(2)   "Systematic Errors in Diffractometer Alignment." R. Jenkins (chair), Philips Electronic Instruments, Inc.

(3)   "Guinier Monochromator/Camera Alignment." A. Brown (chair), Studsvik Energiteknik AB; and J. W. Edmonds (chair), Dow Chemical Company.

(4)   "Introduction to Automated Powder Diffraction (Tutorial)." J. W. Edmonds, Dow Chemical Co.; C. L. Mallory, Signetics Corp.; and R. L. Snyder (chair), Alfred University.

(5)   "Operational Considerations of Wavelength Dispersive X-Ray Spectrometry." D. W. Beard (chair), Siemens Corporation; J. Croke, Philips Electronic Instruments, Inc.; and M. Janiak, Rigaku/USA.

(6) "Strategies of XRF Data Reduction." J. M. Doster, North
     Carolina State University; A. P. Quinn, Corning Glass Corp.;
     and J. C. Russ (chair), North Carolina State University.

(7) "Methods of Sample Preparation; X-Ray Spectrometry." V. E.
     Buhrke (chair), The Buhrke Company; F. Claisse, Claisse
     Scientific Corp.; J. Croke, Philips Electronic Instruments,
     Inc.; J. V. Gilfrich, Naval Research Lab; O. Ingamels, AMAX
     Company; J. Taggart, U.S. Geological Survey; S. Wahlberg,
     U.S. Geological Survey; and M. Yanak, International Nickel
     Research Labs.

(8) "Operational Considerations of Energy Dispersive X-Ray
     Fluorescence." E. Acree (chair), United Scientific Corp.;
     R. A. Vane, United Scientific Corp.; W. Wegscheider, Techni-
     cal University, Graz, Austria; and D. Wherry, Kevex Corpora-
     tion.

The workshop attendance continued to increase this year;
222 workshop attendees out of a total of 364 conference attendees.
We are particularly indebted to the workshop organizers and
instructors who gave unselfishly of their time and talent in
these workshops, and to the JCPDS-International Centre for Diffraction Data for sponsorship.

We are grateful to all who co-chaired the various conference
sessions. These were: C. S. Barrett, D. W. Beard, D. A. Gedcke,
J. V. Gilfrich, R. Jenkins, D. J. Johnson, J. H. Konnert, W. H.
Lemons, D. E. Leyden, J. F. Molina, J. B. Newkirk, P. S. Prevey,
C. O. Ruud, D. K. Smith, D. M. Smith, W. D. Stewart, and R. A.
Young.

We are also grateful to the conference aids who worked long
and unusual hours to ensure that the conference ran smoothly.
These were: Tina Barrett, Bill Bodnar, Barb Cain, Bob Flowers,
Penny Hudson, Rich Miller, Dave Mustoe, Kurt Nordstrom, Steve
Northcott, John Shinton, Yvonne Shinton, and Linda Wallace.

A special word of thanks and appreciation to Mildred Cain,
the conference secretary, for running the whole show.

                                        Paul K. Predecki
                                        For the Conference Committee

Unpublished Papers

The following papers were presented orally only and are not published here for a variety of reasons.

"Algorithms for Automated X-Ray Powder Diffraction Analysis," Chester L. Mallory and Robert L. Snyder, NYS College of Ceramics, Alfred University, Alfred, New York.

"Interactive Profile Analysis," Robert A. Sparks, California Scientific Systems, Sunnyvale, California.

"Control of Systematic Errors in the Computer Controlled Powder Diffractometer," R. Jenkins, Y. Hahm and C. Villamizar, Philips Electronic Instruments, Mahwah, New Jersey; and W. N. Schreiner and C. Surdukowski, Philips Laboratories, Briarcliff Manor, New York.

"EDXRF Rock Analysis Using Background and Interelement Coefficients," William H. Lemons, Wyoming Mineral Corporation, Boulder, Colorado.

"Coal Analysis Using an X-Ray Fluorescent Spectrometer," T. Arai and M. Funahashi, Rigaku Industrial Corporation, Osaka, Japan; and M. J. Janiak, Rigaku/USA, Inc., Danvers, Massachusetts.

"Historical Development of X-Ray Spectrochemical Analysis," L. S. Birks, U. S. Naval Research Laboratory, Washington, D.C.

"On the Use of Secondary Fluorescence Targets and Diffraction Crystals for the Direct Measurement of Si(Li) Response Functions," Marek Lankosz and Robin P. Gardner, North Carolina State University, Raleigh, North Carolina.

"A Microcomputer Based Wavelength Control and Data Acquisition System for Early Model WDXRF Spectrometers," Bruce B. Jablonski and Donald E. Leyden, University of Denver, Denver, Colorado.

"Determination of Trace Metals in Mineral Waters by Energy Dispersive X-Ray Fluorescence," Wolfhard Wegscheider and Kurt Müller, Technical University, Graz, Austria.

"Preconcentration of Nanomolar Amounts of Trace Elements from Natural Waters for X-Ray Energy Spectrometric Analysis Using Pyrollidine Carbodithioic Acid," T. Tisue and C. A. Seils, Argonne National Laboratory, Argonne, Illinois.

"A Versatile, Rapid, Wavelength Dispersive X-Ray Spectrometer: Concept," Alan P. Quinn and William T. Kane, Corning Glass Works, Sullivan Park, Corning, New York.

"A New Low-Cost Bench-Top Simultaneous X-Ray Spectrometer," Sidney
B. Miller, Diano Corporation, Woburn, Massachusetts.

"Small Angle X-Ray Scattering - Determination of the Amorphous
Thickness of Polymer Single Crystals," I. R. Harrison, W. D.
Varnell, and S. Kozmiski, The Pennsylvania State University,
University Park, Pennsylvania.

"Crystallite Sizes of Polyethylene Single Crystals Isothermally
Grown from Solution," J. Lahijani and I. R. Harrison, The
Pennsylvania State University, University Park, Pennsylvania.

"Percent Cristobalite by X-Ray Diffraction," R. L. Hoffmann and
T. A. Johnson, Babcock & Wilcox Company, Lynchburg, Virginia;
and R. L. Snyder, Alfred University, Alfred, New York.

"Phase Separation Techniques for Microsample Analysis," A. M.
Davis, General Electric Company, Schenectady, New York.

"A Method of Correction of Particle-Size Effect in XRF Analysis
of Thin Samples Having Particle-Size Distribution," Barbara
Holyńska and Andrzej Markowicz, Academy of Mining and Metallurgy,
Kraków, Poland.

CONTENTS

PRACTICAL APPLICATIONS OF AUTOMATED ANALYSIS OF DIFFRACTION DATA

XRD MATHEMATICAL METHODS, TECHNIQUES AND INSTRUMENTATION

OTHER XRD APPLICATIONS

XRF APPLICATIONS IN THE MINERALS INDUSTRY

APPLICATION OF THE RIETVELD METHOD FOR STRUCTURE

REFINEMENT WITH POWDER DIFFRACTION DATA

R.A. Young and D.B. Wiles

School of Physics and
Engineering Experiment Station
Georgia Institute of Technology
Atlanta, Georgia   30332

I.   INTRODUCTION

     The object of the Rietveld method is to produce refined values
of crystal structural parameters from powder diffraction data.
Many materials of great interest can not be made available for
study in single crystal form. This may be because it is not possible to prepare a single-crystal form at all (e.g., human tooth
enamel) or because the single-crystal form differs from the polycrystalline form with the properties of interest (e.g., catalysts).
Thus, our basic understanding of the atomic scale mechanisms is
limited on the structural·side by the information that can be deduced from powder diffraction patterns. (Only diffraction and
EXAFS are direct probes of the spatial arrangements of atoms.)
The Rietveld method has greatly extended the amount of structural
detail that we can obtain routinely from powder diffraction patterns.
In this method, structural parameters such as atom coordinate,
thermal motion, and site occupancy parameters are adjusted in a
least-squares refinement procedure until the best fit is obtained
between entire calculated and observed powder diffraction patterns,
as a whole.

     A possible descriptive name for the method would be whole
pattern fitting structure refinement. No effort is made in advance
to allocate intensity observed in the powder diffraction pattern to
particular Bragg reflections; neither is any effort made to resolve
overlapped reflections. Instead, one starts with a reasonably
good model of the crystal structure, and of the various instrumental
and other factors affecting the pattern, and calculates the expected

1

powder pattern as a continuous, or stepwise continuous, plot of intensity vs 2θ.   Parameter shifts which will improve the fit are
then calculated on the basis of the differences between observed
and calculated patterns, a new calculated pattern is obtained, and
the cycle is repeated.   One can readily appreciate that this approach, which makes use of every bit of observed intensity information, is quite different from those which start by decomposing the
powder pattern into estimated Bragg intensities.

The method has been used longer and more widely with neutron
data, but in this paper its more recent successful use with x-ray
data is emphasized.   Earlier methods of using x-ray powder patterns
for structure analysis have been useful and sometimes ingenious.
Kennicott (1963) did least-squares adjustment of structural parameters to give the best fit to integrated powder diffraction intensities which were not all resolved; overlapped reflections were
lumped together.   Similar least-squares procedures with x-ray powder intensities have been used recently by, for example, Gavarri,
Calvarin & Weigel (1975).   A program designed to give fast convergence in such procedures has been described by Pham, Choisnet &
Raveau (1975).   Taupin (1973), Mortier and Costenoble (1973),
Mortier (1980), Huang and Parrish (1975) and Parrish, Huang & Ayers
(1976) have contributed methods for decomposing the whole powder
diffraction pattern into component overlapping peaks without reference to a structural model.   Clearly, such decomposition techniques might be usefully coupled with the programs for least-squares
structure refinement with Bragg intensities.   But if the object is
structure refinement, the Rietveld method is superior in principle
because it uses all of the observed data, including zeroes, with a
minimum of bias.

It is only in recent years that the algorithm and computer
programs necessary for this powerful new method have become available (Rietveld, 1967; 1969).   Rietveld's program and its modifications were first written for the neutron case.   By 1977, Cheetham
and Taylor were able to list 170 structures which had been refined,
with the Rietveld method, from neutron powder diffraction data.
The successful modification of Rietveld's program to work with
x-ray data was reported five years ago (Mackie & Young, 1975) and
the first published papers giving x-ray Rietveld analysis results
appeared only three years ago (Malmros & Thomas, 1977; Young,
Mackie, and von Dreele , 1977; Khattak and Cox, 1977).   The published and soon-to-be-published applications of the Rietveld
method with x-ray data were surveyed as of about June,1978 (Young,
1980).   More than 30 different materials, in 15 space groups,
were listed, including inorganic, organic, polymer, mineral, and
biological materials.

2. THE REFINEMENT PROCESS

2.1 Least-squares Procedure

As in any least-squares refinement procedure, the quantity minimized, called the residual, is the weighted sum of the squares of the differences between observed and calculated values of the observables. Here the observables are the intensity values y_i at each of the $\underline{i\text{th}}$ steps in the pattern, so the residual $\mathcal{R}$ is

$$\mathcal{R} = \sum_i w_i \, (y_i(o) - y_i(c))^2 \tag{1}$$

and the summation is over all of the steps in the pattern. The weight w_i is the reciprocal of the variance, $\sigma_i{}^2$, at the $\underline{i\text{th}}$ step and is usually based on counting statistics for gross and background intensities. Let x_j represent the adjustable parameters. To proceed with the minimization of the residual   (i.e., to produce the calculated pattern that will fit best with that observed), one first forms the normal equations matrix in which the elements are given by

$$M_{jk} = -\sum_i 2w_i \left[(y_i(o) - y_i(c)) \frac{\partial^2 y_i(c)}{\partial x_j \partial x_k} - \frac{\partial y_i(c)}{\partial x_j} \frac{\partial y_i(c)}{\partial x_k} \right] \tag{2}$$

In practice, the first term is usually omitted to simplify the calculations. The shifts, Δx_j, which will best reduce the residual are then given by

$$\Delta x_j = \sum_k M_{jk}^{-1} \frac{\partial \mathcal{R}}{\partial x_k} \tag{3}$$

where M^{-1} is the inverse of the matrix M. The calculated shifts Δx_j are then applied to the adjustable parameters x_j in the model(s), a new set of $y_i(c)$ (i.e., a new calculated pattern) is calculated with the newly adjusted parameters, and the whole procedure is repeated cyclically until some criterion of "completion" is met. For example, the criterion may be that all $\Delta x_j < (1/3)\sigma_j$.

The standard deviations in the adjusted parameters are given by

$$\sigma_j = \left[M_{jj}^{-1} \frac{\sum_i w_i (y_i(o) - y_i(c))^2}{N - P + C} \right]^{\frac{1}{2}} \tag{4}$$

where M_{jj}^{-1} is the diagonal element of the inverse matrix, N is

the number of observations (e.g., the number of y_i's), P is the
number of parameters adjusted, and C is the number of constraints
applied.

2.2 Application to Powder Data

The procedure so far described is much the same as that used
for least-squares structure refinement with single-crystal data,
the only difference being that the y_i's would then be Bragg reflection intensities. But the y_i's here are not Bragg intensities and
are rather complicated because each y_i may contain contributions
from a number of Bragg reflections (Fig. 1) as well as several
other factors:

$$y_i(c) = s \sum_K p_K L_K |F_K|^2 G(\Delta\theta_{iK}) \, P'_K + y_i(b) \tag{5}$$

Where
>   s is a scale factor,
> L_K= Lorentz and polarization factors for the Kth Bragg reflec-
>   tion,
> F_K= structure factor, $\sum_j f_j \, e^{2\Pi i (hx_j+ky_j+\ell z_j)} e^{-M_j}$,
> $f_j, x_j, y_j z_j, M_j$= atomic scattering factor, coordinates, and
>   thermal factor for the jth atom in the unit
>   cell,
> p_K= multiplicity factor,
> P'_K= Preferred orientation function, (e.g., $e^{+P\alpha^2 K}$),
> $\alpha_K\equiv$ angle between reference direction and $\underset{\sim}{d^*_K}$,
> θ_K= Bragg angle for the Kth reflection,
> K= h,k,ℓ, the indices identifying the Bragg reflection for
>   which each of the above is evaluated, and
> $G(\Delta\theta_{iK})\equiv G(2\theta_i-2\theta_K)$ is the reflection profile function con-
>   sisting, in one present version, of a
>   symmetric profile function $g(\Delta\theta_{iK})$ multi-
>   plied by an asymmetry function
> $a(\Delta\theta_{iK})$ given by

$$a(\Delta\theta_{iK}) = 1 - \frac{A(\text{sign } \Delta\theta_{iK}) \, (2 \, \Delta\theta_{iK})^2}{\tan \theta_K}$$

where A is the (adjustable) asymmetry parameter, and "sign" is +
or −.

As is made abundantly clear from the foregoing, there are
many parts to y_i, including contributions from each of the Bragg
reflections, K, occurring within some specified 2θ range of the

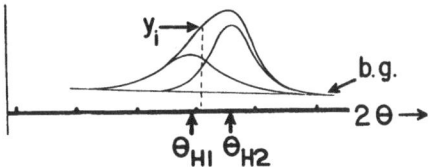

Fig. 1. The intensity y_i includes contributions from more than one Bragg reflection, e.g. those centered at θ_{H_1} and θ_{H_2}, and from the background intensity.

2θ position of y_i. This range is often taken to be 2 or 3 times the FWHM (full-width-at-half maximum) of the individual reflection profile.

The identification of the best reflection profile function is a matter of current research. The reflection profile functions most frequently used by various workers for the x-ray case are given in Table 1. Toraya & Marumo (1980) have also included an asymmetry factor within their modified Lorentzian function. Also given in Table 1 are the functional forms used for the background in those cases in which it was refined simultaneously with the structure. To date, most workers have approximated the actual reflection profile function(s) with Gaussian or Lorentzian forms with H_K (FWHM of the Kth reflection) given by

$$H_K^2 = U \tan^2\theta + V \tan\theta + W$$

where U, V, and W are refinable parameters. Thus, to summarize, the parameters that can be refined simultaneously in a Rietveld analysis include

Structural (for two phases, if needed) parameters
    atom position x_j, y_j, z_j
    atom thermal B_j or β_{jj}
    atom site-occupancy

Instrumental and specimen parameters
    2θ-zero
    overall scale
    overall "B"
    profile breadth (U,V,W)
    profile asymmetry
    background
    preferred orientation
    lattice parameters or wavelength

Table 1
Functions

Symbol	Function	Name
G	$A_1 \exp(-x^2/k_1^2)$	Gaussian
L	$A_2 (1+k_2^2 x^2)^{-1}$	Lorentzian
ML	$A_3 (1+k_3^2 x^2)^{-2}$	Mod 1 Lorentzian
IL	$A_4 (1+k_4^2 x^2)^{-1.5}$	Mod 2 Lorentzian
P	$\dfrac{2\sqrt{(m)}}{\sqrt{\Pi}} \dfrac{\sqrt{2^{(\frac{1}{m})}-1}}{\lceil(m-\frac{1}{2})} \dfrac{1}{k_5} \left[1+4 \dfrac{x^2}{k_5^2} (2^{(\frac{1}{m})}-1)\right]^{-m}$	Pearson VII
V	$A_6 \int_{-\infty}^{\infty} L(x') \, G(x-x') \, dx'$	Voigt
Poly	$\dfrac{A_7}{P_0} (1-\dfrac{x}{Q})$	Polynomial

In the above

$x = 2\theta_i - 2\theta_K$

$A_1 \ldots A_7 \equiv$ required normalization factor

$k_1 \ldots k_5 =$ a constant related to profile breadth H_k

P and Q $\equiv$ polynomials with even exponents, only.

For further details, see Wiles and Young (1981) and Young 1979).

Background representation used by Wiles and Young (1981)

$$y_{ib}(c) = \sum_{n=0}^{5} B_n (2\theta)^n$$

In the 30 structure refinements (by x-ray Rietveld analysis) reviewed recently (Young, 1980) the number of parameters simultaneously
refined ranged from 7 to 51 and the number of data points y_i was
most often in the range between 1500 and 2500.

2.3 Computer Programs

The first computer programs prepared for x-ray Rietveld analysis were generally modifications of the original Rietveld (1969)
program. Recently, several all-new programs have been written.
For the angle dispersive x-ray case, the programs and some of their
features of which we are aware are listed in Table 2. The absence
of a check mark at any point does not necessarily mean that the
program does not have that feature; it only means that we do not
know that it does. In Table 2, "all space groups" means that the program is directly applicable to all with no additional programming
required for any space group. "Single-pass" operation means that
no preparatory programs are needed, one simply enters the program
with the data and exits with the refinement results. "Generalized
coordinates" (Immirzi, 1978) are those which occur as the variables
when one applies constraints in such a way as to reduce the size of
the normal matrix. An example is the use of torsion angles and
distances between molecular units of known structure in a polymer
instead of the coordinates of all atoms present. "R's in shells"
means that the program can provide separate R factors for different
spherical shells in reciprocal space.

The computer central-processor time required per refinement
cycle varies markedly, of course, with the problem, the computer,
and the program. With the DBW 2.9 program, a CDC Cyber 70/74 computer, 2,000 y_i's, and 28 adjustable parameters, the cycle time is
about 60 seconds. For many of the programs, copies are available
on request from the program authors. In particular, the "DBW 2.9"
program (Wiles and Young, 1981) is available without charge other
than possible reimbursement of reproduction and mailing costs.

2.4 Criteria of Fit

An informative visual criterion of fit is given by a composite
plot such as Fig. 2. In the upper portion, the dots with vertical
lines through them show the observed data. The continuous curve
nearly overlaying them is the calculated pattern obtained from the
refined model. The series of short, vertical lines below the patterns indicate the positions of possible Bragg reflections. The
lowest curve shows the difference between observed and calculated
patterns. One can see that the fit is generally very good.

A particular advantage of this type of visual comparison,

TABLE 2

PROGRAMS FOR ANGLE DISPERSIVE CASE

FEATURE	PROGRAM							
	RMT	RHVM	RKC	RWSMT	AI	BH	DBW 2.9	TOR
ALL SPACE GROUPS					✓		✓	✓
SINGLE PASS							✓	✓
MODULAR					✓		✓	
ANSI					✓		✓	
MULTIPHASE				✓			✓	
GEN. COORD.					✓			
CONSTRAINTS		✓			✓	✓	✓	
ANOM. DISPERSION		✓					✓	
R's IN SHELLS					✓			
DATA TYPE	1	2	3	1	2	2	2	2
USED BY OTHERS	✓	✓					✓	
BACKGROUND					✓		✓	✓

Notes:

RMT - modified Rietveld as used by Malmros & Thomas (1977)
RHVM - modified Rietveld described by Young, Mackie & von Dreele (1977)
RKC - modified Rietveld described by Khattak & Cox (1977)
RWSMT - modified Rietveld described by Werner, Salomé, Malmros & Thomas (1979)
AI - written by A. Immirzi (1978, 1980)
BH - written by Baerlocher and Hepp (1980)
DBW 2.9 - written by D.B. Wiles (Wiles & Young 1980)
TOR - written by H. Toraya (Toraya & Marumo, 1980)

1 → Guinier Hägg, α_1 only
2 → Θ-2Θ Diffractometer, α_1, α_2
3 → Θ-2Θ Diffractometer, β only

other than the esthetic pleasure of "seeing" a good fit, is the information it conveys about details of the remaining misfit. Here
there are two features of special interest. First, in the difference curve, note the peaks at ∿44.5° and 65° (2θ) which look like
Bragg reflection peaks. They are. They are due not to the hydroxyapatite specimen but to Aℓ in the holder; a bit of the holder was
allowed to be in the x-ray beam. Because the main phase pattern is
so well fit by the calculated pattern, it is effectively stripped
out of the difference curve and the presence of the second phase
(Aℓ) is thereby made much more obvious. The second feature is the
oscillatory character of the difference curve at the positions of
the main maxima. This shows that the reflection profile function
used did not accurately match the actual one (to see this, subtract
a Lorentzian, for example, from a Gaussian of the same area). This
mismatch contributes to the residual, $\mathcal{R}$, (eqn. 1) but this contribution does not arise from a mis-match of calculated and observed
integrated (Bragg) intensities. In fact, the calculated and observed integrated intensities are actually almost as well fit (in
this case) as they would be if the profile functions were well
matched. Both of these features, the unnoticed second phase and
the profile-function mismatch, enlarge the R_p and R_{wp} values for
the the pattern-fitting in a way that has little or no significance
to the quality of the structure refinement.

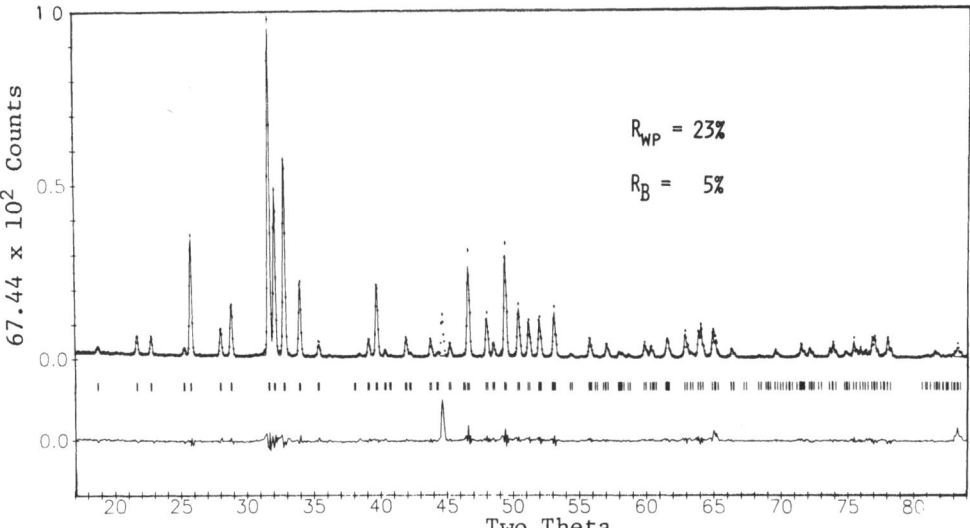

Fig. 2.   X-ray (CuKα) powder diffraction pattern-fitting (refine-
           ment R168) for hydrothermally prepared OHAp (Y155). The
           dots show the observed data, the continuous curve overlay-
           ing them is the calculated pattern, the vertical bars in a
           row beneath these mark the positions of possible Bragg re-
           flections, and the bottom curve shows the difference be-
           tween observed and calculated patterns. Background refined.

The quantitative criteria of fit that are usually used are one
or more of these "R-factors":

$$R_F = \frac{\Sigma \left| \sqrt{I_K(\text{"obs"})} - \sqrt{I_K(\text{calc})} \right|}{\Sigma \sqrt{I_K(\text{"obs"})}} \qquad \text{("R-structure factor")}$$

$$R_B = \frac{\Sigma \left| I_K(\text{"obs"}) - I_K(\text{calc}) \right|}{\Sigma I_K(\text{"obs"})} \qquad \text{("R-Bragg")}$$

$$R_p = \frac{\Sigma \left| y_i(\text{obs}) - \frac{1}{c} y_i(\text{calc}) \right|}{\Sigma y_i(\text{obs})} \qquad \text{("R-pattern")}$$

$$R_{wp} = \left[\frac{\Sigma w_i (y_i(\text{obs}) - \frac{1}{c} y_i(\text{calc}))^2}{\Sigma w_i (y_i(\text{obs}))^2} \right]^{\frac{1}{2}} \qquad \text{("R-weighted-pattern")}$$

Here I_K is the intensity assigned to the Kth Bragg reflection.
In the expressions for R_F the "obs" -- for observed -- is put in
quotation marks because the Bragg intensity I_K is rarely directly
observed; instead, the $I_K(\text{"obs"})$ values are obtained from an allo-
cation of total observed intensity in overlapped reflections to the
individual Bragg reflections according to the ratios of intensities
in the calculated pattern.

Since R_{wp} contains, in its numerator, the quantity being minimized (see eqn. 1), it is the most statistically meaningful of the
four. It is R_p, however, that has been most consistently reported.
In the ∿28 x-ray Rietveld refinements   made with   angle-dispersive
data, reviewed by Young (1980), R_p ranged from 12 to 27% with an
average of about 20%. For the ∿170 neutron Rietveld refinements
reviewed by Cheetham and Taylor (1977), R_p ranged from 5 to 17%
with an average of ∿10%. The smaller R-factors for the neutron
case may reflect the fact that the actual profile function is well
approximated by a Gaussian.

Finally, the usefulness of R_B in assessing structural results
should not be overlooked. It is not as sensitive as are R_p and
R_{wp} to misfit of the background between reflections nor to second
phases nor to the use of calculated reflection profiles which do
not fit the actual shapes of the observed reflection profiles but
do manage to yield good estimates of integrated intensities. Thus,
R_B tends to depend more heavily on the fit of the crystal structural
model and less on non-structural contributions to the measure of

misfit than do R_p and R_{wp}. It must be remembered, however, that since the "observed" integrated intensities going into R_B are really allocations based on the model, they are necessarily biased in favor of the model being used.

One should not expect to compare the Rietveld-analysis R values directly with values quoted in single-crystal work. The Rietveld R_p and R_{wp} values tend to be larger for equally good structural fit because of at least two factors. (1) These R's are based on the summation of the misfit over the entire pattern, and (2) profile function mismatch enlarges the R values even though integrated intensities match. It is probably more appropriate to base comparisons of Rietveld and single-crystal results on the parameter values and their standard deviations, σ_j, although they, too, are affected by any meaningless contributions to R that may occur (eqn. 4).

2.5 Precision

The precision obtained for lattice parameters, refined simultaneously with other parameters, can be a pleasant surprise. Even with data only at angles below $2\theta= 100°$, for a and c in hexagonal crystals we consistently get calculated precision (σ) of about one (0.3 to 1.7) part in 10^4. Khattak and Cox (1977) reported 2 parts in 10^5. In our experience, the actual specimen to specimen reproducibility is somewhat poorer, but it is still gratifying since no special effort is expended to achieve high precision.

The precision in structural parameters, in the 30 odd x-ray Rietveld refinements reviewed by Young (1980), ranged from 10^{-4} to 10^{-2} of the cell edge with an average of about 10^{-3}. This may be compared to an average of $\sim 10^{-4}$ for good single-crystal work. The precision (σ) in site occupancy parameters, in the same review, ranged from 1 to 11%.

3. EXAMPLES

3.1 A Test Case: Fluorapatite

The structure of fluorapatite, unit cell content $Ca_{10}(PO_4)_6F_2$ and space group $P6_3/m$, is well known from single-crystal work (e.g., Sudarsanan, Mackie and Young, 1972). It is indicated in Fig. 3. Fluorapatite can be reproducibly synthesized in powder form. It has enough structural complexity to let the power of the Rietveld method be exhibited without overburdening it. Fig. 4 shows the result of a Rietveld refinement (R152) with crystal-monochromatized CuKα (both α_1 and α_2 x-ray data collected in $0.0375°$ (2θ) steps in the range 17 to $80°(2\theta)$ (1680 y_i's). The individual anisotropic thermal parameters were fixed at their single-crystal values, and a total of 28 parameters were varied. Of these, 19 were structural (12 positional, 6 site occupancy, and 1 thermal). $R_{wp} = 8.2\%$ and

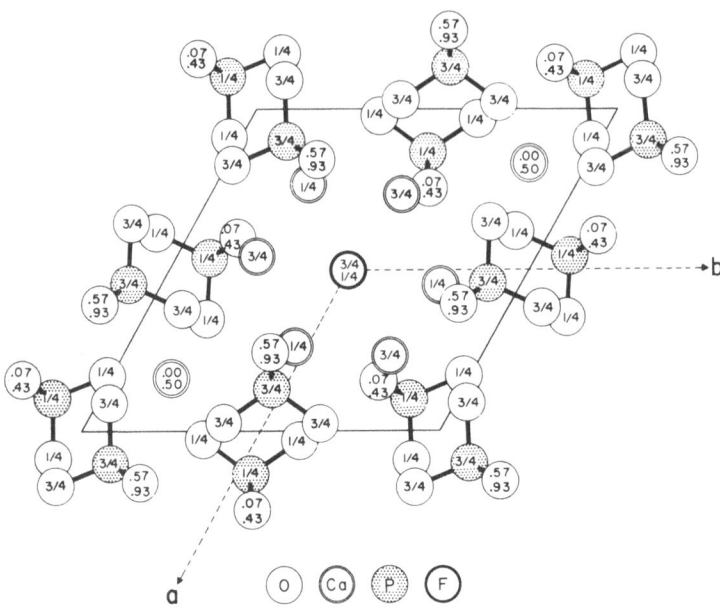

Fig. 3.   The apatite structure projected onto the a,b, plane. The
          numbers inside the atom indicators are z position parameters
          (e.g. height above the paper). Not shown are the "X-" ions,
          e.g. F,OH,Cl, etc, which are located at each corner of the
          inscribed parallelepiped (unit cell boundary). For fluor-
          apatite, X=F and has z=1/4 & 3/4. For hydroxyapatite, X=OH
          with the O at z=0.19 and H at z=0.05, plus the symmetry re-
          lated positions produced by the mirror at z=1/4 and the 6_3
          axis at x=y=0.

R_B= 3.1%. This is the smallest R_{wp} we have obtained in any of our
work to date. The calculated σ_j's are also the smallest, being only
2 or 3 times larger than the single-crystal values, rather than the
usual ∿10 times. Clearly, the fit is excellent. Lattice parameters
are a= 9.3641(3) and c= 6.8810(2). All but one of the position parameters agree with the single-crystal values within three of these
unusually small σ's. The odd one agrees within 5 of these small
σ's. There are no surprises; it is a good test case.

3.2 A Real Case: Human Tooth Enamel

     3.2.1. Specimens and Models. The most dense portion
(sp.g. >2.95) of human tooth enamel (TE) is >99.5% inorganic crystalline material of a single phase that can be approximately
modeled as hydroxyapatite (OHAp), Fig. 3, with many substitutions.
Principal compositional differences are ∿3 wt% CO_3^{2-} distributed between two sites, 3-4 wt% HPO_4^{2-}, 3-4 wt% H_2O, ∿40-50% OH deficiency,
a few tenths wt% each of Na, Cl and sometimes F and Mg, and a great
variety of substitutions in much lesser amounts.

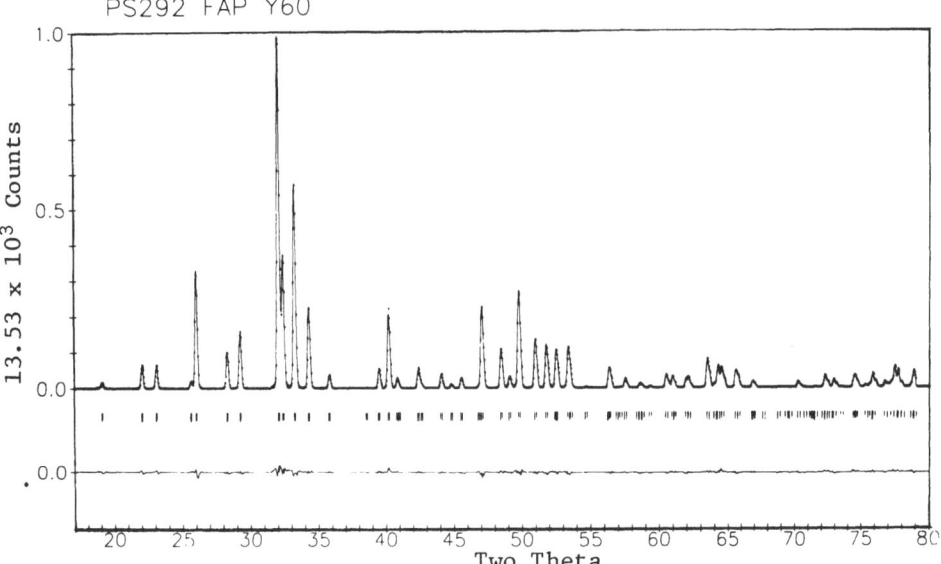

Fig. 4.   X-ray (CuKα) pattern-fitting for fluorapatite, plotted as
for Fig. 2. In this particular refinement (R152), the
background was held fixed at visually estimated values.
Here R_{wp}=8.2% and R_B=3.1%.

For comparison purposes, specimens of OHAp, in powdered form,
were specially prepared by both reflux and hydrothermal methods.

Crystal structure refinements of TE have been carried out with
a model in which Ca, P, and phosphate O atoms were initially
assigned the positions, thermal parameters, and site occupancies
they have in stoichiometric OHAp (Sudarsanan & Young, 1969). The
positions and site occupancies were refined. It was expected that
the presence of CO_3 (B-site) replacing PO_4 might be reflected in
the departure of the refined parameters for PO_4 from those in
stoichiometric OHAp. OH^-, Cl^-, F^-, some of the CO_3^{2-} (A-site) and
possibly H_2O are in the "X-ion channel" along 0,0,z (Fig. 3), the
channel in which OH^- resides in OHAp. Each of these entities except H_2O is known to occur at its own particular z position. They
are too close together to allow simultaneous refinement of their
site occupancy factors. Therefore, the distribution of scattering
density along 0,0,z, was sampled by refinement of the apparent site
occupancy factors of "dummy" oxygen atoms placed at equal intervals
along 0,0,z (Wilson, Sudarsanan and Young, 1977).

Generally, with the 2θ range used, only 3 such dummy X-ions
could be accommodated at once. Therefore, to provide more apparent
detail and also a consistency check, the refinements were carried
out with the dummy atoms first at z= 0.0, 0.10, 0.20 and then at
0.05, 0.15, 0.25, or vice versa. (This procedure was dubbed

"3X-2 refinement".) The sum of the scattering at the three dummy
X-ion positions was the same for both choices of the 3X̄ positions,
as were the other refined parameters.

   3.2.2. Final Refinement Results. In these final refinements,
both background and preferred orientation parameters were refined
simultaneously with all of the other refined parameters. Fig. 2
shows the x-ray pattern-fitting (R168) for the hydrothermally prepared OHAp (Y155), for which R_{wp}= 22.9% and R_B= 5.1%. The R's are
inflated somewhat by the presence of the Aℓ. For a reflux-prepared
OHAp specimen (Y146), refinement (R169) resulted in R_{wp}= 18.9% and
R_B= 4.6%. Fig. 5 shows the x-ray pattern fitting for TE specimen
Y157A. In this refinement (R181), R_{wp}= 19.5% and R_B= 4.9%. Fig. 6
shows the neutron pattern fitting for the same TE material, Y157A.
In that refinement (R173), R_{wp}= 22.2% and R_B= 1.8%.

   The coordinates, lattice parameters, and total X-ion scattering
obtained for, variously, TE, powdered OHAp and single-crystal OHAp
of mineral origin (Sudarsanan and Young, 1969) are compared in
Fig. 7. The vertical bars indicate estimated standard deviations
calculated in the refinements. The plot shows the difference between the value of each parameter obtained by Rietveld refinement
and the corresponding value in the single-crystal work.

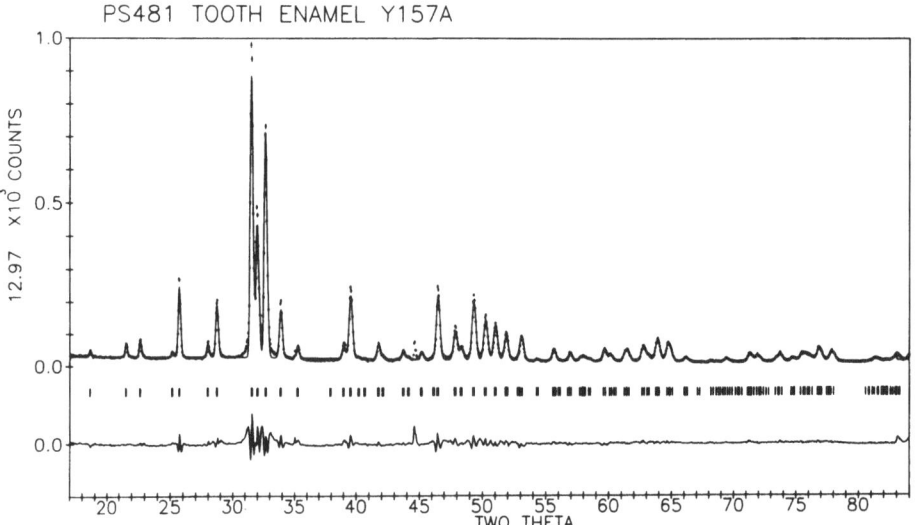

Fig. 5. X-ray (CuKα) pattern-fitting (R181) for TE specimen Y157A
         plotted as for Fig. 2. Here R_{wp}= 19.5% and R_B= 4.9%.
         The background was refined.

NDATA TOOTH ENAMEL Y157A

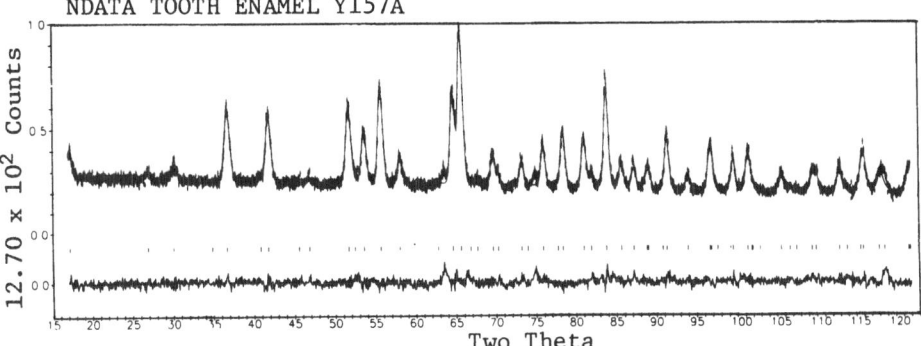

Fig. 6.   Neutron pattern-fitting (R173) for TE specimen Y157A
plotted as for Fig. 2. Here R_{wp}= 22.2% and R_B= 1.8%.
The neutron wavelength was 2.46 Å. The background was
refined.

The enlargement of the a lattice parameters of TE and reflux
OHAp relative to normal OHAp (a= 9.418Å), are thought possibly to
be associated with the incorporation of H_2O or HPO_4^{2-}, or both.
When the specimens have been heated past 400°, the a axes return to
normal OHAp lengths (e.g., Holcomb and Young, 1980).

Let us consider now, the coordinate results. First, we note
that, except possibly for z of O(3), the x-ray results are consistent for the TE specimens. This is true even though two separately
obtained specimens were used and in one case (Y157A+) the x-ray
specimen was made up of equal portions of TE and ground glass.
Second, the x-ray determined coordinate results for powdered synthetic OHAp and for TE are in general agreement; the most significant disagreements occurring for O(3) and, possibly, Ca(2). Third,
the x-ray and neutron results for TE also differ most for O(3) and
Ca(2), the two atoms that constitute the "walls" of the X-ion
channels. Fourth and finally, the consistent disagreement between
all of the powdered specimen results and the single-crystal result
for x of O(1) may indicate that the oddity is in the mineral single
crystal.

These results show that, on the whole, the OHAp model is a
rather good one for TE even though the CO_3^{2-} and H_2O known to be

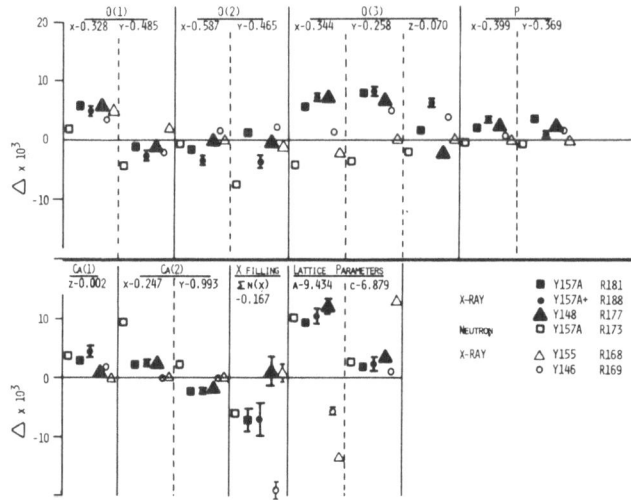

Fig. 7.   Intercomparisons of positional and lattice parameter re-
          sults. The plotted points show the differences between
          the Rietveld-method results and those previously obtained
          for a single crystal of Holly Springs hydroxyapatite
          (Sudarsanan and Young, 1969). Error bars correspond to cal-
          culated σ's. Y157 and Y148 are TE, Y155 and Y146 are OHAp
          prepared hydrothermally and by a reflux method, respectively.

present are not included in it. Possibly, the differences noted in
O(3) and Ca(2) coordinates may constitute some of the clues, alluded
to earlier, about the location of the incorporated CO_3^{2-} and H_2O, presumably in or around the channels. However, that analysis is not
yet done.

    The refined site occupancies for Ca, P and O are compared in
Fig. 8. The site occupancy of O(3) was arbitrarily fixed at unity,
so the refined parameter values are relative to that. The quantity
plotted is the difference between the refined site-occupancy parameter, relative to that of O(3), and its stoichiometric value.
Again, the reproducibility of the results for TE is good and, again,
the OHAp model seems to fit TE rather well on the whole, although
there are detailed differences that may yield additional information
upon further analysis. The site-occupancy for O(1) is consistently
high.

    It is possible that the differences between the x-ray and neu-
tron results for TE may contain clues to H locations, but the observed differences are not dramatic.

    The distribution of scattering density along the 6_3 axis at
0,0,z is shown in Fig. 9. The units are indicated (relative) site
occupancies found for the "dummy" O atoms in the 3X-2 refinements.

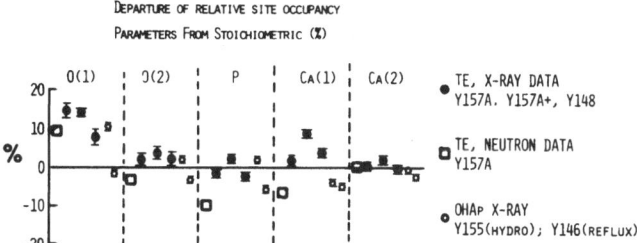

Fig. 8.   Intercomparisons of atomic site occupancies obtained for
          several TE and OHAp specimens.

The x-ray determined X-ion distributions for the two TE specimens
(Fig. 9a) are similar; their apparent differences can be taken as
an indication of probable actual reproducibility.   The neutron-de-
termined distribution (Fig. 9a) is in reasonable agrement with the
x-ray determinations; the negative values in the z= 0.00 to 0.05
range are to be expected because that is the location of the H of
the structural OH^- ions and H has a negative scattering length
$(-0.38 \times 10^{-12}$ cm) for thermal neutrons.   In fact, this x-ray $\underline{vs}$
neutron difference confirms the presence of H in that range of z.

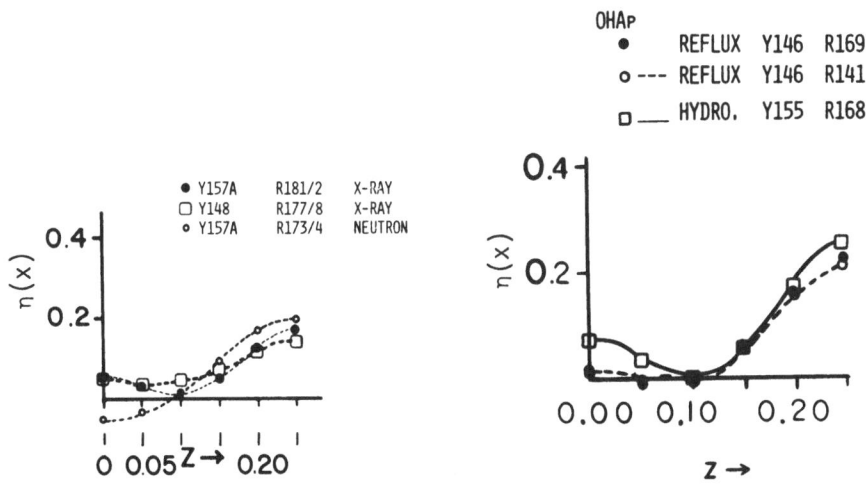

                                                    9a   9b
Fig. 9.   Apparent scattering density along 0,0,$\overline{z}$ as determined with
          3X-2 refinements. (See text.) n(X) is the multiplier
          actually used in the refinement and is linearily related
          to site occupancy. The preferred orientation parameter was
          refined simultaneously with the other parameters.   Fig. 9a
          shows results for TE. Fig. 9b shows results for OHAp.

Comparing the x-ray results for TE in Fig. 9a with those for OHAp in Fig. 9b, one sees very little difference that, at this point, is assuredly significant. There are no surprises -- which, itself, may be considered a disappointment by one expecting to see clearly demonstrated the presence of the molecular H_2O in the X-ion channels where it could be sterically accommodated quite logically. There is a possibility that the preferred-orientation correction has not been properly accomplished in the refinement, and that the actual X-ion distribution in TE therefore may be different from that in Fig. 9a by having larger values in the z= 0.10 - 0.15 region. This possibility is examined in §3.2.3, where it is concluded that Fig. 9a represents the most probably correct result now available. We therefore base our following comments on Fig. 9.

In Fig. 9a, there may be a real TE vs OHAp difference in the X-ion occupancies near z= 0.20 to 0.25. This should be expected, as TE is known (Holcomb & Young, 1980) to be 30-40% deficient in structural OH^- (0 occurs at 0,0,0.19). Whether the higher values indicated for TE at z= 0.0 are significant or not is not yet known. TE does contain enough Cl^- to constitute about one X-ion in 12-15 and it is expected to occur at z= 0.06 - 0.08. Also expected in the X-ion channel is about one CO_3^{2-} for every 6-15 cells (Holcomb & Young, 1980). Thus, there are physical reasons why the X-ion occupancies in TE and OHAp could and should be somewhat different. At this point in the work, however, the reproducibility of detail, particularly of detail in the X-ion distributions, needs to be improved before results such as those in Fig. 9a are subjected to extensive interpretation.

In fact, as is discussed in the next section, the possibility that the X-ion distribution may have a maximum near z= 0.10 - 0.15, as initially thought (Young & Mackie, 1980), has not been entirely ruled out. Such a maximum could indicate the expected presence of H_2O in these X-ion channels.

3.2.3. Preferred orientation effects. Preferred orientation or improper correction for it can have a profound effect on the structural results from Rietveld analysis.

As is well known, preferred orientation produces systematic errors in the measures of Bragg intensities. Unless properly corrected for, it will, therefore, have a profound effect on the structural results derived from powder data. The Rietveld refinement results so far presented in this paper have all been obtained with simultaneous refinement of the preferred orientation parameter P as defined for eqn (5). For the OHAp refinements, P refined to ∿0. For the TE refinements R177, R181, & R188, P refined to 0.25, 0.21, and 0.23, respectively. The data used for R181 were used in

a 3x-2 refinement (R189) differing only in that P was held fixed
at 0.   Then R_{wp}= 22.3%, R_B= 6.6%.   As Table 3 shows, both site occupancies and even some coordinates (e.g., x for O(2) and y for P)
were affected.   The X-ion occupancy distribution shown in Fig. 10
was obtained.   Clearly, if the Fig. 10 results were correct, our
view of the content of the X-ion channels would be markedly changed.
The difference between Fig. 9a and Fig. 10 (for Y157A) is strictly
due to the handling of preferred orientation.   Which is more nearly
correct?

     The reduction of R_{wp} from 22.3% to 19.5% and particularly the
reduction of R_B from 6.6% to 4.9% strongly indicates that the results obtained with P refined are the more nearly correct.   Further,
the site occupancies obtained for the other atoms (Table 3) also
seem better (with P refined) in the sense that they show less marked
unaccounted-for departure from stoichiometric values.   Finally, in
our work with several different powder diffraction data sets collected for both the same and different TE specimens, we have noted
that the reproducibility in both site occupancies and, even, in
atom coordinates is better when P is refined.   Thus, we can conclude that the results in Fig. 9a are better than those in Fig 10.
But, are they correct?

     The powdered TE specimens are composed in part of rod like
particles.   Thus the expected preferred orientation is that for an
acicular habit.   That the TE specimens can take up an acicular preferred orientation is demonstrated by the changing ratio of 00.2 to
2.10 intensity with different specimen loading techniques.   The
preferred orientation function used in the refinements (eqn. 5) is
for a platy habit.   Hence, it can correct only partially, in a
first approximation, for the acicular preferred orientation.   Thus,
while the use of the available preferred orientation function did
improve R factors, reproducibility, and "reasonableness" of the

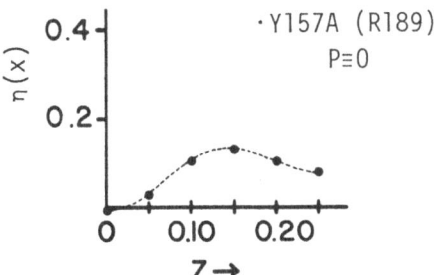

Fig. 10. Apparent scattering density along 0,0,z in TE as indicated
          by Rietveld refinements with the preferred orientation
          parameter fixed at zero.

Table 3

Effect of not refining the preferred orientation parameter

Tooth Enamel Specimen	Data Set	Refinement
Y157A	PS481	R182
Y157A	PS481	R189

O(1)x	O(1)y	O(2)x	O(2)y	O(3)x	O(3)y	O(3)z
0.3341(6)	0.4841(6)	0.5855(8)	0.4666(8)	0.3497(5)	0.2662(5)	0.0721(6)
0.3332(8)	0.4843(8)	0.5896(10)	0.4644(9)	0.3541(5)	0.2723(5)	0.0642(5)

Px	Py	Ca(1)z	Ca(2)x	Ca(2)y	a(Å)	c(Å)
0.4009(4)	0.3726(4)	0.0051(6)	0.2492(3)	0.9909(3)	9.4432(10)	6.9807(9)
0.4021(5)	0.3646(4)	0.0044(6)	0.2490(3)	0.9893(4)	9.4434(12)	6.8808(10)

R_{wp}%	R_B%	O(1)n	O(2)n	Pn	Ca(1)n	Ca(2)n
19.5	4.9	0.575(8)	0.520(6)	0.515(5)	0.363(3)	0.515(4)
22.3	6.6	0.415(5)	0.420(5)	0.440(3)	0.300(2)	0.439(2)

X(00)n	X(05)n	X(10)n	X(15)n	X(20)n	X(25)n	SUM
0.027(6)	0.028(5)	0.007(10)	0.045(10)	0.126(4)	0.086(6)	0.160
-0.006(5)	0.027(4)	0.105(8)	0.133(8)	0.103(4)	0.040(5)	0.201

T	U	V	W	Z	ASYM	PRF
2.18(12)	-0.257(86)	0.451(70)	-0.038(13)	-0.211(5)	0.51(36)	0.21(1)
2.65(14)	-0.376(99)	0.556(80)	-0.059(15)	-0.211(6)	0.56(42)	0.00

SCALE
0.00164(8)
0.00326(11)

n is the site occupancy parameter, which is equal to the site occupancy divided by the site symmetry factor.
X(05)n, for example, is the site occupancy parameter for the dummy O ion placed at 0,0,0.05.
SUM is the sum of the site occupancy parameters for the X ions; multiplying it by 12 yields the number of X-ions per cell expressed as O-ion scatterers.
T is the overall temperature factor in units of $Å^{-2}$.
Z is the 2θ-zero correction.
ASYM is the reflection profile asymmetry parameter referred to as A in the text.
PRF is the preferred orientation parameter referred to as P in the text.

results for the calcium and phosphate ions, it is not necessarily true that the results obtained with it are strictly correct. In particular, the possibility remains open that some of the "hump" character of Fig. 10 may be real and does not appear in Fig. 9a because of an effective overcorrection arising from the use of the wrong preferred orientation function. Unfortunately, the correct orientation function has not yet been developed and implemented.

Clearly, one should collect data on TE specimens free of preferred orientation. For R181 the TE specimen Y157A was mixed with an equal volume of ground glass, but this did not change significantly the apparent preferred orientation. In fact, P went from 0.21(1) to 0.23(2).

In conclusion, then, it is clear that preferred orientation can occur in TE, and that improper correction for it can seriously affect the structural refinement results, but the extent to which these effects have influenced the results in Fig. 9a is not determined. Probabilities seem to suggest that Fig. 9a is more nearly correct than Fig. 10, but the truth, for the distribution of scattering density in the X-ion channels, may well lie between these two representations.

CONCLUSIONS

With good, digitized powder diffraction data, highly useful crystal structure refinements can be made for materials of moderate complexity which could never be made available in single-crystal form. An example is human tooth enamel. Currently available computer programs are easy to use and have most of the desired features. Improvements needed include improved reflection profile functions and incorporation of a preferred-orientation function appropriate to an acicular habit.

ACKNOWLEDGEMENTS

The authors are grateful to Dr. D.E. Cox and Brookhaven National Laboratories for the neutron powder diffraction data used here and to the USPHS for financial support under project NIH-NIDR-DE-01912.

REFERENCES

Bärlocher, Ch., and Hepp, A., A New Pattern Fitting Structure Refinement Program for X-ray Powder Data, Symposium on Accuracy in Powder Diffraction, National Bureau of Standards, Gathiersburg, MD, 165.

Cheetham, A.K., and Taylor, J.C. 1977, Profile Analysis of Powder Neutron Diffraction Data: Its Scope, Limitations, and Applications in Solid State Chemistry, J. Sol. S. Chem., 21:253.

Gavarri, J.R., Calvarin, G., and Weigel, D., 1975, Oxydes de Plomb. II. Etude Structurale a 5 K de la Phase Orthorhombique de l'Oxyde Pb_3O_4, J. Sol. S. Chem., 14:91.

Holcomb, D.W., and Young, R.A., 1980, Thermal Decomposition of Human Tooth Enamel, Calcif. Tiss. Int'l., 31:189.

Huang, T.C., and Parrish, W., 1975, Accurate and Rapid Reduction of Experimental X-ray Data, Appl. Phys. Lett., 27:123.

Immirzi, A., 1978, Profile Fitting Refinements Using Generalized Coordinates, Acta Cryst., A34:S348.

Immirzi, A., 1980, Constrained Powder Profile Refinement Based on Generalized Coordinates. Application to X-ray Data of Isotactic Polypropylene, Acta Cryst., (in press).

Kennicott, P.R., 1963, A Modification of the Busing-Levy Least Squares Program to Account for Overlapped Data, Rep. No. 63-RL (3321G), General Electric Research Laboratories, Schenectady,NY.

Khattak, C.P., and Cox, D.E., 1977, Profile Analysis of X-ray Powder Diffractometer Data: Structural Refinement of $La_{0.75}Sr_{0.25}CrO_3$, J. Appl. Cryst., 10:404.

Mackie, P.E., and Young, R.A., 1975, Profile-Fitting-Structure-Refinement Applied with X-ray Powder Data, Acta Cryst., A31:S198.

Malmros, G. and Thomas, J.O., 1977, Least-squares Structure Refinement Based on Profile Analysis of Powder Film Intensity Data Measured on an Automatic Microdensitometer, J. Appl. Cryst., 10:7.

Mortier, W.J., and Costenoble, M.L., 1973, The Separation of Overlapping Peaks in X-ray Powder Patterns with the Use of an Experimental Profile, J. Appl. Cryst., 6:488.

Mortier, W.J., 1980, Structures from Powder Data: Data Sampling, Refinement and Accuracy, Symposium on Accuracy in Powder Diffraction, National Bureau of Standards, Gaithersburg, MD, 315.

Parrish, W., Huang, T.C., and Ayers, G.L., 1976, Profile Fitting: A Powerful Method of Computer X-ray Instrumentation and Analysis, Trans. Amer. Cryst. Assoc., 12:55.

Pham, C., Choisnet, J., and Raveau, B., 1975, Programme d'affinement par moindres carrés des structures à partir des données des diagrammes X de poudre, Bull. Cl. Sci. Acad. R. Belg., 75:473.

Rietveld, H.M., 1967, Line Profiles of Neutron Powder-Diffraction Peaks for Structure Refinement, Acta Cryst., 22:151.

Rietveld, H.M., 1969, A Profile Refinement Method for Nuclear and Magnetic Structures, J. Appl. Cryst., 2:65.

Sudarsanan, K., and Young, R.A., 1969, Significant Precision in Crystal Structural Details: Holly Springs Hydroxyapatite, Acta Cryst., B25:1534.

Sudarsanán, K., Mackie, P.E., and Young, R.A., 1972, Comparison of
    Synthetic and Mineral Fluorapatite, $Ca_5(PO_4)_3F$, in Crystallo-
    graphic Detail, Mat. Res.   Bull., 7:1331.
Taupin, D., 1973, Automatic Peak Determination in X-ray Powder
    Patterns, J. Appl. Cryst., 6:266.
Toraya, H., and Marumo, F., 1980, Application of Total Pattern-Fit-
    ting, to X-ray Powder Diffraction Data, Report of the
    Laboratory of Engineering Materials, Tokyo Institute of
    Technology,   Number 5, Nagatsuta, Yokohama, Japan, pp.55-64.
Werner, P.E., Salmoné, S., Malmros, G., and Thomas, J.O., 1979,
    Quantitative Analysis of Multicomponent Powders by Full-Pro-
    file Refinement of Guinier-Hägg X-ray Film Data, J. Appl.
    Cryst., (in press).
Wiles, D.B., and Young, R.A., 1981, New Computer Program for Rietveld
    Analysis of X-ray Powder Diffraction Patterns, J. Appl. Cryst.,
    (in press).
Wilson, A.J.C., Sudarsanan, K., and Young, R.A., 1977, The Structures
    of Some Cadmium 'Apatites' $Cd_5(MO_4)_3X$. II. The Distributions
    of the Halogen Atoms in $Cd_5(VO_4)_3I$, $Cd_5(PO_4)_3Br$, $Cd_5(AsO_4)_3Br$,
    $Cd_5(VO_4)_3Br$ and $Cd_5(PO_4)_3Cl$, Acta Cryst., B33:3142.
Young, R.A., Mackie, P.E., and Von Dreele, R.B., 1977, Application
    of the Pattern-Fitting-Structure-Refinement Method to X-ray
    Powder Diffractometer Patterns, J. Appl. Cryst., 10:262.
Young, R.A., 1980, Structural Analysis from X-ray Powder Diffraction
    Patterns with the Rietveld Method, Symposium on Accuracy in
    Powder Diffraction, National Bureau of Standards,
    Gaithersburg, MD, 567:143.
Young, R.A., and Mackie, P.E., 1980, Crystallography of Human
    Tooth Enamel: Initial Structure Refinement, Mat. Res. Bull.,
    15:17.

CRYSTALLINITY, CRYSTALLITE SIZE AND LATTICE PERFECTION

IN FIBROUS POLYMERS

D. J. Johnson

Textile Physics Laboratory
Department of Textile Industries
University of Leeds
Leeds LS2 9JT, UK.

INTRODUCTION

The X-ray diffraction patterns from fibrous polymers are generally characterised by a small number of relatively broad over-lapping peaks often overlaid with a diffuse halo. Single fibres give the same pattern as a parallel bundle of fibres, so that the pattern is equivalent to the rotation pattern of a single crystal. Despite the inherent difficulty of insufficient well-defined reflections, all commercially useful fibres have been indexed, starting with cellulose[1], through nylon 66[2] and polyester[3], to the more recent fibres such as Nomex[4] and Kevlar[5]. Only the structural complexities of the natural fibres of keratin (wool and hair) remain to be fully determined.

Many sophisticated computations have been carried out to refine unit-cell constants; unfortunately, much of this work is of doubtful value to those interested in fibre characterisation. Fibres are semicrystalline materials and their molecules are insufficiently well ordered for precision calculations of atomic positions. Indeed, there can be significant error involved in evaluating unit-cell constants from overlapping peaks, unless precise profile resolution is carried out[6].

Although single-crystal work on fibrous polymers is impossible by conventional X-ray diffraction methods, the development of high-resolution, ultrahigh-vacuum electron microscopes, has given a boost to the analysis of polymers by electron-diffraction and lattice-imaging methods. Perhaps one of the most far reaching discoveries made with transmission electron microscope (TEM) methods, has been that polymer single crystals are composed of chain-folded molecules[7]. In the fibrous form polymers such as the polyamides and the polyesters are considered to be composed of chain-folded regions with paracrystalline order, linked by tie molecules, and with totally disordered molecules in close association. It is this complexity which confounds those concerned with fibre characterisation by X-ray diffraction methods.

In general, fibre scientists in the X-ray analysis field are not primarily concerned with indexing, unless a new polymer is developed or a new polymorphic form of an established polymer is suspected, and are rarely concerned with identification, except for forensic or archaelogical reasons. Their major purpose is that of character-isation by means of such parameters as crystallinity, crystallite size, lattice order, and preferred orientation, parameters which may be affected by changes in processing conditions, or by the local environment. Preferred orientation can be measured by several methods and, for high-modulus fibres in particular, correlates well with Young's modulus. It is a parameter which deserves a detailed study in its own right and this paper will be concerned only with a crit-ical assessment of methods for the measurement of crystallinity, crystallite size, and lattice order.

CRYSTALLINITY

Crystallinity has always been a most elusive parameter to measure quantitatively, and until methods using automatic scanning diffractometers or microdensitometers are perfected, a rapid quanti-tative assessment of the full diffraction pattern will be impossible. Although the human eye can do a remarkable semi-quantitative job in this respect, and, despite the availability of increasingly sophis-ticated computational facilities, we are still far from our goal. It is unfortunate that, although fibre scientists have laboured over the years to measure crystallinity, and although the number of papers on the subject is legion, significant advance has been painfully slow. Some of the work in this area has been detailed in our own publica-tions[8,9,10]; here it will be convenient to discuss two categories of methods for assessing crystallinity, the relative and the absolute.

Relative Methods

All methods of crystallinity measurement should be based on Vainshtein's law of conservation of intensity, which states that total scatter in reciprocal space from equivalent regions with per-fect lattice order (crystalline), imperfect lattice order (para-crystalline), and complete disorder (amorphous), is identical. For a randomized system this law can be expressed as

$$\int_0^\infty I(s)\ dv_s = 4\pi \int_0^\infty I(s)s^2 ds = 4\pi \int_0^\infty \overline{f^2} s^2 ds$$

$$\overline{f^2} = \sum N_i f_i / \sum N_i$$

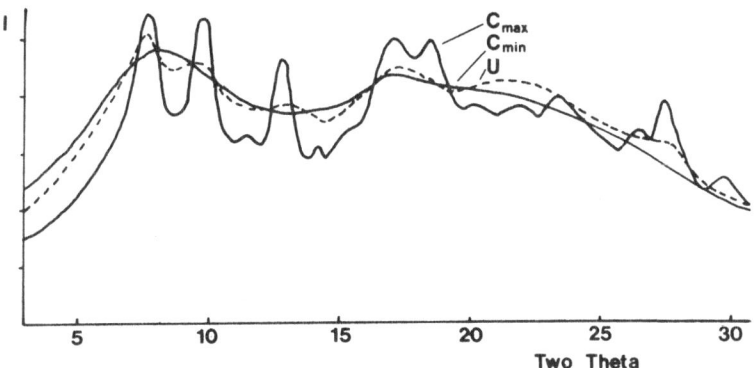

Fig. 1.   Normalised traces of cellulose triacetate.
          C_{max} 100% standard; C_{min} 0% standard;
          U unknown specimen;   I - arbitrary units.

where f_i is the atomic scattering factor of an atom of type i, N_i
is the number of atoms of type i, and s is the reciprocal lattice
vector s = $2\sin\theta/\lambda$.
     An excellent ranking method which makes use of this law, is
the Correlation Crystallinity Index (CCI) method, first introduced
by Wakelin et al.[11].   Standard specimens of a material are used to
produce diffraction traces ranked as 100% (C_{max}) and 0% (C_{min});
specimens of unknown crystallinity (U) are randomized, the traces
normalised, and the ordinate differences U - C_{min} and C_{max} - C_{min}
correlated to find a regression line whose slope is the CCI.   Fig. 1
shows how normalised traces from cellulose triacetate specimens
have been established[12].   The method is executed computationally and
is very straightforward once appropriate standards have been chosen.
Thermoplastic fibres can be induced to provide a wide range of
crystallinity, but the provision of standard samples for high modulus
fibres is not so simple.

Absolute Methods

     The classical method of Hermans and Weidinger[13] gives a measure
of so-called "absolute" crystallinity in terms of the ratio of
integrated intensity under the peaks to the integrated intensity
under the complete trace, Fig. 2.   This can be expressed as

$$x_{cr} = \frac{\int_0^\infty I_{cr}(2\theta)d(2\theta)}{\int_0^\infty I_{tot}(2\theta)d(2\theta)} \qquad \ldots\ldots \text{(i)}$$

The arbitrary separation of the scatter under the peaks, supposedly

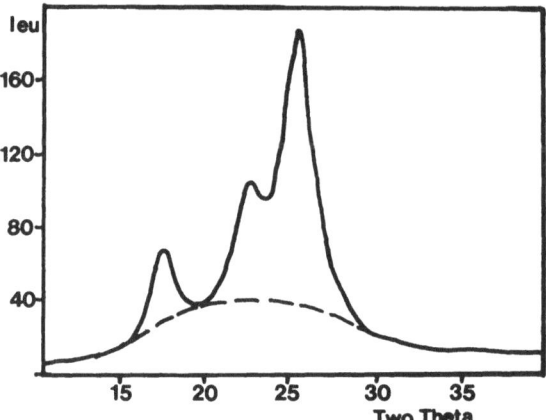

Fig. 2.   Arbitrary separation of scatter under the crystalline
          peaks of a polyester specimen.

due to crystalline regions, from the scatter in the background due
to the non-crystalline regions, is a major difficulty. Very many
methods for doing this have been suggested, all of which neglect
peak overlap, and thus give different results with ostensibly the
same material.

Our own "absolute" peak-area method[10,14] employs a peak-
resolution procedure, resolving peaks in terms of best-fit profiles,
and consigning the remainder of the scatter to the background. Each
peak is considered to have a profile of the form

$$f_t G_t \ + \ (1 - f_t C_t)$$

where G_t is the Gaussian function

$$A_t \ \exp \left\{ -\ln 2 \left[\frac{2(X - P_t)}{W_t} \right]^2 \right\}$$

and C_t is the Cauchy function

$$A_t \ / \ \left\{ 1 \ + \left[2(X - P_t) \ / \ W_t \right]^2 \right\}$$

The peaks are defined by the parameters A_t the peak height, W_t the
width of the peak at half height and P_t the position; f_t is the profile function parameter which effectively describes the tail region
of the profile, see Fig. 3, and can sensibly vary from about −0.5
to +1.0. The background generally has the polynomial form

$$a \ + \ bX \ + \ cX^2 \ + \ dX^3$$

where X may be in either the two theta or the s scale. It is

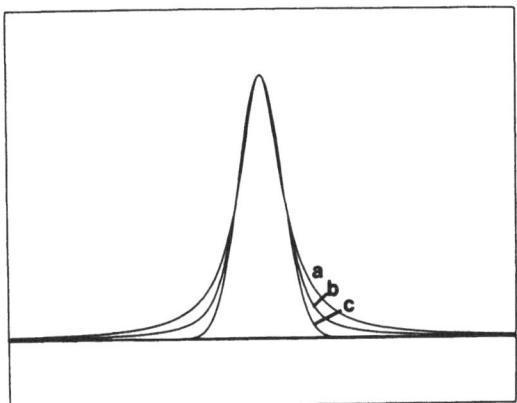

Fig. 3.   Schematic profiles with function parameters   a - 0,
           b - 0.5, c - 1.0.

possible to include functions for intermediate crystallinity or broad
paracrystalline peaks. Asymmetry in the profile can be allowed by
including both left and right hand W and f parameters. Several
examples of the application of this profile resolution method have
been described at length in a recent paper[6].

An alternative method for measuring "absolute" crystallinity
was proposed by Ruland[15], in which a correction factor based on a
disorder parameter is incorporated into the standard peak area
equation (i). Ruland's measurement of the crystalline fraction can
be expressed as

$$X_{cr} = \frac{\int_{s_o}^{s_p} s^2 I_{cr}(s)ds}{\int_{s_o}^{s_p} s^2 I_{tot}(s)ds} \qquad K \qquad \ldots\ldots (ii)$$

where the correction factor K is given by

$$\frac{\int_{s_o}^{s_p} s^2 \overline{f^2} ds}{\int_{s_o}^{s_p} s^2 \overline{f^2}\, D\, ds}$$

D is the isotropic disorder represented by $D = \exp(-ks^2)$ with k
incorporating lattice distortions of the first and second kind and
thermal distortion. Because x_{cr} must be a constant, the correction
factor K is evaluated for different intervals of s by changing the

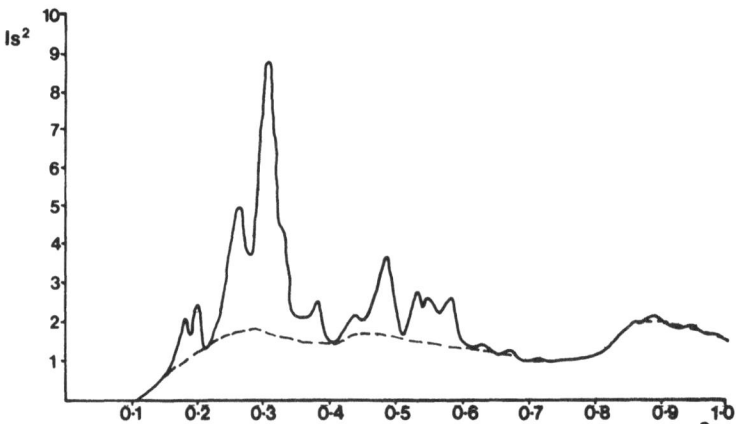

Fig. 4. Schematic representation of separation of crystalline and
         non-crystalline scatter by Ruland's method.

upper limit s_p and the parameter k until appropriate values are found
to make x_{cr} a constant. Fig. 4 shows a typical separation of crystalline and non-crystalline scatter; this arbitrary procedure is the
most difficult problem with this method, since it must affect both k
and x_{cr}. A mathematical solution, using a standard non-crystalline
sample as in the CCI method, has been applied by Sotton et al.[16,17]
who have made extensive use of Ruland's method in their investigations of fibrous polymers.

CRYSTALLITE SIZE

     Crystallite size evaluation is usually carried out by means of
the peak-broadening formula

$$L_{hkl} \;=\; K/ds \;=\; \lambda K/\cos\theta d(2\theta) \qquad \ldots\ldots \text{(iii)}$$

where L_{hkl} is a weight-average measurement of size normal to the
planes (hkl), and ds or d(2θ) is the breadth (integral breadth or
half width) of the peak hkl. K is the Scherrer parameter, often
taken as 1 for integral breadth and 0.89 for half width, but best
considered as

$$K \;=\; \frac{\text{true crystallite size}}{\text{apparent crystallite size}}$$

With K = 1, an apparent crystallite size is measured, which can be
converted to a true size if the actual value of K is known. Recent
investigations of optical transforms from computer-drawn lattice
fringes have shown that K is a function of both crystallite size and
lattice distortion[18].

     Since the half width or integral breadth must be found before
crystallite size can be evaluated, and because there is considerable
overlapping of peaks in diffraction patterns from fibrous polymers,

peak-resolution methods are required to obtain reliable results.
Indeed, even with peak resolution, inappropriate choice of the peak
profile can lead to errors of up to 100% in the crystallite size
measurement[9].

Following peak resolution, it is necessary to correct the profile for instrumental broadening. This is carried out either by a
Stokes deconvolution or the Jones approximations. With peaks of
profile function parameter f, the Jones correction becomes

$$\beta = f(B^2 - b^2)^{\frac{1}{2}} + (1 - f)(B - b)$$

where β is the corrected breadth, B the observed breadth, and b the
instrumental breadth. In most cases the Stokes method is preferred
since it also provides Fourier coefficients which can be used in a
Warren and Averbach type analysis for measurement of a number average
crystallite size together with a measure of lattice distortion.

LATTICE PERFECTION

The evaluation of lattice perfection or, as is most usual,
lattice disorder, is inextricably linked with most methods for the
evaluation of crystallite size, although Ruland's method is an
exception. The effect of disorder in a lattice is to create another
line-broadening factor so that a deconvolution method for separating
size and distortion effects is necessary. If Δs_s defines the size-
broadening component as given in (iii), the lattice disorder component (actually a lattice strain) is often defined as

$$\Delta s_d = 2es = 4e \sin\theta/\lambda \qquad \ldots\ldots \text{(iv)}$$

where $e = \Delta d/d$ can be considered as a lattice strain giving a maximum
distortion Δd in a lattice of spacing d. A Jones type relation for
the assumption that both the line-broadening components have Gaussian
profiles, gives

$$(\Delta s_1)^2 = (K/L_{001})^2 + 4e^2 s_o^2 l^2 \qquad \ldots\ldots \text{(v)}$$

and for the assumption that both components have Cauchy profiles,

$$\Delta s_1 = K/L_{001} + 2es_o l \qquad \ldots\ldots \text{(vi)}$$

for 001 reflections with $s_o = 1/d(001)$. For mixed line broadening
profiles the following expression has been suggested

$$\Delta s_1 = K/L_{001} + 2\pi^2\sigma^2 s_o^2 l^2 \qquad \ldots\ldots \text{(vii)}$$

where σ, the root mean square lattice distortion, is usually taken
as 0.8e.

Another relationship, based on Hosemann's theory of paracrystallinity
is

$$(\Delta s_1)^2 \;=\; (K/L_{001})^2 \;=\; \pi^4 \sigma^4 s_o^2 l^4 \qquad \dots \quad (viii)$$

All these methods require the presence of several orders of a
reflection so that successive breadths Δs_1 can be obtained and
appropriate straight-line plots made from which crystallite size
and distortion can be evaluated[6,10].

An alternative method due to Warren and Averbach uses the
Fourier coefficients of successive orders of a reflection in terms of
the relationship

$$\ln F(t) \;=\; \ln F_s(t) - 2\pi^2 s_o^2 l^2 t^2 \sigma^2 \qquad \dots \quad (ix)$$

where t is a measure of length normal to the diffraction planes.
The slope of $\ln F(t)$ versus $s_o^2 l^2$ is

$$\frac{dF_s(t)}{dt}_{(t=0)} \;=\; \frac{1}{\bar{N}} \qquad \dots \quad (x)$$

where $\bar{N}$ is the number average crystallite size. In practice, this
procedure is very difficult to follow with polymers, even when enough
orders of a reflection are available[9,19].

RESULTS

PPT Poly(p-phenylene terephthalamide) fibres

A typical equatorial scan of a PPT sample after correction and
normalisation to electron units by our standard method[6] followed by
resolution of the overlapping 110 and 200 peaks is shown in Fig. 5,
with the resolved peak parameters given in Table 1. The total peak-
area crystallinity found by this method, i.e. within the two theta
limits $10°$ to $50°$ is 64.2%. The background non-crystalline scatter
is 35.8% with the polynomial function $0.534 + 2.07X - 0.087X^2 + 0.009X^3$. This value of crystallinity has no meaning except in
relative terms; it is taken from a series in which specimens were

Table 1.   Resolved peak parameters for PPT specimen.

Parameter		110	200
f		0.1	0.2
A		94.3	122.3
W		1.90	1.86
P		19.77	22.21
Area		28.8%	35.4%
SD	1.24	Total area	64.2%

f - profile function parameter, A - peak height (eu),
W - peak width (two theta), P - peak position (two theta).

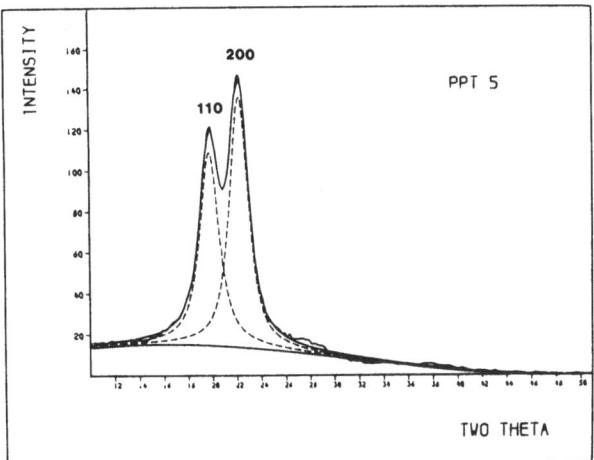

Fig. 5. Equatorial trace of a PPT specimen; corrected, normalised, and resolved into two peaks and a background.

being assessed for the effect of spin-stretch factor on crystallinity, crystallite size and tensile properties.

The resolved peak profiles were corrected using Stokes method after the instrumental broadening was found using a hexamethylene tetramine specimen[19]. The results are given in Table 2, apparent crystallite sizes being calculated according to equation (iii) with K = 1.

PET Poly(ethylene terephthalate) fibres

The resolution of a polyester specimen PET 8 is illustrated in Figs. 6a and b, with the resolved peak parameters given in Table 3. Clearly the fit with the crystalline peaks 010, 110 and 100, Fig. 6a, is unsatisfactory, but with the addition of a paracrystalline (non indexable) peak Fig. 6b, much better resolution is achieved. This is also indicated by the lower value of the root mean square deviation. Apparent crystallite sizes after a Stokes deconvolution are included.

Table 2. Apparent crystallite size in PPT specimen.

	110		200	
	d(2θ)	L nm	d(2θ)	L nm
Uncorrected width	1.90	4.7	1.86	4.8
Corrected width	1.73	5.2	1.70	5.3
Corrected integral breadth	2.49	3.6	2.37	3.8

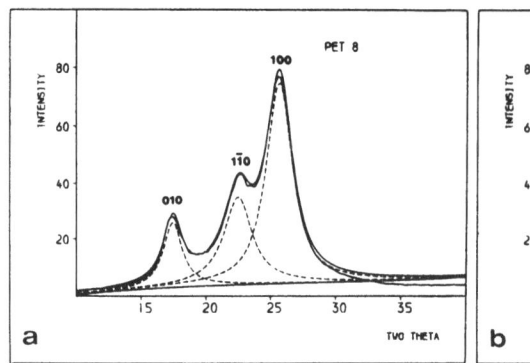 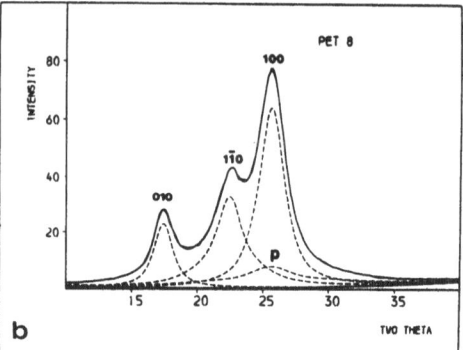

Fig. 6.   Equatorial trace of PET specimens, corrected, normalised, and resolved, (a) into three crystalline peaks, (b) into three crystalline peaks and one paracrystalline peak.

Table 3.   Resolution of specimen PET 8 into three crystalline peaks, and three crystalline peaks plus a para-crystalline peak

Peak	f	A	P	W	L_{hko}	% Area
0$\overline{1}$0	0.0	22.7	17.46	1.75	5.2	10.7
1$\overline{1}$0	0.0	31.0	22.44	2.63	3.6	21.8
100	0.1	70.4	25.65	2.48	3.8	45.0
SD	1.80				Total Cryst.	77.5
0$\overline{1}$0	0.1	22.6	17.43	1.77	5.2	11.3
1$\overline{1}$0	-0.3	32.3	22.46	2.64	3.6	27.2
100	0.2	63.5	25.64	2.40	3.8	42.3
Para	-0.9	7.44	25.59	5.49	-	13.6
SD	0.36				Total Cryst.	80.8

f - profile function parameter,   A - peak height (eu),
W - peak width (two theta), P - peak position (two theta),
L - apparent crystallite size (nm),   SD - root mean square standard deviation.

Size and Distortion

The emphasis in our work on peak resolution has been on crystallite size rather than crystallinity measurement, with lattice distortion measured whenever possible, which was not the case with the two specimens above because of the lack of higher order equatorial reflections. We may note that Sotton et al.[16] applied Ruland's method to the PET 8 specimen and measured the distortion as 2.5%.

Comparison of the various methods used for estimating crystallite size and lattice distortion can only be made if a true crystallite size is known. Two methods have been applied to utilise

Table 4.   Size and distortion parameters for simulated and
            real 001 profiles

Method	Equation	ACS sim nm	ACS real nm	K sim	K real	Distortion sim %	Distortion real %
Gaussian	(v)	6.2	4.9	0.6	0.9	1.6	1.5
Cauchy	(vi)	10.5	6.3	0.4	0.7	3.6	1.1
Mixed	(vii)	3.8	4.4	1.0	1.0	7.0	3.6
Hosemann	(viii)	4.4	4.2	0.9	1.1	5.5	2.4
W and A	(xi)	4.7	3.8	0.9	1.2	2.2	2.5
True Value		4.0	4.5*	1.0	1.0	5.0	-

* EM measurement   ACS - apparent crystallite size
                         K - Scherrer parameter

true size: (i) by means of a computer simulated distribution of
size, (ii) by direct measurement on lattice-fringe images[20]. Lattice
distortion can only be simulated, it cannot be evaluated absolutely.
Computed profiles were used to simulate the three orders of 001 from
a high-modulus carbon fibre, real examples of which were resolved
from overlapping hk0 peaks and corrected for instrumental broadening.
The results of analyses by five methods of the simulated and the real
profiles are given in Table 4. It must be noted that correlations
according to the specified equations were poor and there is considerable error in all evaluations. The mixed profile method gave the
most reliable crystallite size measurement, whereas the Hosemann
method gives the best estimate of distortion.

DISCUSSION

     All methods for measuring crystallinity have inherent limita-
tions and are best regarded as giving relative rather than absolute
values. Very few comparisons of crystallinity evaluation have been
made. Sotton et al.[16] obtained a good correlation between the CCI
method and Ruland's method for several samples of polyester. For the
specimen PET 8 they found $x_{cr} = 0.24$ by Ruland's method, 0.36 by the
CCI method, and 0.47 by a peak-area method without resolution. We
have obtained a value of 0.80 by the peak-area method with
resolution. We must conclude that any of these methods is useful as
a relative method, but the correlation crystallinity method is the
most straightforward in application, if standard samples can be
obtained for C_{max} and C_{min}.

     Apparent crystallite size can be measured after peak resolution,
but the value obtained will depend on the profile function parameter,
the size itself, and the lattice distortion. The true crystallite

size will depend upon the appropriate Scherrer parameter (K).
Separation of size and distortion broadening, where successive
orders are available, is possible by several methods, none of which
is considered ideal. Tests with real and simulated lattices suggest
that the mixed profile method gives the best results for crystallite
size, but overestimates lattice distortion. The method based on
Hosemann's theory of paracrystallinity gives the best result for
distortion but the measurement of lattice perfection is generally
unsatisfactory.

ACKNOWLEDGEMENTS

     The author would like to thank S.C. Bennett, P.J. Frost,
A.M. Hindeleh, and P.E. Montague for their assistance, and M. Sotton
of L'Institut Textile de France for provision of the polyester
specimen.

REFERENCES

1. K.H. Meyer and H. Mark, Ber.Dtsch.Chem. Ges 61:1939(1928).
2. C.W. Bunn and E.V. Garner, Proc.Roy.Soc.A, 189:39(1947).
3. R.de P. Daubeny, C.W. Bunn and C.J. Brown, Proc.Roy.Soc.A,
        266:531(1954).
4. H. Kakida, Y. Chatini and H. Tadokoro, J.Polymer Sci., Polymer
        Phys.Ed., 14:427(1976).
5. M.G. Northolt, Eur.Polymer J., 10:799(1974).
6. D.J. Johnson, in: "Diffraction Methods for Structural Deter-
        mination of Fibrous Polymers", ACS Symposium Series, to
        be published.
7. A. Keller, Rep.Prog.Phys., 31:623(1968).
8. A.M. Hindeleh and D.J. Johnson, Polymer, 13:27(1972).
9. A.M. Hindeleh and D.J. Johnson, Polymer, 13:423(1972).
10. A.M. Hindeleh and D.J. Johnson, Polymer, 19:27(1978).
11. J.H. Wakelin, H.S. Virgin and E. Crystal, J.Appl.Phys.,
        30:1654(1959).
12. A.M. Hindeleh and D.J. Johnson, Polymer, 11:666(1970).
13. P.H. Hermans and A. Weidinger, Text.Res.J., 31:558(1961).
14. A.M. Hindeleh and D.J. Johnson, J.Phys.D., 4:259(1971).
15. W. Ruland, Acta Cryst., 14:1180(1961).
16. M. Sotton, A. Arniaud and C. Rabourdin, J.Appl.Polymer Sci.,
        22:2585(1978).
17. M. Sotton, in: "Diffraction Methods for Structural Determina-
        tion of Fibrous Polymers", ACS Symposium Series, to be
        published.
18. A.M. Hindeleh and D.J. Johnson, Polymer, to be published.
19. D.R. Buchanan and R.L. Miller, J.Appl.Phys., 37:4003(1966).
20. S.C. Bennett, D.J. Johnson and R. Murray, Carbon, 14:117(1976).

QUANTITATIVE ANALYSIS OF DUST SAMPLES FROM OCCUPATIONAL

ENVIRONMENTS USING COMPUTER-AUTOMATED X-RAY DIFFRACTION

M. T. Abell, D. D. Dollberg, J. V. Crable

National Institute for Occupational Safety and Health
4676 Columbia Parkway
Cincinnati, Ohio 45226

The goal of the occupational health professions is "... to assure so far as possible every working man and woman in the Nation safe and healthful working conditions and to preserve our human resources" (Occupational Safety and Health Act, PL91-596). When the health of workers is threatened by chemical agents in the air, occupational health chemistry has a role to play. Analyses of workplace air are performed by the thousands each year by industry and by state and federal agencies. In the federal government, these analyses, as well as occupational health work in general, are performed by three agencies: the National Institute for Occupational Safety and Health (NIOSH, DHHS), which does occupational health research; the Occupational Safety and Health Administration (OSHA, DOL), which enforces safety and health standards in the general workplace; and the Mine Safety and Health Administration (MSHA, DOL), which enforces similar standards in mining operations.

The quantitative analysis of substances collected from air is important not only for the original determination of the toxicity of a substance (toxicology, epidemiology), but also for measuring the protection afforded workers (control technology, enforcement). Methods of chemical analysis are available for approximately 500 substances of occupational health interest (1) and several of these utilize X-ray powder diffraction (XRD). The small but growing list of XRD analytes includes free silica dust (usually quartz or cristobalite) which is a widespread occupational hazard, as well as other dusts for which quantitation of chemical compounds is required. Such compounds include asbestos (2), boron carbide, titanium diboride, and the oxides of lead, mercury, zirconium,

and zinc. The need to distinquish these compounds from others
containing the same elements makes XRD increasingly more
important in occupational health analysès. The use of XRD for
the analysis of dusts containing compounds of occupational
health interest has been reviewed in several general papers
(3,4).

XRD ANALYSIS PROCEDURE

It is useful to briefly summarize the sampling and analysis
requirements for occupational assessment of airborne dusts by
X-ray diffraction in order to illustrate the need for computer
automation. To assess occupational exposures to dust a
"breathing zone, respirable air sample" must be collected.
This is accomplished by outfitting a worker with a
battery-powered pump which pulls air through a sampling train
which is usually attached at the collar. The sampling train
consists of a cassette containing a membrane filter, preceeded
in the airstream by a cyclone, a device which passes only those
particles whose size would permit them to penetrate to the
lungs. The cyclone does not pass particles which are larger
than 10 micrometers in diameter, a favorable condition for XRD
measurement. Sample collection time is usually a full eight
hours, but the low flow rate required for the proper operation
of the cyclone keeps the volume of sampled air down to
approximately 0.75 m^3. In the case of quartz, a sample of
this size typically produces a filter loading less than 150 μg
of quartz in a milligram or two of dust and the amounts are
often much lower. Quartz levels as low as 25 μg need to be
quantitated since the permissible exposure limit proposed by
NIOSH is 50 μg/m^3.

In the laboratory, the membrane filter is destroyed by
ashing, and the remaining dust is dispersed in alcohol with the
help of ultrasonic agitation. The suspended dust is then
redeposited as a uniform layer on a silver membrane filter
which is put onto an XRD holder for analysis. Aliquots of
standard suspensions are deposited on silver filters for
standards. Although other filters may be used, silver filters
have the advantage of low background throughout the 2θ scan
region. In addition, the silver peak intensity can be used for
matrix absorption corrections (substrate standard method).

The XRD analysis of personal air samples is a
time-consuming process because of the small amount of analyte.
Measurement of integrated peak intensities and background
intensities requires about twenty-five minutes for a quartz
profile if measurements are to be made at the 20 microgram
level. Similarly, measurement of the secondary peak will
double this time and measurement of the integrated intensity of

the silver peak adds several minutes, so that an analysis could
require about one hour of instrument time. The silver peak is
measured because it can be used to correct for absorption in
those samples which are not infinitely thin. One of the
procedures used when absorption is suspected is to turn the
silver filter over and measure the silver peak intensity from
the clean surface. If this intensity is more than that found
on the dusty side, absorption is occurring in the dust and a
correction factor must be applied to the weight of analyte
calculated. The absorption correction is $(-R \ln T)/(1-T^R)$,
where T is the ratio of the silver peak intensities (dust side
to clean side) and R is the ratio of the sines of the θ angles
(silver angle to analyte angle) (5,6).

NEED FOR AUTOMATION

The above procedure can benefit from computer automation of
the X-ray diffractometer in several ways, and these benefits
are the subject of this paper. The most immediate benefit of
any automated instrument is that it frees an analyst from
routine tasks. This is important in occupational health
chemistry because of the large number of samples. Last year
OSHA analyzed 6000 silica samples, MSHA analyzed 20,000 (peak
heights were used for rapid checks), and NIOSH analyzed or had
analyzed 2400 on a routine basis and about 2000 other samples
for research.

NIOSH and MSHA anticipated this work load in 1972 and each
ordered automated equipment consisting of two diffractometers
and a spectrometer running under the control of one computer.
These hybrid systems were advanced in concept, but because of
reliability problems, were later replaced by instruments then
available from a major X-ray instrument manufacturer. At this
time, OSHA has six automated diffractometers, MSHA and NIOSH
each have two, and a number of other laboratories running
occupational health samples have one or more. The use of one
of the newer units in an industrial laboratory has been
reported by Bumsted (7).

ADVANTAGES OF COMPUTER AUTOMATION

The advantages of the computer-controlled X-ray
diffractometer may be put into four categories: economy or
efficiency, improved data quality, flexibility through
reprogramming, and the ability to extend the automation of the
analysis into the decision and computation spheres. The first
two categories emphasize the automatic nature of these instruments which can continually cycle through samples while the
last two depend on the computer basis of this automation. The

categories are arbitrary and overlap but do serve to point out
the important areas where improvements have and will be made.

Economy

   The savings in time realized with an automated
diffractometer analyzing occupational health samples can be
appreciated by examining the usual laboratory procedure. The
time required to prepare a sample is little more than five
minutes apiece in batches of 10 or more. The instrumental
time, as explained above, can be an hour. An analyst using a
manual instrument would analyze only eight samples in a day
with sufficient time between steps to prepare samples and
analyze data but little time for anything else of consequence.
With an automated instrument, the analyst could spend several
hours preparing samples while the instrument analyzes 24
samples in a day. Because the automated instrument can run 24
hours a day unattended while the manual instrument runs only
while attended, it is apparent that the automated instrument
has a 3:1 advantage. In money terms, the cost of operating
each type of instrument is about the same since the higher
purchase and maintenance cost of the automated instrument is
balanced by the lower cost of personnel (about one-fourth).
Thus, the cost per sample for the manual instrument is several
times that of the automated instrument if sufficient samples
are available to keep the automated instrument busy.

Quality of Data

   In the preceeding comparison of manual and automatic
diffractometers, it was assumed that the operator would
manipulate the manual instrument to match what is done in the
automated instrument. This involves collecting total counts
while step scanning the peak and subtracting background counts
collected on each side of the peak for half the scan time.
This is the proper procedure, but it is not likely to be
followed in a manual operation. The quality of the data
suffers if the procedure is not repeated exactly for each
sample and suffers more if peak heights are used. The
automatic instrument not only follows the procedure exactly but
also is usually programmed to perform a calibration check of
the angle and intensity after every sample or two on a
reference specimen. In the example of input parameters given
in Figure 1 this information is requested in question 9. As a
result, the 2 θ angle of the reference specimen peak is checked
and the intensity of the peak is stored. Since the intensity
of this peak is quite constant, it can be used to normalize all
the data (question 7) thus compensating for slow drifts in

```
(0)  RUN ID: TRIDYMITE (SAMPLES 1 to 11)
     AND QUARTZ (SAMPLES 12 TO 16)
(1)  QUANTITATIVE MODE   FIRST SAMPLE=1
     LAST SAMPLE= 11
(2)  PRESENT TWO-THETA =    .00?  NO
     PRESENT TWO-THETA = 8.00
     SUMMARY DATA MODE
(4)  TWO-THETA LIMITS
     LOWER-LIMIT   UPPER-LIMIT   TIME
       1= 20.0        21.0       10.0
       2= 37.03       39.03       0.5
       3=
(5)  TWO-THETA INCREMENT= .02
(7)  DATA NORMALIZATION? YES EXTERNAL? YES
(8)  CHART FS= 5000
     DATA SCALE FACTOR= 10000
(9)  TWO-THETA CALIBRATION? YES
     PEAK TWO-THETA= 26.66
     AT SAMPLE INTERVALS= 2
```

Fig 1.   Parameter input sequence of automated diffractometer
         with a choice of step time for each phase

X-ray tube output and other variables. The two theta check and
normalization step improve the quality of the data.

The advantages of economy and improved data quality are
related in one sense; they are the result of removing the human
operator from a repetitive routine which is prone to error, and
this is what is usually expected from an automated instrument.
In addition to the regular workload of filter samples from the
workplace, NIOSH has performed several studies of the XRD
method which serve as examples of work that would be nearly
impossible without a computer-automated diffractometer.

A first example concerns the precision of the XRD method
for silica. The method is less precise than many other kinds
of analyses and this lack of precision explains why research
projects on the method require that large numbers of samples be
run. In fact, actually determining the precision for a given
amount of analyte requires a large number of samples, certainly
more than ten. Five hundred standards were prepared and
analyzed (integrated, normalized peak areas) to produce the
data graphed in Figure 2. The relation shown there between
precision and amount of analyte (quartz) is not surprising but
until this large amount of data was collected, the precision at
these levels could not be predicted with much certainty.

Another study that involved a large amount of data was done
in an attempt to find out how best to do an absorption
correction using silver filters (8). For an absorption
correction, two measurements of silver integrated intensity
must be made, one with and one without the dust in place.
There are three possible procedures for making these
measurements and they are illustrated in Figure 3. In the

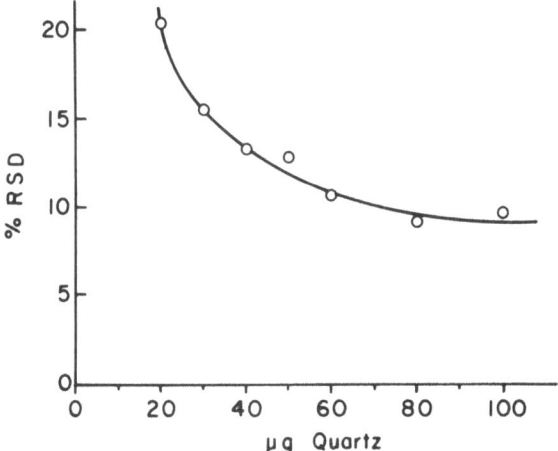

Fig. 2.   Precision of XRD quartz analyses as a function of mass
          of quartz for a very large number of standards
          (Min–U–Sil 10)

averaging procedure, a number of clean filters are used to
establish the "without dust" intensity (I°_{Ag}), and the
intensity though the dust (I_{Ag}) is measured for each sample
filter. In the FR (front-reverse) procedure, the deposit and
reverse surfaces of the one sample filter are used to obtain
these intensities. And in the FF (front-front) procedure the
measurements are made on the same surface of the sample filter
before and after the sample dust is deposited. After hundreds

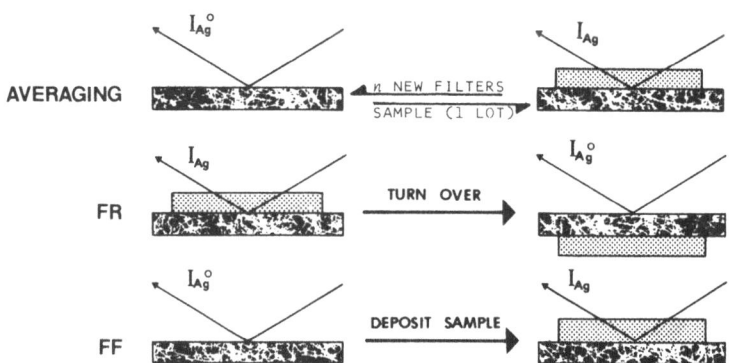

Fig. 3.   Three possible procedures for measuring silver
          intensity both with (I_{Ag}) and without (I°_{Ag}) the
          sample dust in place.

of measurements on 0.45 μm pore size silver filters, it was
found that although the FR procedure exhibited a smaller
variation of 3% between sides, the averaging procedure was also
acceptable, giving about 4% variation when using the average of
6 or more filters from a lot. Convenience, then, may dictate
the choice of procedures; FF, the least convenient method, was
not as thoroughly investigated, but appeared comparable to the
FR procedure.

Flexibility

Automated equipment is very useful for performing routine
tasks, but if the automation is based in mechanical systems
with timers and cams and so on, the routine is rigid. Computer
automation can be tailored through programming to an individual
user's needs. This should be considered when purchasing an
instrument because some instrument suppliers, unfortunately,
have policies regarding computer software that make changes
difficult if not impossible. Ideally, the provided software
should meet most of the users' needs and also be easy for user
to change.

An example of the need for this flexibility is provided by
the analytical method presented above in which very small
analyte peaks are to be measured at the same time as large
silver peaks. Because most chemists would not face this
situation, the instrument supplier allowed only one choice of
step time. Scanning as slowly on silver peaks as on analyte
peaks would overload the scaler. So when the instrument was
reprogrammed (9) the possiblity of different step times for
each phase was provided as can be seen in question 4,
Figure 1. Question 6 originally allowed input of one step time
and that question was eliminated. Typically, as shown in the
figure, step times of 10 and 0.5 seconds are used for analyte
and silver step times, respectively.

Figure 4 is an example of instrument results in
quantitative mode and shows how the net counts (third column)
for silver (phase 2) is greater than that for the analyte
(phase 1) in spite of the shorter step time. The figure also
shows the 2θ calibration check data in the two columns on the
right (unlabeled). Perusal of these columns is a good check on
instrument performance and the data in the net counts (third)
column has been normalized relative to the calibration
intensity.

Finally, the figure also shows a short block of input
between two sets of results. The input parameters for the new
analysis were supplied automatically from a prerecorded tape
cassette. By programming the computer to turn on the tape

```
TRIDYMITE (SAMPLES 1 to 11) AND QUARTZ (SAMPLES 12 TO 16) JAN 31 78
TWO-THETA CALIBRATION
     PEAK TWO-THETA- 26.66    PEAK INTENSITY- 27231
TRIDYMITE (SAMPLES 1 TO 11) AND QUARTZ (SAMPLES 12 TO 16) JAN 31 78
   SAMPLE    PHASE    NET-COUNTS    BACKGROUND    COUNT ERROR(%)
      1         1         3601         46526          3.25       27231   26.66
      1         2        61005         11931           .22       27231   26.66
      2         1         3499         45367          3.30       27231   26.66
      2         2        79693         11329           .22       27231   26.66
      3         1         2857         42575          3.89       27383   26.65
      3         2        77617         10785           .22       27383   26.65
      .         .            .            .             .           .       .
      .         .            .            .             .           .       .
      .         .            .            .             .           .       .
     11         1         5766         45238          2.05       27484   26.66
     11         2        72101         10926           .23       27484   26.66
   ENTER COMMAND!
(1) QUANTITATIVE MODE   FIRST SAMPLE- 12   LAST SAMPLE- 16
   ENTER COMMAND!
(4) TWO-THETA LIMITS
     LOWER-LIMIT   UPPER-LIMIT   TIME
       1- 25.8        27.1       10.0
       2-
(5) TWO-THETA INCREMENT- 0.02
TRIDYMITE (SAMPLES 1 TO 11) AND QUARTZ (SAMPLES 12 TO 16) JAN 31 78
   SAMPLE    PHASE    NET-COUNTS    BACKGROUND    COUNT ERROR(%)
     12         1        17858         87255           .96       27290   26.66
     13         1        24751         65092           .65       27290   26.66
     14         1        16374         59933           .89       27620   26.66
     15         1        13028         62254          1.11       27620   26.66
     16         1         7903         61798          1.75       27425   26.66
```

Fig. 4. X-ray diffraction results for two types of sample with
         a block of automatic parameter input between.

transport for a block of input, it is possible to leave the
instrument with two or more types of samples in the sample
magazine and each will be analyzed using the correct conditions
as stored on the tape. This example of flexibility is also an
example of efficiency, since a second set of samples can be
analyzed without waiting for the operator.

    Another feature incorporated along with the programming
changes already mentioned was the selective recording of the
output data on magnetic tape cassette. This stored data can be
conveniently retrieved for additional work as explained in the
next section.

Extending the Automation

    Because large amounts of data can be fairly easily
transferred from one computer to another, one need not be
limited to the power of the instrument's computer. In the
examples being given here, the instrument's computer is very
limited but data temporarily transferred to magnetic tape
cassette can be sent via phone lines to a large computer. At
NIOSH, this is done when preparing calibration curve data
(Figure 5a) and also when computing the quantity of analyte on
sample filters including absorption corrections (Figure 5b).
The program can also format final analytical reports and is an

MINUSIL-10 CALIBRATION CURVE 20-250 UG/FILTER

CALCULATION OF STANDARD DEVIATION

CONC.	AV. CNT.	SIGMA	RSD
20.00	4503.40	986.24	21.90
50.00	11235.00	1751.56	15.59
100.00	23445.00	4329.66	18.47
150.00	33563.40	3327.57	9.91
200.00	43701.40	3448.77	7.85
250.00	58796.40	1431.35	2.43

LINEAR LEAST SQUARES ANALYSIS

SLOPE = 226.10 STD. DEV.= 7.31
Y INTERCEPT= -143.16 STD. DEV.= 77.53
LINEAR CORRELATION COEFFICIENT= 0.9973

X(I)	Y(I)	YCALC(I)
20.00	4503.398	4377.
50.00	11235.000	11163.
100.00	23445.000	22469.
150.00	33563.398	33773.
200.00	43701.398	45075.
250.00	58796.398	56383.

a

NIOSH X-RAY DIFFRACTION ANALYSIS SYSTEM

P.A.T SAMPLES ROUND 46 1/23/78
J. PALASSIS

PHASE 1

LAB. NO	SAMP. NO.	UG	RSD
1	461	106.69	1.16
2	411	82.86	7.85
3	462	61.73	2.15
4	463	75.91	2.32
5	464	28.65	5.10

b

Fig. 5. Results of calculations done on a remote computer: a)
calibration curve data, b) sample results.

example of how the computer can be extended into a particular
analytical situation (9).

Going farther, some laboratories doing routine analyses of
occupational health samples have written decision-making
programs which automate the sequence of steps by which a sample
is taken from beginning to end of the quantitative analysis.
Such an analysis proceeds in the following way. The first step
is always the same and will usually be a fairly fast scan of
the analyte's primary peak. If this peak is nonexistent or
below a certain level, no further analysis is necessary unless
another analyte is to be determined. But if the peak is large
enough to indicate the analyte is at hazardous levels, further
verification is needed. These first few steps are shown in the
flow chart example of Figure 6, in which diamonds indicate
decision points. The decision criteria inside the diamonds are
very objective and can be quickly tested in a computer
program. Although this task seems more sophisticated and
"human" than most performed by the computer in XRD work, it is
the repetitive nature of the work that makes it possible for a
human to program a computer to do it. The final outcome is
that the computer once again saves time when it follows the
steps at which its programmer said, "Don't bother with the rest
of the analysis on this sample."

The first two examples of how the computer can be further
extended into an analysis could be implemented on modest
computers. A program that would tax the limits of the
minicomputers available on the newest XRD instruments is a
deconvolution program written specifically for quantitative

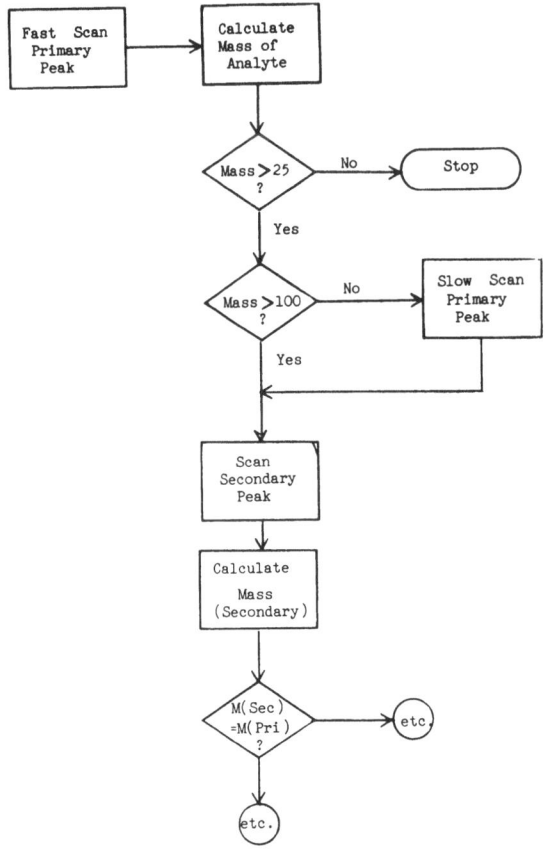

Fig. 6. Flow chart for routine XRD quantitative analysis
         program.

analysis of occupational health samples (10). Here again, a
procedure has been worked out for transferring XRD line profile
data to a large remote computer with relative ease. The
purpose of the deconvolution program is to separate interfering
peaks from analyte peaks; an example showing the resolution of
zircon (interference) and quartz (analyte) peaks is in
Figure 7. The success of these computations for the low
intensity data from occupational health samples has not yet
been fully tested. This last example of the use of computers
in XRD is based on sheer computational power and this use,
instead of saving time, _per se_, actually extends the usefulness
of the XRD method.

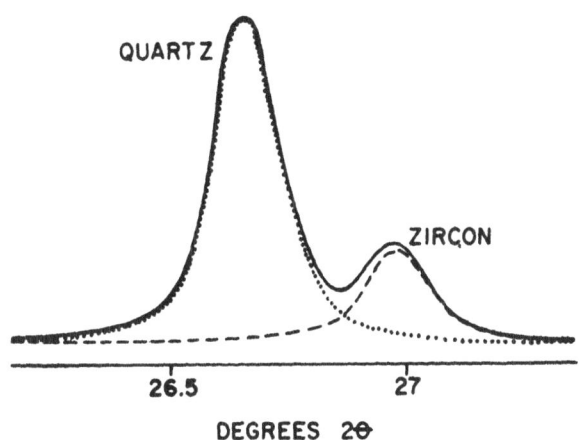

Fig. 7. Printout of smoothed diffraction data (____) and
         resolved peaks of quartz (···) and zircon (---).

CONCLUSION

Occupational health laboratories have taken good advantage
of computer-automated X-ray powder diffractometers. While
these laboratories do not directly participate in instrument
development, they occasionally underwrite such activity and
they are a strong stimulus in the marketplace for commercial
instrument producers. The improvement in efficiency and data
quality with the available instruments is very important for
the large volume of occupational health samples, but the
special character of these samples has also prompted some
changes in the operating software which have yielded additional
improvements. The "special character" of these samples refers
particularly to the minute amounts of sample dust which must be
collected and analyzed on filters. One development which would
improve the XRD method for occupational samples, although not
involving the use of computer, is use of the collection filter
for analysis instead of redepositing the dust on a second
filter. Human involvement in the analysis would again be
minimized. Beyond this, further improvements in the
quantitative analysis of occupational health samples are likely
to depend on the expanded use of computers for calculations and
decision making.

REFERENCES

1. NIOSH Manual of Analytical Methods., Taylor, D. G.; Crable, J. V., Ed.; NIOSH Publication Nos. 77-157 (Vols. 1-3), 78-175 (Vol. 4), 79-141 (Vol. 5).

2. Lange, B. A.; Haartz, J. C., Determination of Microgram Quantities of Asbestos by X-Ray Diffraction: Chrysotile in Thin Dust Layers of Matrix Material, Anal. Chem., 51, 520 (1979).

3. Nenadic, C. M.; Crable, J. V., Application of X-Ray Diffraction to Analytical Problems of Occupational Health, Amer. Ind. Hyg. Assoc. J., 32, 529 (1971).

4. Dollberg, D. D.; Abell, M. T.; Lange, B. A., Occupational Health Analytical Chemistry: Quantitation Using X-Ray Powder Diffraction, Analytical Techniques in Occupational Health Chemistry, Dollberg, D. D.; Verstuyft, A. W., Ed., ACS Symposium Series 120, 1980 pp 43-66.

5. Williams, P. P., Direct Quantitative Diffractometric Analysis, Anal. Chem., 31, 1842 (1959).

6. Leroux, J.; Davey, A. B. C.; Paillard, A., Proposed Methodology for The Evaluation of Silicosis Hazards, Amer. Ind. Hyg. Assoc. J., 34, 409 (1973).

7. Bumsted, H. E., Free Silica Determination with an Automated X-Ray Diffractometer, presented at the Amer. Ind. Hyg. Conf., Houston (Abstract 135).

8. Abell, M. T.; Dollberg, D. D.; Lange, B. A.; Hornung, R. W.; Haartz, J. C., Absorption Corrections in X-Ray Diffraction Dust Analyses: Procedures Employing Silver Filters, Electron Microscopy and X-Ray Applications, Vol. II, Russel, P.; Hutchings, A., Ed., Ann Arbor Science Publishers, Inc., Ann Arbor, Michigan (1980).

9. Abell, M. T.; Dollberg, D. D., A Quantitative Analysis Software Package for Use with the APD 3500, Norelco Reporter, 25 (2), 21 (1978).

10. Slaughter, M. S. Deconvolution of X-Ray Spectra Produced by Mineral Samples, Electron Microscopy and X-Ray Applications, Vol. II, Russel, P.; Hutchings, A., Ed., Ann Arbor Science Publishers, Inc., Ann Arbor, Michigan (1980).

POROTECTOSILICATE STRUCTURE DETERMINATION FROM MODEL BUILDING

George T. Kokotailo and John L. Schlenker

Mobil Research and Development Corporation
Research Department
Paulsboro, N.J. 08066

ABSTRACT

Porotectosilicates are a class of siliceous crystalline materials which includes both zeolites and materials which resemble zeolites in crystal structure, but may or may not have ion exchange
capability. The framework structures of these porotectosilicates
are comprised of "T" atoms tetrahedrally coordinated to oxygen,
where "T" can be Al, Si or any other element capable of isomorphous
substitution for silicon. The occurrence of small crystals and
the additional problems introduced by the presence of stacking
faults and crystal twinning make structure determination of porotectosilicates by conventional approaches difficult.

The industrial significance of these materials has led to the
development of a technique which permits the determination of
their structure. The method involves the construction of appropriate hypothetical models, a DLS refinement followed by computation of a Smith plot for comparison with the experimental powder
pattern. Model crystal structures may now be refined using the
Rietveld technique. It is expected that this technique will contribute significantly to the solution of porotectosilicate structures which are difficult--if not impossible--to establish by
other techniques.

INTRODUCTION

Porotectosilicates encompass that class of siliceous crystalline materials which includes both zeolites and those materials

which resemble zeolites in structure but may or may not have ion
exchange capabilities. The basic unit in these framework structures consists of a T-atom, (Si,Al) or any other element capable
of isomorphous substitution for Si, tetrahedrally coordinated to
four oxygen atoms. The tetrahedra are linked through common oxygen atoms to form loops and chains which are in turn linked to
form three-dimensional frameworks. The excess negative charge of
the framework is neutralized by positively charged cations located
in the cavities and channels of these structures.

The importance of porotectosilicates in the petroleum and
chemical industries can scarcely be overestimated. The first application was the use of rare earth exchanged synthetic faujasite
zeolites as cracking catalysts.[1-3]   The new pentasil (ZSM-5/11)
catalysts selectively convert methanol to high quality gasoline.[4]
This class of zeolites has made possible the development of commercially significant petroleum and chemical processes, such as
distillate dewaxing, ethylbenzene synthesis, xylene isomerization
and toluene disproportionation.[4,5] The adsorption and rates of
diffusion of molecules into a zeolite are controlled by the size
and shape of the channel system, and by the nature and location of
the cations. Stability and catalytic properties are also a function of structural characteristics.

Determining structural features is essential for the understanding of the physical and catalytic properties of these materials, but small crystallite size, faulting, and twinning make it
difficult, if not impossible, to use single crystal methods.
This report will discuss the use of model building and powder pattern simulation for the solution of porotectosilicate structures.

Representation of Structures

Porotectosilicate frameworks may be depicted in a number of
ways. Although a tabulation of the space group, lattice parameters, atomic coordinates, interatomic distances, and angles constitutes a complete description of the structure, pictorial depic-
tion via framework models is more illustrative. Models, such as
that of mordenite, where the atoms are represented by spheres, were
introduced by Bragg and Brown.[7] Atomic radii are readily evident
in such a representation of a framework structure.

Pauling[8] used polyhedral models consisting of linked polyhedra representing a basic unit or repeating group of atoms. A
framework representation in which atoms are at the nodes and con-
nections between nodes represent bonds was used by Wells.[9] Meier
utilized physical models in which "tetrahedral stars," representing
T-atoms, are connected via tubing (representing bonds). When built

to the scale 1Å = 1cm, this type of model permits rapid and accurate estimation of unit cell parameters and atomic coordinates.
Such models are useful for determining the symmetry of a structure,
and permit rapid recognition of basic building units. A drawing of
this type of framework model as well as the others, is shown in
Fig. 1.

Trial model building is not a random process. All available
information, lattice parameters, and symmetry information from
diffraction studies, estimates of channel dimensions from IR and
diffusion studies, measured density, and ring ellipticity from diffusion rates, must be incorporated in the hypothetical model. The
next step is the simulation of a powder pattern for the model.

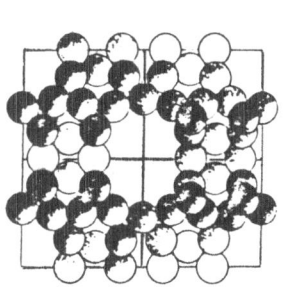

Mordenite Spheres

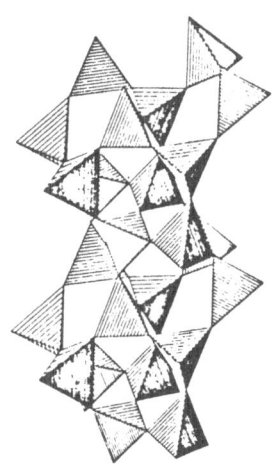

Mordenite Polyhedra

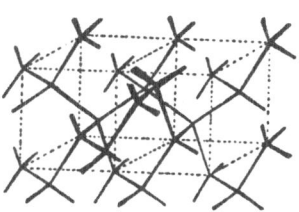

Quartz 3D Net

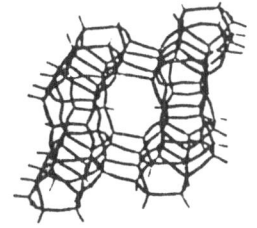

Dachiardite
Tetrahedral Connected

Fig. 1. Structure Representations

Simulation of Patterns

For porotectosilicates, interatomic distances and bond angles can be predicted within fairly narrow limits.

If the lattice parameters of a model are known, the atomic coordinates of the individual atoms may be adjusted so that the interatomic distances correspond as closely as possible to the predicted distances. These positional parameters can be computed from the prescribed interatomic distances D_j° by a least squares procedure which minimizes the residual function

$$\rho_w = \sum_{\substack{j \\ (m,n)}} w_j^2 (D_j^{\circ} - D_j^{m,n})^2$$

where w_j is the weight ascribed to the interatomic distance of type j. This method of refinement--referred to as a DLS refinement--was described by Meier and Villiger.[10] This DLS refinement method gives idealized model structures using prescribed inter-atomic distances and unit cell constants for a given space group. The weight w_j of each error equation is based on bonding considerations or observed variations in bond-length values.[11] The simulation results vary depending on the weights. In silicates, it was found that the Si-O bond was related to the Si-O-Si bond angle by the following relation:[12]

$$d(Si-O) = 1.527 + .068 \ [-Sec(Si-O-Si)]$$

This relation is used to weigh the Si-O bond distance as a function of bond angle.

Only approximate atomic coordinates are required to initiate the least-squares refinement. Convergence is usually rapid for chemically reasonable structures. A final "R-factor" is provided which can be used to estimate the "goodness" or chemical reasonableness of the structure.

The Pentasil Family of Structures

ZSM-5 and ZSM-11 are members of a family of highly siliceous porotectosilicates[13,14] which possess unusual catalytic properties and high thermal stability. Knowledge of the structures of these materials was essential in order to ascertain the details of the channel systems. Since large fault-free crystals were not available, powder diffraction and model building was used to establish the frameworks. As stated, model building is not a random process. All the information available must be used. In the case of ZSM-5, X-ray and electron diffraction, adsorption and diffusion data established that the cell was orthorhombic with a = 20.07, b = 19.97, c = 13.4Å and space group Pnma, and that the channel system consisted of ten-membered rings. Using this information, a model was assembled which satisfied all these conditions. The framework structure can be described in terms of a characteristic layer as shown in Fig. 2. The heavy dark outline outlines the secondary building units from which the characteristic layers can be built. If these layers are linked such that they are related by an inversion, the required three dimensional framework results (Fig. 3). A DLS refinement established the positional parameters. A Smith plot[15] of these data agreed very well with the X-ray diffraction pattern of ZSM-5, Fig. 4. Single crystal data refinement revealed the correctness of the structure.[16]

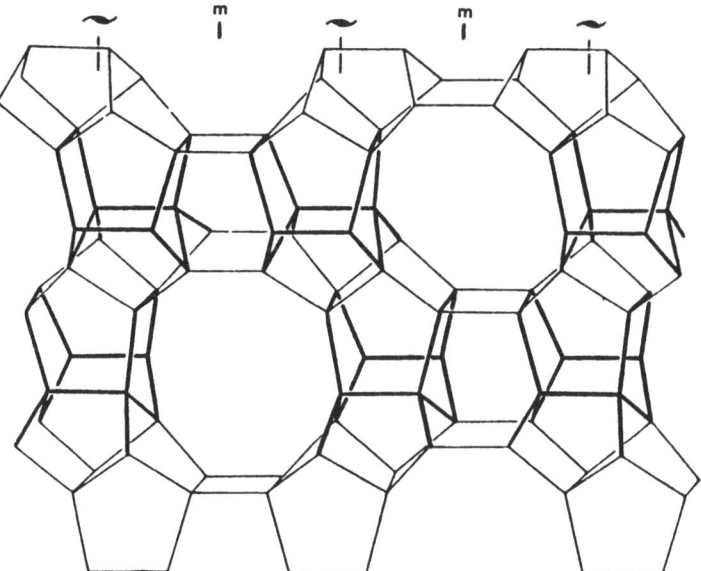

Fig. 2. Characteristic Pentasil Layer

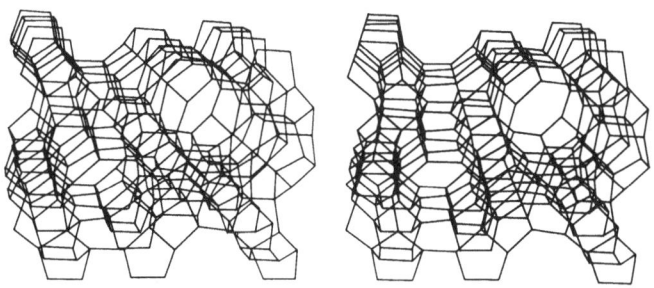

Fig. 3. Framework Structure of ZSM-5

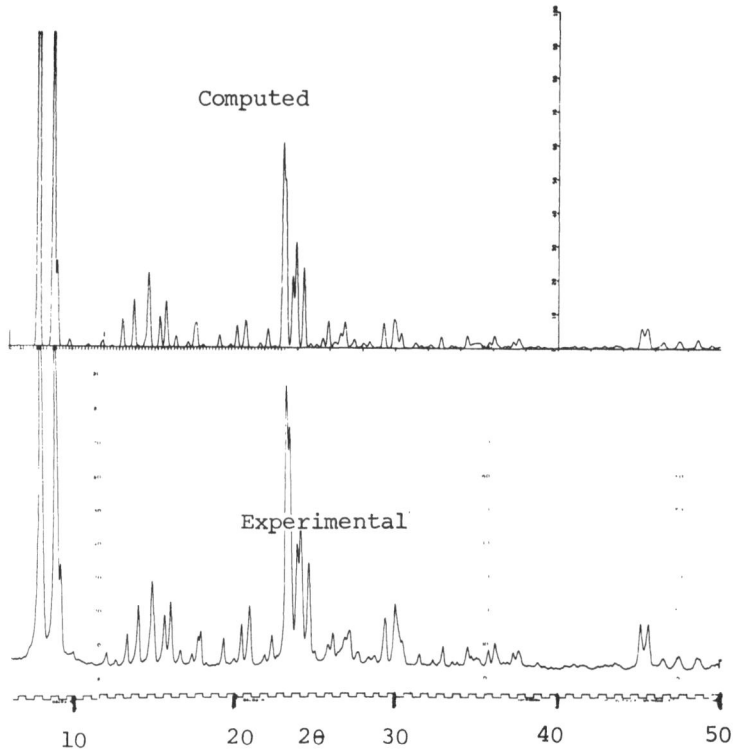

Fig. 4.   Comparison of the Computed and Experimental X-ray
          Diffraction Patterns of ZSM-5

    If the characteristic layers in Fig. 2 are linked such that
neighboring layers are related by a reflection, another framework
structure results (Fig. 5). A DLS refinement determined the positional parameters and a Smith plot of these data matched the X-ray
diffraction pattern of ZSM-11, Fig. 6, establishing its structure.[17]

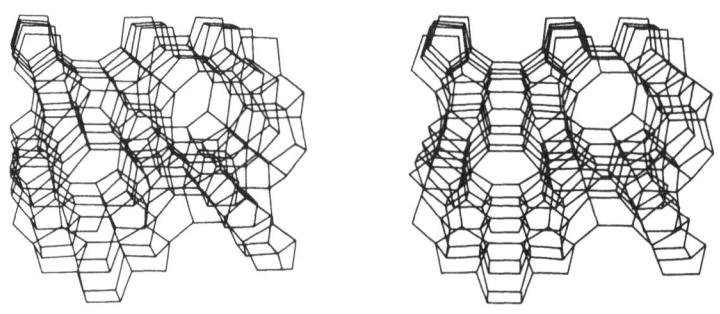

Fig. 5.   Framework Structure of ZSM-11

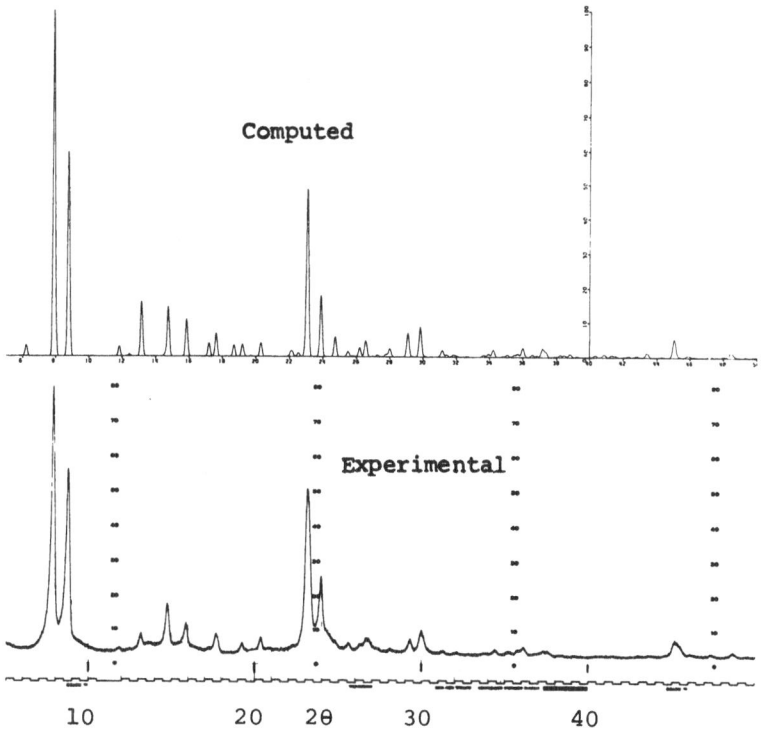

Fig. 6. Comparison of Computed and Experimental X-ray
        Diffraction Patterns of ZSM-11

In ZSM-11, the sequence of layers is related by a reflection, σ,
and in ZSM-5 it is related by an inversion i, where:

$$\sigma: A \rightarrow A' \quad i; A \rightarrow B'$$

These are shown schematically in Fig. 7a and 7c.

        Structures with other possible combinations are easily derived.
A simple combination σiσi or AA'AB' is shown in Fig. 7b. The sequence σσii or AA'AB represents another simple example. The number
of possible combinations is essentially infinite and is limited
only by the particular repeat distance chosen in each case. The
ZSM-5 and ZSM-11 structures are the end members embracing a substantially infinite series of intermediate structures.[18] All intermediate structures are derivable from ZSM-11 by applying the i
transformation and from ZSM-5 by applying the σ transformation.
This series of structures is so closely related that a simple generic
name, pentasil, is proposed to encompass all family members.

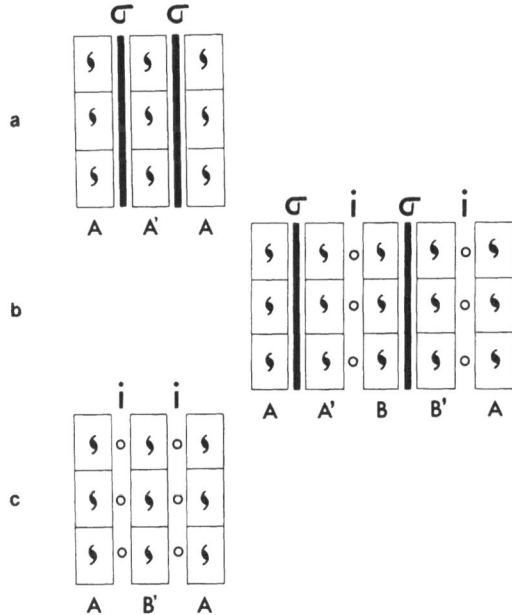

Fig. 7. Stacking Sequence of Layers in Pentasil

Stacking Faults in Erionite-Offretite System

The structures of erionite[19] and offretite[20] have been determined. Both structures can be represented by layers of cancrinite
cages. In offretite, the layers are oriented so that there are
open 12 MR channels parallel to the c-axis (Figs. 8 and 9). If
alternate layers are rotated 60° (Figs. 10 and 11), the 12 MR channels are closed and the erionite structure results. Zeolite
Linde T[21] is related to both offretite and erionite in that random layers are rotated 60°. The presence of stacking faults in
offretite has the effect of causing streaking in electron diffraction patterns along reciprocal lattice rows parallel to c.* [22]

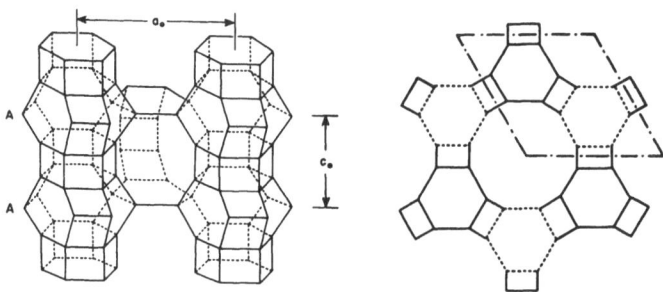

Fig. 8. View of Offretite   Fig. 9. 001 Projection of the
         Framework   Offretite Framework

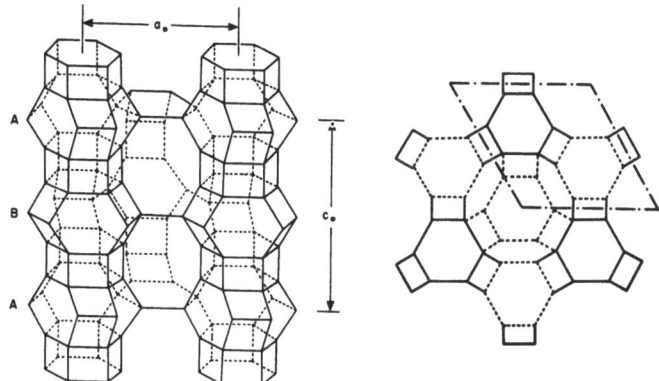

Fig. 10. View of Erionite   Fig. 11. 001 Projection of
         Framework   Erionite Framework

     Contrast lines in transmission electron micrographs of erionite
from Pershing County, Nevada, indicate the presence of stacking
faults.[22]

     In order to determine the effect of stacking faults on diffrac-
tion patterns of erionite or offretite, a method of superimposing
the structure of a double unit cell of offretite on erionite and
assigning an occupancy of 1 to all atoms that coincide and an occupancy of 1/2 for split atoms (1-6 in Fig. 11), was developed. This
is a model of a randomly disordered erionite-offretite structure
with 50% of each component. Smith plots of erionite-offretite using
single crystal data[19,20] and a 25-75% disordered structure using
this split atom assignment are shown in Fig. 12. The odd ℓ lines
(starred) in the erionite pattern decrease in intensity in the 25-
75 disordered structure pattern, and disappear in the offretite pattern. Linde T zeolite is a disordered offretite structure. Smith
plots of a 25-75 and 50-50 disordered offretite-erionite structure
are compared with the experimental X-ray diffraction patterns of
Linde T; the agreement with the 25-75 disordered structure is fairly
good. The odd ℓ lines in the Linde T pattern are broad because of
random stacking which has the effect of giving rise to X-ray reflections corresponding to $d_\infty = \infty$ or $c^* = o$. The intensity of these
lines is the area under the peak. This method can now be used to
estimate the concentration of stacking faults. The method is applicable to other porotectosilicates with stacking faults.

Contribution of Water and Cations to Powder Pattern

     The structure of the natural zeolite ferrierite was determined
by Vaughn.[23] A view along direction [001] of the framework will
clearly show the 10 MR channels and also the chains of 5 MR which,

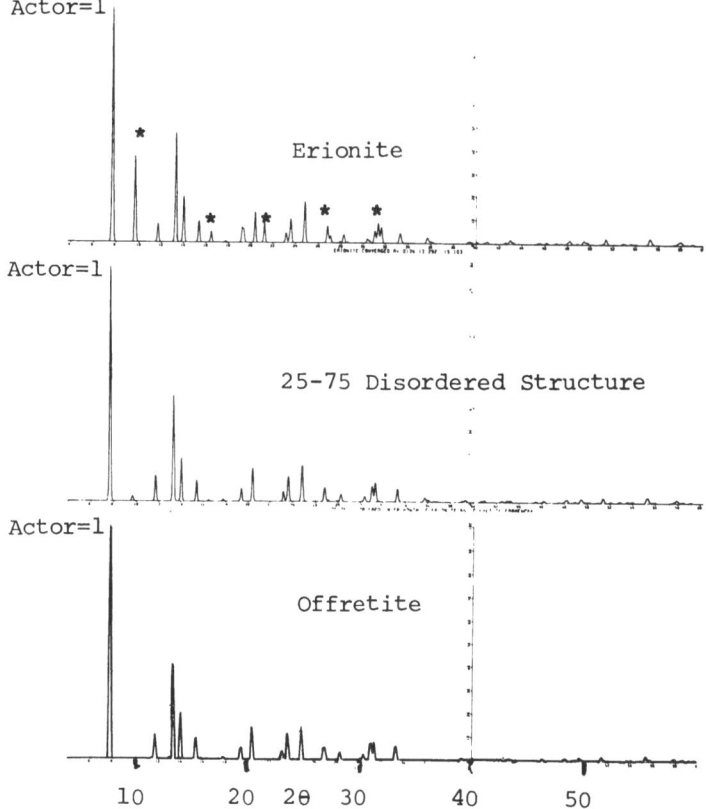

Fig. 12.   Smith Plots of Framework Structures

when linked together, form the channel system. A view along [010]
depicting the position of the cations is shown in Fig. 13.[24] In
hydrated ferrierite $Mg(H_2O)_6^{++}$ cation complexes are situated in
cavities with 8 MR openings, as indicated. The Na^+ and K^+ cations
have not been located.

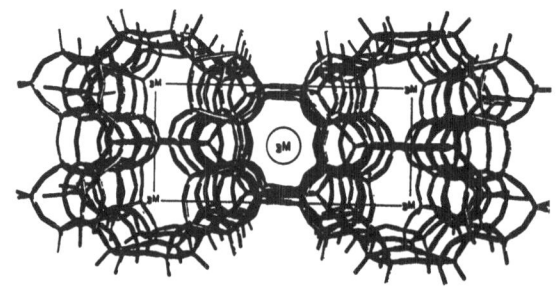

Fig. 13.   Framework Structure of Ferrierite

Differences in the relative intensities of the lines in the
X-ray diffraction patterns of samples from various locations led
us to an analysis of the various contributions to the diffracted
intensities. A Smith plot of Vaughn's data, when compared to a
powder pattern of Staples,[25] reveals differences in relative intensity. The lower intensity of especially the 110-line is evident
in the Staples data. In order to resolve these discrepancies,
Smith plots of the contribution of the cation complexes were obtained, Fig. 14. It is compared to the Smith plot of the Vaughn
data and also the Vaughn data with the Mg complexes removed. It is
evident that the higher intensity of the (110) peak in the Vaughn
pattern is due to the Mg complexes. By obtaining plots of structures with partial occupancy, a better agreement of powder data can
be obtained.

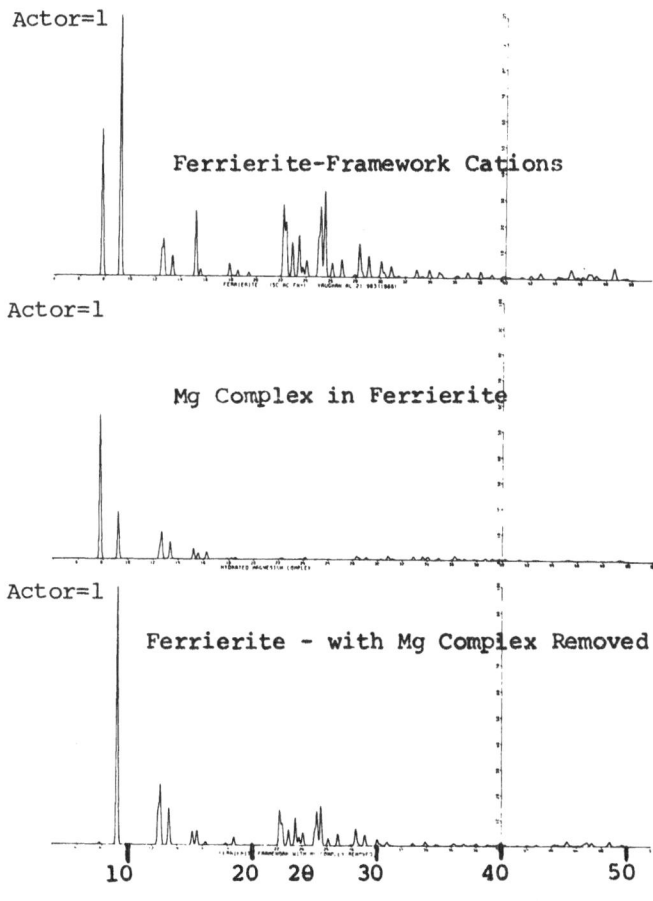

Fig. 14.   Simulated Plots of Ferrierite

Pseudosymmetry

Some crystal structures can be solved using symmetries higher than the true symmetry and reduction to the true symmetry is not trivial. Fourier and direct methods often fail in determining the true structure. Meier[26] actually developed the DLS program to cope with this problem.

The structure of Linde A was determined by Reed and Breck[27] from powder data. They found the space group to be $Pm\overline{3}m$ with a = 12.3Å and composition $Na_{12}Al_{12}Si_{12}O_{48} \cdot 27H_2O$. Since the Si/Al ratio in NaA is very close to 1 and if Lowenstein's rule is followed, there should be an ordered arrangement of Si and Al atoms leading to a doubling of the cell parameter[28] and also certain additional weak b reflections. The space group is now $Fm\overline{3}c$ and a = 24.61Å. Since the difference between the true and pseudosymmetric substructure is very small, the determination of the phases of the b-reflections was not possible. On DLS, it was possible to show that Si and Al atoms are indeed ordered in space group $Fm\overline{3}c$.[29,30]

The DLS technique has been used to study several other pseudosymmetric structures such as cubic SiP_2O_7[31] and tridymite.[32]

ACKNOWLEDGEMENT

The help of W. M. Meier, S. L. Lawton, S. Sawruk, and A. C. Rohrman, Jr. is gratefully acknowledged.

REFERENCES

1. P. B. Weisz and V. J. Frilette, J. Phys. Chem. 64:392 (1960).
2. S. C. Eastwood, R. D. Drew, and F. D. Hartzell, Oil and Gas Journal 60:152 (1962).
3. C. J. Plank, E. J. Rosinski, and W. P. Hawthorne, Ind. Eng. Chem. Prod. Res. Dev. 3:165 (1964).
4. S. L. Meisel, J. P. McCullough, C. H. Lechthaler, and P. B. Weisz, Chem. Tech. 6:86 (1978).
5. N. Y. Chen, R. L. Gorring, H. R. Ireland, and T. R. Stein, Oil and Gas J. 75:165 (1977).
6. N. Y. Chen, W. E. Garwood, W. O. Haag, and A. B. Schwartz, Symposium on Advances in Catalytic Chemistry, Oct. 1977, Snowbird, Utah.
7. W. L. Bragg and G. B. Brown, Z. Kristallogr. 65:528 (1926).
8. L. Pauling and J. H. Sturdivant, Z. Kristallogr. 68:239 (1928).
9. A. F. Wells, Acta Cryst. 7:535 (1954).
10. W. M. Meier and H. Villiger, Z. Kristallogr. 129:411 (1969).
11. W. H. Baur, Phys. and Chem. Minerals 2:3 (1977).

12. G. V. Gibbs, E. P. Meagher, J. V. Smith, and J. J. Pluth, Fourth International Conference on Molecular Sieves, April, 1977, Chicago, Illinois.
13. U.S. Patent 3,702,886.
14. U.S. Patent 3,709,979.
15. D. K. Smith, A Revised Program for Calculating X-ray Powder Diffraction Patterns, UCRL 50264, Lawrence Radiation Laboratory, University of California, Livermore, CA (1967); Norelco Reporter 15:57 (1968).
16. G. T. Kokotailo, S. L. Lawton, D. H. Olson, and W. M. Meier, Nature 272:437 (1978).
17. G. T. Kokotailo, P. Chu, S. L. Lawton, and W. M. Meier, Nature 275:119 (1978).
18. G. T. Kokotailo and W. M. Meier, Proceedings of the Conference on the Properties and Applications of Zeolites, Soc. Chem. Ind. London, April 1979.
19. L. W. Staples and J. A. Gard, Mineral Mag. 32:261 (1959).
20. J. M. Bennett and J. A. Gard, Nature 214:1005 (1967).
21. U. S. Patent 2,950,952
22. G. T. Kokotailo, S. Sawruk and S. L. Lawton, Am. Mineral. 57:439 (1972).
23. P. A. Vaughn, Acta Cryst. 21:983 (1966).
24. D. W. Breck, Zeolite Molecular Sieves, John Wiley & Sons, New York (1974).
25. L. W. Staples, Am. Mineral. 40:1095 (1955).
26. W. M. Meier, Adv. Chem. Series 121:39 (1973).
27. T. B. Reed and D. W. Breck, J. Amer. Chem. Soc. 78:5972 (1956).
28. R. M. Barrer and W. M. Meier, Trans. Farad. Soc. 54:1074 (1958).
29. V. Gramlich, Dissertation ETH Zurich, 1971.
30. V. Gramlich and W. M. Meier, Z. Kristallogr. 133:134 (1971).
31. E. Tillmanns and W. H. Baur, J. Sol. State Chem. 7:69 (1973).
32. W. A. Dollase and W. H. Baur, Am. Mineral. 61:971 (1976).

ANALYSIS AND INTERPRETATION OF DIFFRACTION DATA FROM AMORPHOUS

MATERIALS

J. H. Konnert, P. D'Antonio and J. Karle

Laboratory for the Structure of Matter
Naval Research Laboratory
Washington, D. C.   20375

INTRODUCTION

Amorphous materials give rise to rather diffuse diffraction patterns. In contrast to diffraction patterns from polycrystalline materials which are characterized by a large number of sharp diffraction rings, patterns from amorphous materials are composed of relatively few broadened features. Such a diffuse pattern, however, contains much structural information in the form of an interatomic distance distribution that may be computed directly from the measured diffraction pattern. This radial distribution function (RDF) provides information concerning bonded distances, the types of atomic groupings and the extent of ordering in the sample. Special care, however, must be taken during the analysis to avoid introducing spurious details into the RDF that may be confused with, or mask real structural features. Such false detail may arise both from data collection and data reduction procedures. The availability of modern computers has greatly facilitated the calculation of accurate RDF's by readily permitting the introduction of physical criteria into the data reduction procedure that must be satisfied by the RDF.

The interpretation of an RDF generally entails the calculation of a corresponding function for a proposed model and a comparison of the result with the experimental curve.

The first three sections of this paper concern techniques for the calculation of accurate RDF's from diffraction data; the fourth section deals with the calculation of these functions for known bonding topologies; and the last describes several applications.

63

GENERAL THEORY

     In this section the components of a diffraction pattern are
described, the portion that contains the structural information
is identified and the functional relationship between the intensity
and the RDF is defined.

     The total diffracted intensity from an amorphous material that
has been corrected for effects such as polarization and absorption
is designated I_t and is composed of the interatomic interference
scattering, I, the coherent atomic scattering, I_c, and the
incoherent atomic scattering, I_i. A total intensity curve for
silica glass is illustrated in Fig. 1. The related equations are

$$I_t(s) = I(s) + I_c(s) + I_i(s) = I(s) + I_b(s) \qquad (1)$$

$$\text{and} \quad i(s) = [I_t(s) - I_b(s)] / \Sigma f^2(s) = I(s)/\Sigma f^2(s) \qquad (2)$$

where $s = 4\pi \sin \theta/\lambda$, 2θ is the angle between the incident and
diffracted beams and λ is the wavelength. The summation in Eq. 2
is taken over the squares of the atomic scattering factors, f,
for the atoms comprising a unit of chemical composition. Division
of I by Σf^2 yields an intensity function that corresponds approximately to the scattering from vibrating point atoms.

     The radial distribution function of interatomic distances,
$G(r)$, is obtained by taking the Fourier sine transform of $si(s)$.

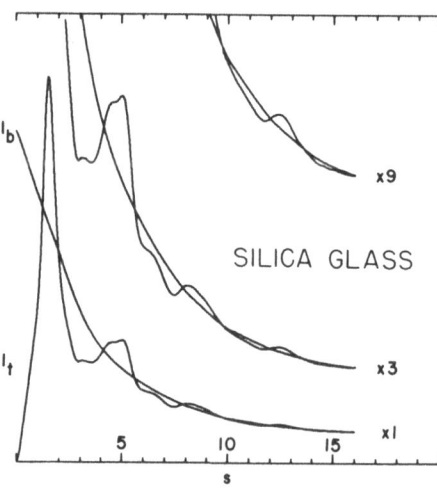

$$G(r) = 4\pi r(\rho(r) - \rho_o)$$

$$= 2/\pi \int_0^\infty si(s)\sin srds \quad (3)$$

where $4\pi r^2 \rho(r)$ represents the
probability, weighted by the
product of the scattering
factors of atoms i and j and
divided by Σf^2, of finding
atoms j in the sample separated
from atoms i by a distance
between r and r+dr. Bulk
density is represented by ρ_o.
The objective of the data
reduction procedure is to
isolate $i(s)$ from I_t accurately
and perform the integration
in Eq. 3 without introducing
spurious detail into $G(r)$
because of the termination of
the integration at finite s.

Fig. 1. The total x-ray diffraction pattern for silica glass,
I_t, and the background intensity,
I_b.

For application of the data reduction procedure to be described later, it is necessary to express G(r) in terms of the interatomic distances and the bulk density. With the assumption of
harmonic motion between pairs of atoms, G(r) may be expressed in
the following manner

$$rG(r) = \frac{2r}{\pi} \sum_{i,j} \frac{N_{ij}}{r_{ij}} \int_0^\infty \frac{f_i f_j}{\sum f^2} \exp(-l_{ij}^2 s^2/2) \sin sr_{ij} \sin srds - 4\pi r^2 \rho_0' \frac{(\sum f_{(s=0)})^2_{uc}}{\sum_{uc} f^2_{(s=0)}} \quad (4)$$

where r_{ij} is the distance between the ith and jth atoms, N_{ij} is
the coordination number, l_{ij} is the disorder parameter, uc is the
unit of composition (SiO_2 for the example employed), and ρ_0'
is the bulk density in uc's per cubic Angstrom. The first term
on the right hand side of Eq. 4 represents $4\pi r^2 \rho(r)$, and the
second term represents $4\pi r^2 \rho_0$. Further details concerning the
general theory may be found in the early work in this field[1,2] or
in textbooks.[3,4]

DATA REDUCTION

The objective of obtaining an RDF with minimal error by
isolating i(s) accurately from the total intensity and minimizing
termination error may be achieved by requiring I_b and the
resultant RDF to satisfy a variety of physical and mathematical
criteria. These are:

1. The inner region of the RDF should be featureless where
distances are known not to exist. In this connection, it will be
seen that the correction for the termination error involves
removal of the first few distances from the RDF. The requirement
for a featureless RDF in its inner region is thus extended into
the region involving the distances that are removed.

2. The outer region of the RDF should show a uniform distribution of distances. Termination effects and errors in the
experimental data cause spurious features in this region and
require correctional procedures.

3. The scaling of the data should be consistent with
density measurements, known coordination numbers or both.

4. The background scattering, I_b, should be a smoothly
changing function whose shape is compatible with that expected
from theory.

A number of authors have developed procedures for introducing these criteria into an analysis for the purpose of enhancing the reliability of the RDF's.[4,5,6,7,8] The procedure to be
described here[7,9] permits very rapid convergence of the data
reduction by expressing the RDF explicitly as a function of parameters defining the background intensity and short distances.

The separation of that portion of the intensity, $i(s)$, containing the structural information requires that the background be represented by a smooth, flexible function. This flexibility is necessary because the exact shape of I_b must be determined by the procedure. Coherent and incoherent atomic scattering factors, for example, are available only for isolated atoms. Additionally, experimental conditions often result in a small amount of extraneous scattering that must be accounted for by I_b.

It has proved useful to express I_b in the form of overlapping exponential functions:

$$I_b(s) = \sum_n W_n(s) \exp(a_n + b_n s^{cn}) + d \cdot s^e \qquad (5)$$

Each exponential function overlaps to the midpoint of the adjacent junction, and functions are constrained to be equal at points of abuttment. In the analysis of x-ray diffraction data, d is set equal to zero, and, in the analysis of neutron diffraction data, d and e are chosen so that I_b is amenable to fitting with the exponential functions. For silica glass, the shortest distances which are approximately 1.6A contribute to $si(s)$ a sine wave with a periodicity of about 4 s units. It is important to avoid introducing an incorrect I_b with this periodicity or less. Thus, in the initial stages of refinement each exponential should span more than 4 s units.

The treatment of the termination of the integral in Eq. 3 is based on the recognition that generally only the shortest distances make a significant contribution to the experimental intensity function, $i(s)$, beyond the measured range of scattering angle due to their small thermal or disorder parameters. The intensity contributions from the larger distances are damped out at the upper limit of experimental data, s_{max}. The contributions from the shortest distances may be removed from $i(s)$ so that s_{max} can accurately replace infinity as the upper limit of integration in Eq. 3. Such an intensity function may be written as a function of parameters defining I_b and the short distance parameters. The Fourier transform of this function, $i'(s)$, may be expressed as

$$rG'(r) = 4\pi r^2 [\rho'(r) - \rho_o] = \frac{2r}{\pi} \int_o^{S_{max}} si'(s) \exp(-\alpha s^2) \sin srds \qquad (6)$$

$$i'(s) = i(s) - \sum_{sd} N_{ij} f_i f_j \exp(-1_{ij}^2 s^2/2) \cdot \sin sr_{ij}/(sr_{ij} \Sigma f^2) \qquad (7)$$

$$i(s) = \{I(s) - A[\sum_n W_n(s) \exp(a_n + b_n s^{cn}) + d \cdot s^e]\}/\Sigma f^2(s) \qquad (8)$$

$$\rho_o = \rho'_o (\sum_{uc} f_{(s=o)})^2/\sum_{uc} f^2_{(s=o)} \qquad (9)$$

$$I(s) = I(s) \text{ measured}/K \qquad (10)$$

where K places the intensity on an absolute scale, sd refers to
the shortest distances, α is an artificial damping factor that may
be given a small value, if necessary, to remove residual termination effects, $\rho'(r)$ equals $\rho(r)$ minus contributions from the
shortest distances that have been removed from $i(s)$, and A scales
I_b.

LEAST-SQUARES REFINEMENT OF RDF

The formulation of the calculation is:

$$\text{minimize} \left\{ \sum_{p<r<q} [G'(r) + 4\pi r\rho_o]^2 + w \sum_{u<r<v} [G'(r)]^2 \right\} \qquad (11)$$

where p and q delineate the inner region, u and v delineate the
outer region of the RDF and w fixes the relative weights of the
two regions. $G'(r)$ is defined in Eqs. 6-10 in terms of the parameters that are to be optimized by the use of Eq. 10. The first
sum expresses the condition that the inner region of the RDF
should equal $-\rho_o$. The second sum expresses the condition that the
outer region display a uniform distribution of distances. Normally
it is necessary to minimize only the first sum while monitoring
both the inner and the outer portion of the RDF. Optimal values
for the a, b, c and A defining I_b and for K, r_{ij}, l_{ij}, and N_{ij}
are obtained from the calculation of Eq. 10.

It was found for SiO_2 that it was necessary to remove the
three shortest distance types, which are comprised of the covalent
bonded Si-O, the O-O within the tetrahedra, and the Si-Si of
adjacent tetrahedra, in order to rid the integral in Eq. 3 of
termination errors. The sequence of the refinement for silica
glass is illustrated in Fig. 2. The inner region of the RDF is
illustrated at several stages of the refinement along with the
corresponding intensity functions, $si(s)$ and $si'(s)$. Table 1
lists the short distance parameters at various stages. I_b for
cycle O was obtained by scaling the theoretical background shape
such that the positive and negative areas of $si(s)$ were equal.
Cycles 1, 3 and 5 refined the short distance parameters, K, A and
ρ_o. Cycles 2 and 4 refined the I_b shape parameters. The ρ_o
refined to 0.0212 uc/A^3 vs. the measured value of 0.02205 uc/A^3.
The resulting $r^2G(r)$ is illustrated in Fig. 4.

Table 1. Initial and Final Short Distance
Parameters for the Silica Glass Refinement

		r_{ij}	l_{ij}
Si-O	Cycle 0	1.608A	0.03A
	Cycle 5	1.610	0.06
O-O	Cycle 0	2.626	0.10
	Cycle 5	2.639	0.10
Si-Si	Cycle 0	3.060	0.10
	Cycle 5	3.083	0.09

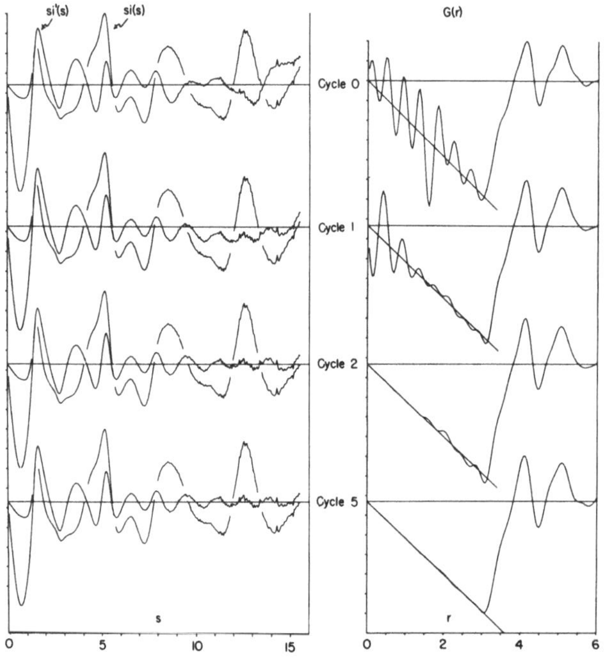

Fig. 2. The intensity functions si(s) and si'(s) and the RDF's
associated with si'(s) at various stages of the refinement.

CALCULATION OF RDF's FOR KNOWN BONDING TOPOLOGY

Figure 3 illustrates the 240 atom[10] asymmetric unit for tridymite, a crystalline polymorph of SiO_2. Each of the 240 atoms
in the asymmetric unit has a unique environment. With the
assumption of gaussian motion along the interatomic vectors, an
RDF for such a crystalline material may be calculated from the
environment of the atoms of a single asymmetric unit as follows:[11]

$$rG(r) = \frac{1}{\sqrt{2\pi}} \sum_i \sum_j W_i W_j \frac{f_i f_j}{\sum f^2} \exp\left[-(r-r_{ij})^2/2\sigma_{ij}^2\right] \quad (12)$$

where i is summed over the atoms in the asymmetric unit, j is summed over the atoms in sufficient asymmetric units and unit cells to attain the maximum r_{ij} considered, r_{ij} is the distance between atoms i and j, and σ_{ij} is the associated rms amplitude for vibration or disorder. For the calculation of an RDF for a nonperiodic model, i and j are summed over all of the atoms in the model.

In order to approximate the RDF, $\rho(r,\infty)$, for an infinite sample, one may divide the RDF for the finite sample by the RDF, $\varepsilon(R,r)$, for a uniform density solid of the same shape as the model

$$\rho(r,\infty) = \rho(r)/\varepsilon(R,r) \tag{13}$$

For a sphere[12]

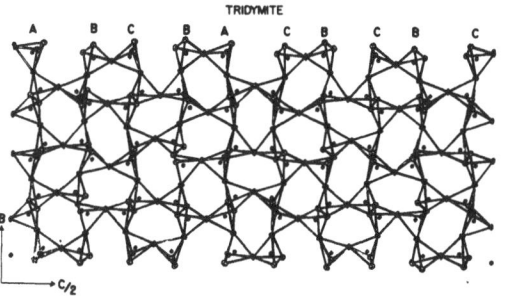

$$\varepsilon(r,R) = 1 - (3/2)r/R + \frac{1}{2}(r/R)^3, r \leq R,$$

where R is the radius of the sphere. Often a $\rho(r,\infty)$ is calculated for a crystalline material or nonperiodic model in which the positions of the maxima in the outer portion of the model RDF are in agreement with those for the amorphous material, but the amplitudes are greater. This effect may be approximately corrected by modifying $\rho(r,\infty)$ with various (r,t) representing different correlation lengths, t.

Fig. 3. The 240 atom asymmetric unit for tridymite, a crystalline polymorph of SiO_2.

$$G(r) = 4\pi r[\rho(r,\infty) - \rho_o] \cdot \int_0^\infty P(t)\, \varepsilon(r,t)\,dt \tag{14}$$

P(t) represents the probability of occurrence of an ordered region of radius t.

INTERPRETATION OF RDF'S

Fig. 4 illustrates the RDF's computed from the bonding topology of tridymite and from a 1412 atom nonperiodic model along with that for silica glass. For these calculations the probability for various correlation lengths in Eq. 13 was represented by two narrow Gaussian distributions centered at 12 and 20A with weights 3/4 and 1/4 respectively.

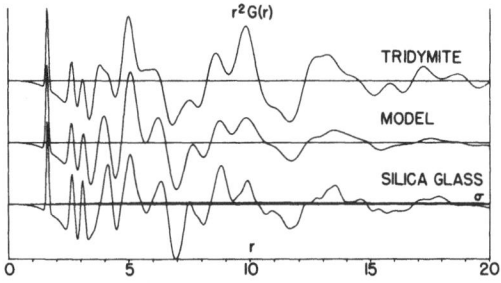

Fig. 4. RDF's for silica glass, tridymite and a 1412 atom model

The nonperiodic model was formed from a model built for amorphous silicon[13] by placing oxygen atoms between silicon atoms and relaxing the model in order that the shortest Si-O, O-O and Si-Si distances assumed nearly ideal values.

The tridymite structure is composed wholly of six-membered rings of silicate tetrahedra, whereas the model contains five and six-membered rings in a ratio of 0.37. The correspondence between the theoretical RDF's and the experimental one suggests the presence of many six-membered rings of silicate tetrahedra in silica glass.[14]

Diffraction data were collected for an amorphous carbonaceous material of large surface area.[15] The experimental intensity is illustrated as the upper curve in Fig. 5. Analysis of the data produced the RDF illustrated in Fig. 6. The oscillations of the interatomic distance distribution about the dotted envelope rather than the horizontal at small r evidences the porocity of the material. This porocity results in short range electron density fluctuations that produce the extensive small angle scattering which blends into the high angle scattering. The Fourier sine transform of the dotted envelope yields the associated small angle scattering. When this scattering is subtracted from the experimental curve, the lower intensity function in Fig. 5 results. A very broad peak that might be associated with a broad 002 reflection from graphite is now revealed. The breadth indicates that only several graphite layers at most are stacked upon one

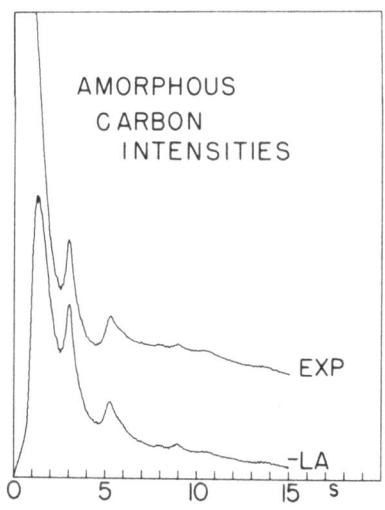

Fig. 5. Total diffraction pattern for an amorphous carbonaceous material, EXP, and EXP minus the low angle scattering, -LA.

another on the average. Indeed, the experimental RDF illustrated in Fig. 7 with the small angle scattering effects removed is in very good agreement with the RDF for a single graphite layer. In addition, the features in the experimental curve at large r are

shifted to progressively smaller
r values than the corresponding features in the curve for a
graphite layer. This appears
to be evidence for the bending
of the graphite layers.

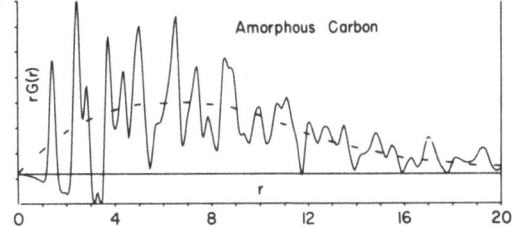

Fig. 6. RDF for an amorphous
carbonaceous material.

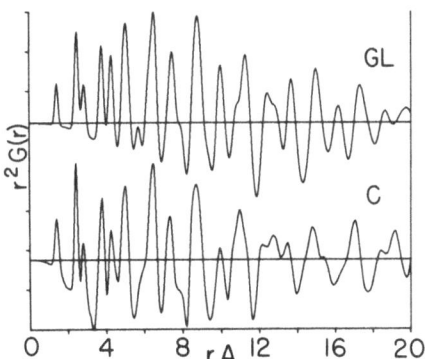

Fig. 7. RDF for an amorphous carbonaceous
material, C, and a single graphite layer, GL.

CONCLUSION

Modern computing capability permits the convenient computation
of accurate RDF's for amorphous materials. Comparison of RDF's
obtained from amorphous materials with RDF's calculated from known
bonding topologies may often indicate the geometry of structural
groupings of atoms in an amorphous material extending over 20A or
more.

REFERENCES

1. P. Debye, Zerstreuung von Röntgenstrahlen, Ann. Phys. 46:809
   (1915).
2. F. Zernicke and J. A. Prins, Die Beugung von Röntgenstrahlen
   in Flüssigkeiten als Effekt der Molekülanordnung, Z. Phys.
   41:184 (1927).
3. A. Guinier, "X-ray Diffraction," W. H. Freeman, San Francisco
   (1963).
4. B. E. Warren, "X-ray Diffraction," Addison-Wesley, Reading,
   Mass. (1969).

5. R. Kaplow, S. L. Strong and B. L. Averback, "Radial Density Functions for Liquid Mercury and Lead," Phys. Rev. 138A:1336 (1965).

6. A. J. Leadbetter and A. C. Wright, "Diffraction Studies of Glass Structure," J. Non-Cryst. Solids 7:141 (1972).

7. J. H. Konnert and J. Karle, "The Computation of Radial Distribution Functions for Glassy Materials," Acta Cryst. A29:702 (1973).

8. V. I. Korsunsky and Y. I. Naberukhin, "An Analytical Method of Computation of Radial Distribution Functions at Large Distances for Liquids and Amorphous Substances," Acta Cryst. A36:33 (1980).

9. P. D'Antonio and J. H. Konnert, "Computer Program for Radial Distribution Analysis of X-ray, Neutron and Electron Diffraction Data," J. Appl. Cryst. 12:634 (1979).

10. J. H. Konnert and D. E. Appleman, "The Crystal Structure of Low Tridymite," Acta Cryst. B34:391 (1978).

11. P. D'Antonio and J. H. Konnert, "Computer Program for Radial Distribution Calculation from Known Bonding Topologies," J. Appl. Cryst. in press.

12. L. H. Germer and A. H. White, "Electron Diffraction Studies of Thin Films," Phys. Rev. 60:447 (1941).

13. M. G. Duffy, D. S. Boudreaux and D. E. Polk, "Systematic Generation of Random Networks," J. Non-Cryst. Solids 15:435 (1974).

14. P. D'Antonio, J. H. Konnert and J. Karle, "Radial Distribution Functions for Silica Glass and Two Different Model Bonding Topologies," submitted to J. Non-Cryst. Solids.

15. P. D'Antonio and J. H. Konnert, "Simultaneous Analysis of High and Low Angle Diffraction Data from Amorphous Carbon," ACA Abstracts, Eufaula, PB20:26 (1980).

X-RAY POWDER DIFFRACTION IN EUROPE

J. L. de Vries

N. V. Philips

Eindhoven, The Netherlands

INTRODUCTION

Talking about X-Ray Diffraction in Europe, I'll limit myself to personal experiences. I started with XRD in Amsterdam University, namely structure analysis, but as well powder diffraction for identification of minerals, determination of crystallinity of stretched gelatine and many more such applications. Exclusively, cameras with photographic recording were used, and they are still used in Europe, to a great extent; maybe in 30-40% of all cases. We used a mechanical rectifier controlled by an adjustable spark gap. This system worked fine, till our laboratory cat decided to sniff at the sparks! That was the end of our generator and the cat. The powder samples in our D.S. cameras, mostly glued onto glass fibers were rotated during exposure. As we could not afford small motors, the rotation was done by blowing air against a windmill system mounted on the sample holder. A typical Dutch solution.

After I joined Philips in 1954, we started working with the goniometer diffractometer which at that time came into use in Europe. Although a system was published by Trost in 1940, due to war conditions commercial production only started in the early fifties by many firms, e.g. Labor Berthold, Seimens in Germany, Hilger in UK, Secasi, C.G.R. in France and Philips. Not many of those firms are still active in this field.

The equipment in 1954 consisted of a goniometer; basically,

the same as used today, a stabilized generator with sealed off
tubes, but the electronics were very much simpler, consisting only
of a power supply for the GM detector, indication tubes, and strip
chart recorder. Some years later the scintillation counter and
proportional counter came into use, enabling Pulse Height Selection to reduce the background. This equipment was already tran-
sistorized and therefore, more compact. The only improvements made
in modern equipment are higher ratings of generators and tubes,
adding of a crystal monochromator and automation by μ processor or
minicomputer. But the basic geometry remains unchanged.

Looking at the considerable possibilities of XRD, at the very
wide variety of applications, it may be asked why has XRD, compared
with XRFS, and IR, not become more popular today? One of the reasons must be the high price of diffractometers, or the inconven-
ience of handling films. It is a difficult technique; qualita-
tively due to the enormous number of compounds with millions of
possible lines, quantitatively due to lack of precision and accuracy.

In nostalgic reminiscing, I would like to discuss some problems we encountered. In interpreting diffraction lines, one is often hindered by extra lines. The anodes commonly used only give
$\alpha 1$, $\alpha 2$ and β lines, but Mo does show $K\beta_5$ lines! Contamination
lines e.g. $WL\alpha$ between Cu $K\beta$ and $K\alpha$ may give wrong impressions.
There may be stray lines, due to wrong alignment of the goniometer
slits.   High background at high θ angles may be caused by direct radiation through the tube head, entering the detector. Reflections against the sample holder may give extra diffraction
lines. Some spurious lines find their origin in the properties of
the counter medium and the P.H.S. Wavelengths of the white continuum are also reflected against the lattice planes, but in general
they do not pass the P.H.S. If however, their energy is such that
their escape peaks can pass thru the P.H.S. window, extra broad
peaks seem to appear. Those energies must be higher than the sum
of the excitation energy e.g. 30 kV for Xe^+ and the lower window,
6 kV for Cu. Figure 1 gives some examples of extra peaks in an
NaCl matrix at 30, 40, 50, 60 kV.

Of course, in modern times, the monochromator eliminates these
extra lines. Often a filter is used to remove $K\beta$ lines and some
background. The optimum place of this filter has been a topic of
discussion for many years. Generally speaking, the filter should
always be placed between sample and detector unless the sample
contains the filter material in high concentrations.

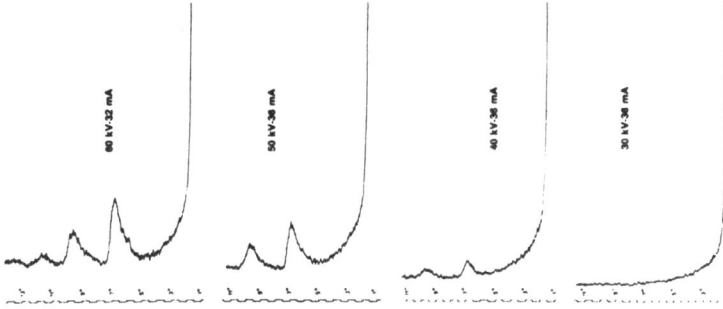

Figure 1

Reflection in the conventional parafocussing geometry can
only occur against crystal planes parellel to the surface. For
reproducible measurements, the partical orientation should be random and there should be sufficient particles to allow good statistical probabilities. Large crystallites may have a dominant influence, however, Figure 2 shows lines of a $Pb(NO_3)_2$ powder taken in
1957.

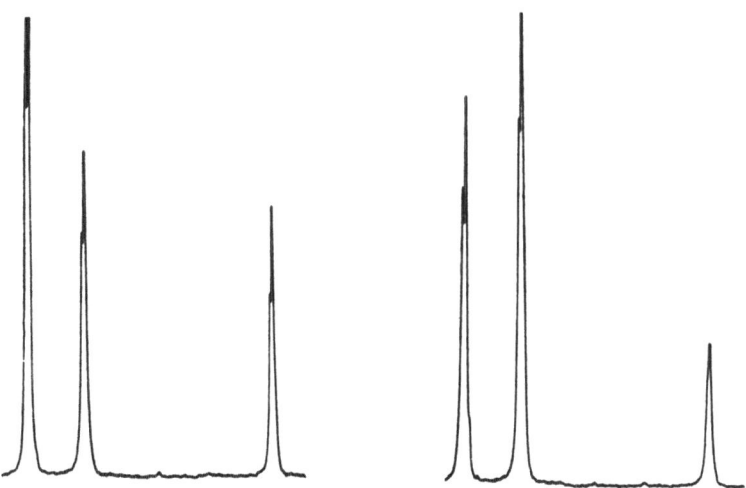

Fig. 2. Stationary Sample

The scan was repeated in two different orientations of the sample
holder, the line might even shift! Rotating the sample improves
the reproducibility as shown in Figure 3.

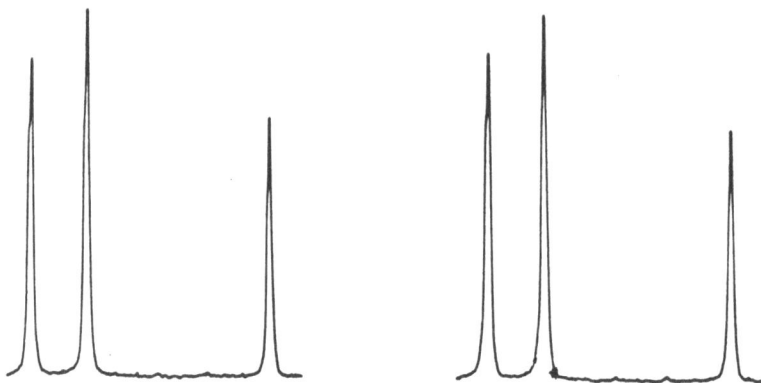

Fig. 3. Rotating Sample

Even in a completely homogeneous sample the limited penetration
depth thus the limited number of particles contributing, may give
intensity fluctuations. A simple calculation shows that for particles of 5μm cube size the s.d. of an intensity measurement might
be 4% for a pure compound or 12% for a 10% volume concentration;
even with rotating samples, these figures are 1.5% and 5% for 10μm.
That is one of the reasons why it is extrememly difficult to get
reproducible intensity measurements from specimens in transmission,
using Guinier geometry or on a diffractometer with focalizer.

    Often one can vary parameters to get optimal results, this
should be done based on statistics. Often peak intensity increase
is accompanied by higher background. If one divides counting times
on peak and background optimally, one should maximize $\sqrt{n_1} - \sqrt{n_2}$.

This follows from the equation:

r.s.d. $= \dfrac{100}{\sqrt{T} \, (\sqrt{n_1} - \sqrt{n_2})}$ %, where n_1 and n_2 are countrates on
peak and background and T is the total counting time.
$T = t_1 + t_2$ and $t_1/t_2 = \sqrt{n_1/n_2}$.

Table 1

	Iron oxide powder sample							
	Cu radiation				Co radiation (filter on the divergence slit)			
Peak	43.5°				51°			
Background		42°				49°		
	n_1	n_2	$\sqrt{n_1} - \sqrt{n_2}$	$\dfrac{n_1 - n_2}{n_2}$		n_2	$\sqrt{n_1} - \sqrt{n_2}$	$\dfrac{n_1 - n_2}{n_2}$
Scintillation counter with filter	1060	832	3.5	0.2	1152	480	12	1.4
Scintillation counter without filter	5630	5000	4	0.1	2020	864	15	1.3
Proportional counter with filter	288	160	4.5	0.8	770	224	13	2.5
Proportional counter without filter	1250	800	7	0.5	1600	704	14	1.3

Table 1 shows some results in choosing between Cu and Co radiation and S.C. or P.C. for quantitative analysis $\sqrt{n_1} - \sqrt{n_2}$ should be as high as possible, whereas for qualitative analysis, the highest value of $(n_1 - n_2)/n_2$ should be best.

Application We estimate that there are approximately 6000 installations in Europe, of which 80-90% are used for powder work. Approximately 40% are still using cameras, for a large part focussing cameras. We also estimate that approximately 60% of these systems are used in Universities and research institutes, the others in more industrial laboratories. The number of instruments actually used for process control is rather small. In which fields are they used?

Table 2
Estimated Distribution of X.R.D. Instruments Over
Various Application Fields

MINERALOGY	15%	GENERAL PHYSICS	20%
GEOLOGY	10	GLASS/CERAMICS	5
METALS	20	ENVIRONMENTAL	3
CHEMISTRY/PHARMAC.	20	OTHERS	7

Table 2 gives some indication, but a survey of papers published gives another impression as shown in Table 3. Their references
are taken from my personal files, which have pretty constant but
limited coverage. The total number of papers is actually slightly
decreasing. It is interesting to note that when we use a certain
method we don't publish anymore. Most of these installations do
not work full-time, however.

Table 3

NUMBER OF PAPERS PUBLISHED

	1956	-58	-62	-66	-70	-74	-78	TOTAL
DIFFRACTOMETER	4	6	9	7	31	48	52	157
HIGH/LOW T, P	37	46	73	59	40	41	26	322
MONOCHROMATOR	12	9	11	12	13	15	9	81
S.A.S.	16	28	15	35	19	16	28	157
TOPOGRAPHY	-	29	64	65	40	31	19	248
METALS STRESS	30	63	71	85	60	32	11	352
TEXTURE	-	13	22	20	22	15	12	104
PHASE ANAL	1	5	9	12	17	15	4	63
MINERALS	14	22	13	14	9	12	8	92
QUANTITATIVE	-	6	8	18	13	12	9	66
FIBRES	20	65	20	9	16	13	5	148
POLLUTION	3	1	1	1	5	7	14	32
CERAMICS	7	19	18	15	15	9	5	88
	144	321	334	352	300	266	202	1910

The application fields vary enormously; besides standard identification, quantitative analysis is mainly used for metals, pig-
ments, pharmaceuticals and polymers. One of the first applications
I heard of was the ageing of bread (Katz 1930's). I would like to
mention some of the first applications which I did.

There are many common types of aspirin drug commonly used in
Europe, mainly consisting of phenacetin, aspirin, caffeine and
carriers. It was possible to determine aspirin quantitatively by
peak intensity count. Calibration was done by adding successively
known amount of the analyte, the method of "spiking". Even with
the very simple equipment we used in 1955, it was possible to perform quantitative analysis. Figure 4 shows a diffraction pattern
of the blue pigment copper phtalocyanine.

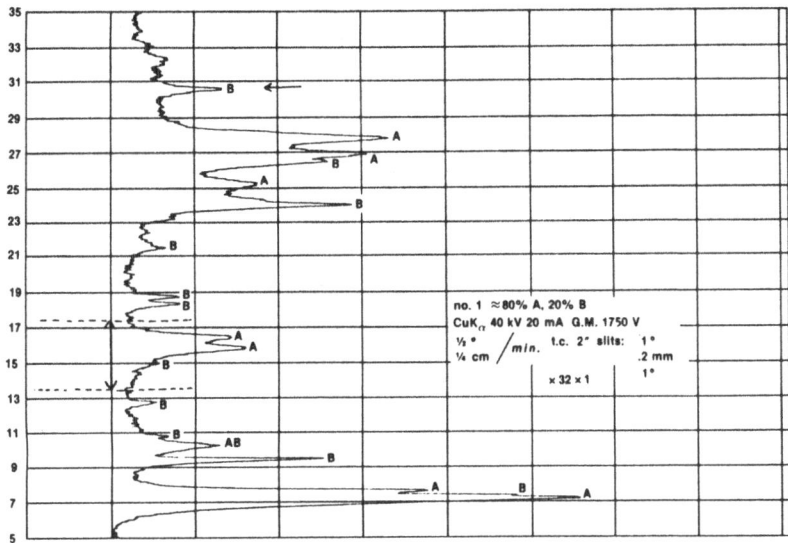

Figure 4

Two modifications can be distinguished: a poorly crystallized component A, showing only broad lines and another modification, B.
The ratio A:B must be determined by combining a peak count for B
and an integrated intensity scan for A, it was possible to set up
a straight calibration line using 4 standard samples as shown in
Figure 5.

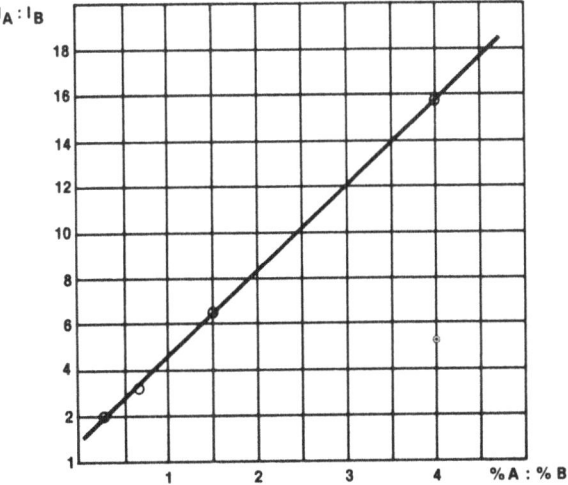

Fig. 5   Calibration curve for A and B

Quantitative analysis is now performed routinely in Europe.
For absorption corrections various methods are applied, e.g. internal standards, reflection against sample support. Recently calculation of matrix corrections is gaining in importance. In favorable cases no matrix corrections are necessary and relative intensities can be calculated. A well known example is the determination of residual austenite in ferrite. Although widely publicised,
this method is only slowly being applied in actual process control.

In our laboratory we also performed studies finding optimal
conditions concerning: anode material, intensity available, resolution needed and mode of analysis. By repeated scans over peak
areas, the measured intensity can be integrated to achieve sufficient statistical precision. The background can best be determined
on two stationary 2θ positions. The total available measuring time
is best used if the number of scans approximates the square root
of the ratio of integrated intensity to background. In this way,
it was possible to go down to approximately 0.5% austenite. Figure
6 shows some scans over relevant peaks for varying concentrations.

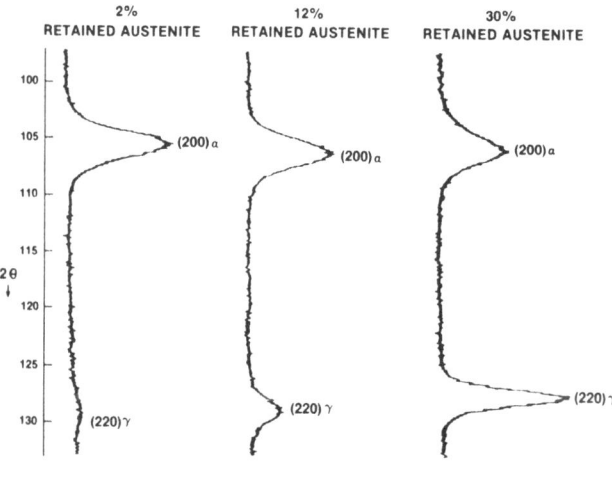

Figure 6

Another favorable case is the determination of the anatase/rutile
ratio. Already in 1961 a fixed two-channel system was developed
in Germany. Ratio measurements were done automatically within one
minute, and more than 30,000 analyses were performed annually.

Quite some studies are published on <u>stress analysis</u>, but the
actual number of equipment in Europe is rather limited, may be less
than 50; mainly in Germany and U.K. Texture determination in met-

als, minerals and polymers is a growing field. To find the texture
in aluminum cups for deep drawing, I used in 1955 a method by comparing line intensities in a textured and random orientated sample.
If sufficient lines can be measured and evaluated by computer calculations, this method of inversed/pole figures gives off results
complementary to the conventional techniques and coming more into
use.

What improvements can be expected? Mainly in ease of operation, automation for qualitative and quantitative analysis. New
detectors have also come into use. The Si(Li) detector has been
used to better utilize the PHS, but the cooling system needed makes
a larger scanning range difficult. Fixed angle diffraction, using
the white spectrum and energy resolution was much published in
Europe in the early seventies, but was never actually used routinely.

Position sensitive detectors are being used widely in Europe
since 1978, mainly for small-angle scattering studies. Although
most systems are based on the linear high resistance wire, gas
detectors and other systems are now being studied also. Another
growing field of interest is the combination of XRD with other
methods of analysis. An example is the addition of an attachment
for differential Thermal Analysis on an X-ray diffractometer in
Germany in the sixties for the analysis of clay minerals. Another
case is the determination of total Ca by X-ray spectrometry added
to the analysis of phases by X.R.D. Examples of this combination
are found in the aluminum and cement industries. Summarizing, we
can conclude that X.R.D. is finally entering the field of routine
analysis, thanks mainly to the growing degree of automation, thus
making easy operation possible.

Finally, I would like to express my great gratitude for the
honors bestowed upon me. I would like also to thank my many col-
laborators who made all this work possible and my two great teach-
ers, Prof. McGillavry formerly of Amsterdam University and Bill
Parrish, formerly of Philips Laboratories, Inc., Irvington.

A HANAWALT TYPE PHASE IDENTIFICATION

PROCEDURE FOR A MINICOMPUTER

Robert L. Snyder

New York State College of Ceramics

Alfred University, Alfred, N.Y. 14802

INTRODUCTION

The use of computers to aid in the identification of phases
from their powder diffraction patterns was pioneered in the mid
1960's by Frevel, Nichols and Johnson (1-3). Today's most widely
used Johnson algorithm conducts a reverse sequential search by
comparing each reference pattern in the JCPDS powder diffraction
file (PDF) to the unknown pattern. A figure of merit is computed
for each match and the patterns with the best figures of merit
are listed at the end of the search. The Nichols approach is a
reverse search of a singly inverted reference file. An inverted
file is one which stores the reference patterns according to the
d value of the lowest angle 100% intensity line (d_1). This type
of file is analogous to the Hanawalt search books distributed by
the JCPDS for manual searching. When an inverted file is stored
in a random format, along with suitable disk directory files,
only reference patterns containing d_1 values of interest need be
read in the search.

Both the Johnson and Nichols algorithms use the full PDF
which today contains about 35000 patterns of highly variable
quality (4). The Frevel algorithm is the only one, to date,
which attempts to deal with the problem of poor quality reference
patterns. This approach also uses a singly inverted reference
file; however, the file is restricted to three hundred commonly
identified phases in unknowns (5). The use of this drastically
restricted file enabled this algorithm to be the first to be
converted to run on a laboratory sized minicomputer. This program has been recently generalized by Edmonds. Johnson has also
converted his algorithm for use on a minicomputer but due to the
exhaustive search approach it also must use a greatly reduced

Frevel type of reference file.

Recent developments in computer technology have brought powerful minicomputers with large amounts of mass storage into a price range that most laboratories can afford. This in turn has increased the desirability of a full PDF minicomputer search-match system. For reasons of either mass storage capacity or computational speed none of the existing algorithms are directly convertable to a minicomputer environment. The development of the full PDF search-match system described here was done under the sponsorship of the Siemens Corporation and is currently part of the software supporting their D500 powder diffraction system.

DESIGN CONSTRAINTS

For the implementation of a full file search-match system on a modern minicomputer a number of desirable or essential features must be incorporated. The following design constraints, once established, dictated the form of the search algorithm.
1) The entire data base should be contained in less than five megabytes of disk storage. This allows for implementation on all but the very smallest of current laboratory minicomputer configurations. To meet this constraint a binary compression format was devised for each of the active patterns in the PDF. All d values are converted to integer d-codes by dividing them into 1000. Thus the integers 1 to 2048 will represent d values ranging from 1000 to 0.488. This integer conversion of the d values introduces an average $\Delta 2\theta$ round off error of 0.025°, for Cu radiation d values greater than 1.0, with a maximum round off of 0.05°. Since the average $\Delta 2\theta$ for the cubic patterns in the PDF is 0.1° (4), the maximum round off error of 0.05° does not significantly degrade the reference data. The advantage is that the integers 1 to 2048 can be stored in 11 binary bits.

Using 11 bits to store the integer d-code leaves five bits which can be used to encode the intensity value and result in the storing of each d-I pair in one 16 bit minicomputer word. Intensities are therefore stored on a scale of 0 to 30. On removing the trailing blanks from the formula the above measures allow for the compression of the 15 megabyte PDF into about 2.5 megabytes.
2) Due to the slow input/output speeds of minicomputers the entire data base should not be searched. This constraint dictates the use of an inverted file with a Hanawalt search strategy. However, a random file structure would violate the file size constraint discussed above. The solution of this dilemma is to create a pseudo-random or indexed sequential file structure. This type of file is inverted by sorting the patterns according to their d_1 values and then the sorted, binary compressed patterns are written sequentially to a disk file. A single disk access will put the program to within 256 words of the

pattern or d_1 range of patterns desired. A double buffered assembly language subroutine performs this function with a negligible amount of time wasted looking at patterns outside the desired d_1 range.

3) Since floating point arithmetic is very slow on most mini-computers it should be avoided. This design contraint is met by first converting the input d values of the unknown pattern into integer d-codes. The entire reverse search is then conducted using only integer arithmetic.

4) Each yearly update to the PDF should be added to the search system with minimum effort. To meet this design constraint each of the sets of the PDF are independently inverted and stored. This produces 28 data files each with an associated d_1 and PDF number directory file.

5) Due to the large volume of patterns in the PDF, and the low probability of most of them being found in common unknowns, a strategy which searches the most likely references first will greatly minimize search time. Following the work of Frevel (5) a file containing approximately 300 of the most common phases in unknowns was created in exactly the same format as described above. This file called the MICRO file is searched first. Any correctly matched phases are quantitatively subtracted from the unknown and only the residual pattern is passed on for further searching. The list of 2500 phases which the JCPDS has designated as frequently encountered have been gathered into a second file called MINI. This file is searched in the second phase of a search. The full 28 set MAXI file is only searched if a residual pattern remains after the MICRO and MINI file searches have been completed.

THE SEARCH ALGORITHM

The hierarchical Hanawalt search proceeds as follows:

1) The d_1 disk block directory of the MICRO file is read into memory.

2) A binary search procedure is used to locate the disk blocks containing the reference patterns whose d_1 values lie within a $\pm 0.1^{\circ}$ 2θ Cu error window around the d_1 of the unknown.

3) Each reference pattern in the correct d_1 range must pass the following three tests:
   a. Its subfile code (e.g. inorganic, mineral, etc.) must agree with those specified by the user.
   b. All diffraction lines with $I \geq 50$ must at least be present in the unknown.
   c. If the user specified chemical constraints, they must be met.

4) For those patterns which pass the tests a figure of merit (FOM) is. computed as described in the next section.

5) The single reference pattern with the highest FOM for d_1 is saved for the MATCH procedure.

6) Steps 2 through 5 are repeated for lines d_2 and d_3.

7) The MATCH routine is where one or more of the three saved
   patterns may be quantitatively subtracted from the unknown.
   If a match was found the rescaled residual pattern with a
   new d_1 value is returned and the search procedure is repeated
   starting at step 2. If no match is found steps 2 through 7
   are repeated with an error window of $\pm 0.2°$ 2θ Cu. If this
   also fails to produce a match the entire procedure is repeated
   using the next three most intense lines of the unknown (d_4
   through d_6) and then again for d_7 through d_9.
8) When no (further) matches are found on searching the 9 highest
   intensity lines in the (residual) unknown pattern the MICRO
   file is abandoned. Steps 1 through 7 are then repeated on
   the MINI file. If any residual pattern remains after the MINI
   file search, steps 1 through 6 are repeated for each of the
   28 files in the MAXI file. After the 28 set MAXI file search
   is complete 7 is executed, and the process is repeated for
   lines d_4 through d_9.

   The search program accepts data from either the d-I file output by the automated data reduction (ADR) program (6) or from
manual entry. It rejects any $K_{\alpha 2}$ lines and only searches patterns
in the user selected subfiles. Chemical constraints may be applied.

THE MATCH ALGORITHM

   The match routine is entered to evaluate the three patterns
found by the search. Any pattern with a figure of merit less than
10 is rejected. If chemistry checking is in effect this minimum
acceptable FOM value is lowered to 7.0. The acceptance of an
incorrect pattern at this stage will so distort the residual pattern as to make any further correct matches unlikely.

   The Figure of Merit contains only three terms:

$$FOM = d_R \times I_R^2 \times d_U$$

d_R = percent of reference lines which match the unknown
     and which have I>I of the lowest matched line.

I_R = percent of the reference intensity matched.

d_U = percent of the unknown lines matched.

   The d_R term does not take into account the closeness of the
d agreement. If an unknown line falls within the rather wide
error window of the reference line it is called a match. The
goodness of fit is only considered when the windows of multiple
lines overlap, and then only to match the correct reference and
unknown intensities. Since the average $\Delta 2\theta$ Cu for the 2000
indexable cubic PDF patterns is $0.1°$ (4) and over 1000 are not
indexable within a $\pm 0.5°$ error window, any term in the FOM which
considers goodness of fit beyond the match/no match criterion is
not justified.

The I_R term is essential to the Hanawalt strategy. The entire file structure and search procedure is based on the assumption that the intensities are not distorted beyond recognition. This term is given added weight by squaring it. The last term d_U forces the search to proceed until it has located one of the major phases in the unknown before any reference pattern subtraction occurs. An example of this effect can be seen in sample 7 discussed below.

If one or more of the three patterns found in the search have acceptable FOM's, the one with the highest FOM will be subtracted from the unknown first. Before subtractraction a more accurate intensity scale factor is computed by performing an intensity weighted linear regression between the I_{ref} and I_{unk} values. The scale factor obtained is used to calculate I_{unk} values from the I_{ref} values. All lines which calculate lower than observed by 5 or more units are rejected as being possibly overlapped by a second phase. A second intensity weighted least squares is computed on the non-overlapped lines to obtain the final scale factor. If a second reference pattern also passes the acceptance tests it will also be rescaled and subtracted. The residual pattern, if one remains, is rescaled, resorted and passed back to the search routine for continued searching.

RESULTS

We have to date, passed over one hundred experimental patterns through this search procedure. Without chemical information being specified the program has correctly identified every one or two phase unknown given it. This includes unknowns containing such complex phases as mullite, spodumine, pedalite and numerous other low symmetry and mildly solid solution prone materials. As the number of unknown phases increases to four or more the probability of finding an incorrect phase increases. The desirability of using chemical constraints increases with the number of phases in the unknown. The average time it takes to perform a three line (d_1 to d_3) search in each of the three steps of the hierachy is: MICRO file - 5 sec., MINI file - 27 sec., and 4.22 min. for the MAXI file.

A standardized test of the algorithm can be obtained by using the Jenkins and Hubbard (7) round robin samples. The first six of these patterns were synthesized while the seventh was distributed as an experimental powder. The results for these samples are shown in Table 1.

The right side of the table shows the actual weight percents in each sample versus the normalized scale factors found by the program. When matrix absorption effects can be ignored, as in the artificial samples (RR 1B through RR 6B), the scale factors give the weight fraction quite accurately. However, in most real specimens where matrix effects are strong, as in the experimental pattern (RR 7B) there is significant disagreement.

TABLE 1. Results of Search Procedure for the Jenkins and Hubbard Round Robin (RR) Samples

PHASES PRESENT	HIERARCHICAL SEARCH			MAXI FILE SEARCH ONLY	SCALE FACTOR	
	MICRO	MINI	MAXI		OBS	ACTUAL
RR No. 1B						
Fe_3S_4	0 (0)	1 (1)	– (–)	1 (1)	.42	.40
Fe_2CuO_4	0 (0)	2 (2)	– (–)	2 (2)	.30	.30
CuS	0 (0)	2 (2)	– (–)	2 (2)	.18	.20
FeS_2	0 (0)	3 (3)	– (–)	3 (3)	.10	.10
Time (Min)	.30	1.05		8.09		
Passes	4	3	– (–)	3		
RR No. 2B						
$PbBr_2$	1 (1)	–	– (–)	1 (1)	.42	.40
$PbOSO_4$	0 (0)	1 (1)	– (–)	2 (2)	.19	.20
$Pb_3O_2Cl_2$	0 (0)	1 (1)	– (–)	2 (2)	.29	.30
$PbCl_2$	0 (0)	2 (2)	– (–)	3 (3)	.09	.10
Time (Min)	.47	1.45		15.48		
Passes	5	2	– (–)	3		
RR No. 3B						
$BaSO_4$	0 (0)	1 (1)	– (–)	1 (1)	.33	.30
$BaCO_3$	0 (1)	2 (–)	– (–)	2 (1)	.41	.40
$SrSO_4$	0 (0)	3 (2)	– (–)	3 (2)	.14	.15
$CaSO_4$	0 (0)	4 (3)	– (–)	4 (3)	.12	.15
Time (Min)	3 (.4)	2.1 (1.9)		18.4 (14.6)		
Passes	4 (5)	4 (3)	– (–)	4 (3)		
RR No. 4B						
Calcite	1 (1)	– (–)	0 (0)	1 (1)	.28	.30
Argonite	1 (1)	– (–)	0 (0)	1 (1)	.29	.30
Vaterite	0 (0)	0 (4)	0 (0)	0 (5)	.21	.20
$ZnCO_3$	0 (0)	0 (3)	0 (0)	0 (4)	.21	.20
$CoCO_3$*		2*		3*		
Time (Min)	.5 (5)	3.2 (3.1)	16.3 (11.9)	32.9 (36.6)		
Passes	5 (5)	6 (8)	4 (4)	7 (9)		
RR No. 5B						
$Ca_3Si_2O_7$	0 (0)	1 (1)	– (–)	1 (1)	.39	.40
Fe_2SiO_4	0 (0)	2 (2)	– (–)	2 (2)	.30	.30
SiO_2	0 (0)	3 (3)	– (–)	3 (3)	.20	.20
$MgSiO_3$	0 (0)	5 (5)	– (–)	5 (–)	.12	.10
$MgNiSi_2O_6$*				4*		
$Mg_{.88}SiO_3 \cdot 12FeSiO_3$				(4*)		
Time (Min)	.35 (.35)	35 (35)	– (–)	34.5 22.7		
Passes	4 (4)	5 (5)	– (–)	6 4		
RR No. 6B						
$m-C_6H_4N_2O_4$	0 (0)	1 (1)	0 (0)	1 (1)	.44	.40
$2,4-C_6H_4N_2O_5$	0 (0)	0 (0)	1 (1)	2 (2)	.39	.40
$2,6-C_6H_4N_2O_5$	0 (0)	0 (0)	0 (0)	3	.17	.20
C_5H_8N*			2* (2*)	(3*)		
$C_{19}H_{30}O_4$*				(1*)		
Time (Min)	.23 (23)	1.4 (1.4)	26.5 (26.5)	24.4 (29.3)		
Passes	4 (4)	5 (5)	6 (6)	6 (7)		
RR No. 7B						
$BaCl_2 \cdot 2H_2O$	0 (0)	3 (1)	– (–)	3 (1)	.30	.50
KI	0 (1)	4 (–)	– (–)	4 (1)	.38	.15
ZnO	0 (1)	4 (–)	– (–)	4 (2)	.32	.20
Si	– (–)	– (–)	– (–)			.15
Time (Min)	.40 (.42)	3.1 (3.1)	13.6 12.5	31.9 (20.8)		
Passes	4 (5)	8 4	(4)	8 (6)		

*Incorrect phase
The numbers in parentheses refer to a run with fluorescence chemistry specified.
The numbers listed for each phase are the search pass in which the phase was found.

As shown in the table, Round Robin samples 1B, 2B and 3B, each containing four phases, were fully and correctly analyzed in about two minutes using the hierarchical search. The values listed in parentheses refer to the same problem run with X-ray fluorescence chemistry specified (i.e. all elements with Z<Si and specific elements with Z>Si treated as present). For samples 1B to 3B the specification of chemistry merely resulted in slightly faster search times. The results are correct with or without chemistry. The same correct results are obtained if only the full MAXI file is searched; however, this increases the search time to about fifteen minutes.

Sample 4B is interesting because the vaterite pattern (13-192) used to synthesize the sample has been deleted from the PDF, due to significant errors in it, and replaced by 25-127. The discrepancies in the old vaterite pattern and the similarities between $ZnCO_3$ and $CoCO_3$ cause the program, when run without chemistry, to incorrectly identify the latter phase. The distortion of the residual pattern due to the original errors in pattern 13-192 and the subtraction of the incorrect $CoCO_3$ causes the program to fail to find the current vaterite pattern. When chemical constraints are applied, all correct phases are identified. However, errors in the original vaterite patterns leave a residual pattern which the program exhaustively searches, wasting some search time.

The $MgSiO_3$ phase in sample 5B had a systematic error introduced into the d values. The hierarchical search identifies the other three phases first and then fails to identify any phase with the initial error window of $\pm$ .1° 2θ. On automatically opening the error window to $\pm$ .2° 2θ on the next pass, however, it correctly finds this phase whether or not chemistry was specified. When only the MAXI file is searched two other phases similar to $MgSiO_3$ are found within the initial error window.

Since only fluorescence chemistry was specified with all elements below Si treated as positive, the single effect on the organic sample RR 6B was to lower the acceptable FOM from 10 to 7. The hierarchical search finds two correct phases and then a false phase (C_5H_8N). The subtraction of the incorrect phase prevents the identification of the remaining correct phase. The MAXI file search without chemistry finds all three correct phases. However, the lower acceptable FOM when chemistry is specified causes two incorrect phases to be found along with two correct phases. It is recommended that for C, H, O and N-containing organics that chemistry not be used as a constraint.

Sample 7B was run and analyzed using our automated diffraction system (6). This system is adjusted to leave all small questionable peaks in the output file for manual deletion upon evaluation. However, for the case shown in the table, all false peaks were left on the file and it was passed without editing to the search program to evaluate a fully automatic analysis. The sample contained Si as an internal standard and though the search listed this phase in each pass it does not subtract it because

it only has two matching lines (3 are required for a match by
default). The three unknown phases are correctly identified in
this fully automated analysis whether or not chemistry or
hierarchical searching is employed.

SUMMARY OF THE ROUND ROBIN ANALYSES

Of the twenty six unknown phases in the round robin samples,
at least one run of the search program produced a completely
correct analysis. Examining the effects of hierarchical versus
a MAXI file only strategy and the effects of using chemistry we
see that:

	Correct Phases Found	Incorrect Phases Found
hierarchical no chemistry	23	3
hierarchical with chemistry	25	1
MAXI file no chemistry	24	2
MAXI file with chemistry	24	3

The preferred approach is clearly to do a hierarchical search
with fluorescence chemistry as a constraint. The average hierarchical search time when no false peaks are present is about four
minutes. For a MAXI file search only, or when false residual
peaks remain, forcing the program to go to a full MAXI file
search, the average time rises to about 25 minutes.

REFERENCES

1. Frevel, L.K., "Computational Aids for Identifying Crystalline
   Phases by Powder Diffraction," Anal. Chem. 37:471-482 (1965).
2. Nichols, M.C., "A Fortran II Program for the Identification of
   X-ray Powder Diffraction Patterns," UCRL-70078, Lawrence
   Livermore Laboratory, Oct. 1966.
3. Johnson, G.G. and Vand, V., "A Computerized Powder Diffraction
   Identification System," Ind. Eng. Chem. 59:19-31 (1967).
4. Snyder, R.L., Johnson, Q.C., Kahara, E., Smith, G.S. and
   Nichols, M.C., "An Analysis of the Powder Diffraction File,"
   UCRL-52505, Lawrence Livermore Laboratory, June 1978.
5. Frevel, L.K., Adams, C.E. and Ruhberg, L.R., "A Fast Search
   Program for Powder Diffraction Analysis," J. Appl. Cryst.
   9:300-305 (1976).
6. Mallory, C.L. and Snyder, R.L., "The Alfred University X-ray
   Powder Diffraction Automation System," N.Y.S. College of
   Ceramics Technical Paper 144 (1979).
7. Jenkins, R. and Hubbard, C.R., "A Preliminary Report on the
   Design and Results of the Second Round Robin to Evaluate
   Search/Match Methods for Quantitative Powder Diffractometry,"
   Adv. X-ray Anal. 22:133-142 (1979).

QUALITATIVE PHASE ANALYSIS USING AN X-RAY POWDER DIFFRACTOMETER

W. N. Schreiner, C. Surdukowski, *R. Jenkins

Philips Laboratories
Briarcliff Manor, N.Y. 10510
*Philips Electronic Instruments Inc.
Mahwah, N.J. 07430

During the past three years we have undertaken the development of a complete X-Ray Powder Diffraction facility with the goal of fully integrating experimental and analytical procedures. Such an approach potentially offers substantially improved performance over previously existing systems by virtue of its internal self-consistency and it opens the possibility of significantly extending analytic procedures for both qualitative and quantitative analyses. Our work to date has resulted in improved performance and significant extensions in both areas, and today I will report on those advances in the area of qualitative analysis.

Two years ago we began a study of the systematic and random errors associated with a Bragg-Brentano Parafocusing X-Ray Powder Diffractometer. Current results from our work in this area will be presented shortly in another paper (1). Although it would be nice to understand these errors from a physical point of view, this is not essential as long as one is able to model them empirically. The principal errors have been studied in the past (2), and we have found most of them to be reasonably well obeyed in practice, although significant differences at low and high angles have been observed. Once the form of the systematic errors has been established, one is able to properly treat them in the data analysis.

Last year we developed peak hunting software employing a 2nd derivative algorithm (3). The most unique features of the implementation were the use of a variable width window characteristic of local curvature and the proper statistical treatment of the raw count data within the window. The result was an algorithm capable of reliably detecting lines whose peaks are as small as 1-σ above background and of separating α1/α2 lines down to about 26° 2θ. This sensitivity assures the experimenter of extracting near maximum information from his diffractogram. At his option, the experimenter could trade the sensitivity for shorter scan times, if this were important in his application. More significantly, however, the detection of smaller peaks can be used to enhance the recognition of phases with weak lines and to provide sufficient information to permit positive identification of minor phases with relative concentrations below 5%.

Last year we also packed the entire JCPDS data base, then consisting of about 34000 compounds, onto a single 5 mb disk (4). With a Tektronix graphics terminal, a user could then simultaneously display his diffractogram and up to three JCPDS patterns. A cross-hair cursor could be used to examine specific line matches or read out coordinates of interest from the diffractogram. Such capability allowed the experimenter to rapidly compare his unknown diffractogram with a series of reference patterns and greatly extended and facilitated the manual matching process.

This year we present the culmination of our efforts in the form of SANDMAN, a fully integrated, probability-based search/match/identify program intended for qualitative analysis. Results obtained with SANDMAN have been found to be consistently superior to those obtained from the commonly used search/match programs, and SANDMAN truly rivals the pattern recognition capabilities of a human being. The reasons for SANDMAN's excellent performance lie partly in the integrated approach taken toward the entire X-ray powder diffraction problem for the parafocusing geometry, partly in the proper packing of the data base, partly in several major innovations in the areas of searching, matching and scoring, partly in the development and use of simulation techniques, and partly in extensions beyond simply searching and matching, extensions which became possible only as a result of the other improvements.

SANDMAN may be run automatically in a batch mode as one component of the qualitative analysis package of the APD-3600 (5), or as an interactive stand-alone program for off-line analysis. As an interactive program, SANDMAN has a command structure very similar to the PDSM component of CIS (6). However, some CIS commands are notably absent, such as the d- and I-error windows. SANDMAN automatically compensates for a reasonable amount of the frequently encountered systematic errors leaving only the need to account for random errors. This obviates, and prevents possible misuse of, user-specified error windows. SANDMAN commands can be grouped under 6 headings as seen in Fig. 1. Data may be entered manually or read from d/I lists generated by the 2nd derivative peak hunting algorithm, APDPEAK (3). Parameter commands are only needed if data is entered manually; otherwise these values are set automatically, based on the scanning parameters. The search may be limited by selection of specific subfiles. One of these options permits selection of special

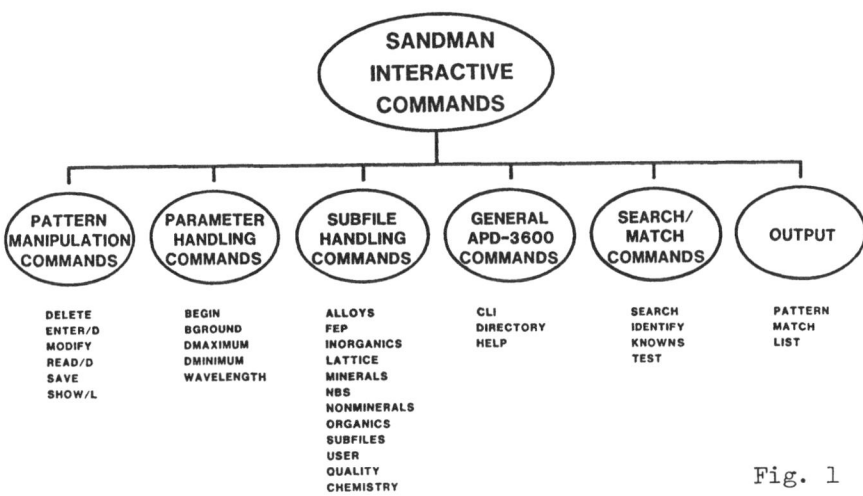

SANDMAN INTERACTIVE COMMANDS					
PATTERN MANIPULATION COMMANDS	PARAMETER HANDLING COMMANDS	SUBFILE HANDLING COMMANDS	GENERAL APD-3600 COMMANDS	SEARCH/ MATCH COMMANDS	OUTPUT
DELETE	BEGIN	ALLOYS	CLI	SEARCH	PATTERN
ENTER/D	BGROUND	FEP	DIRECTORY	IDENTIFY	MATCH
MODIFY	DMAXIMUM	INORGANICS	HELP	KNOWNS	LIST
READ/D	DMINIMUM	LATTICE		TEST	
SAVE	WAVELENGTH	MINERALS			
SHOW/L		NBS			
		NONMINERALS			
		ORGANICS			
		SUBFILES			
		USER			
		QUALITY			
		CHEMISTRY			

Fig. 1

user-defined disk-resident subfiles tailored to meet specific needs.
The 'SEARCH' command invokes the full search/match process much as
in the CIS system. The 'IDENTIFY' option, however, goes one step
further: It picks out those patterns from the match list which, in
combination, best account for all the user lines. The 'KNOWNS' command is used to enumerate any patterns known to be present before
the search, such as internal standards. "TEST' allows the user to
score a single pattern even if it did not appear in the match list.
'MATCH' allows the user to simultaneously score several patterns and
to obtain a detailed list of the line-match assignments.

In batch mode, SANDMAN provides the ultimate ease of use within
the framework of an automated powder diffractometer. One merely spe-
cifies the d/I data file to be analyzed and the program does the
rest. The entire on-line data base is always searched. Subfile restrictions are indicated in the match score table, but they do not
eliminate phases from consideration.

Fig. 2 shows SANDMAN output for one of the recent round-robin
patterns which causes CIS some problems. After the initial search
and match, CIS finds four correct phases (or equivalents thereof) in
positions 1,2,5 and 9. After successively subtracting these phases,
the fourth phase no longer appears, primarily due to problems associated with subtraction as a technique for phase isolation. SANDMAN
finds all four phases on the top. Here the entire inorganic data
base was searched with no chemistry or subfile restrictions. Similar results are obtained for all the other round-robin examples. In
fact, we generally find that synthetic patterns,like the round-robin
patterns,which contain random but not systematic errors present
little or no difficulty for SANDMAN.

```
SANDMAN:  MATCH SCORE TABLE - REV 0.04              7/25/80  10:34:59
                                                    RR2B.DI

  65  USER LINES              7.4345 > D >  1.8739          BKGND = 1.3
  33  LINES SEARCHED WITH  I > 10.0      563 SEC SEARCH,  130 SEC MATCH
25867.  INITIAL PATTERNS           17450 INSPECTED       3800 SCORED
MATCH SCORE TABLE CONTAINS TOP 126 ENTRIES OF  513  ABOVE SEARCH CUTOFF
     *****  REMAINING PHASES MAY NOT BE IN DATA BASE *****
```

	MLMS	HT/MS/BK/UM				EQ.D.E	I%	JCPDS# DM	FORMULA
1	17.2	28	0	1	10	-10	110 +	5- 608 *I	PB BR2
2	12.9	20	0	1	22	10	76 +	18- 702 *M	PB2 (S O4) O
3	11.3	26	1	4	17	20	56 +	25- 448 *I	PB3 O2 CL2
4	4.8	21	2	5	45	-50	58 +	26-1150 *M	PB CL2
5	4.1	8	0	3	0	100	28	23- 838 I	BA V2 O6
6	2.4	15	2	1	10	-100	33	14- 568 I	ND CL2
7	2.4	5	0	2	8	-20	46	23-1144 I	LA4 MO O9
8	2.0	8	1	1	29	-130	76	12- 666 *I	BA2 PB O4
9	1.7	7	1	0	15	-10	43	26-1453 IM	(PD, AG)4 TE
10	1.6	9	2	2	5	-10	47	25- 951 *I	TI IN (S O4)2
11	1.4	6	1	1	8	-160	28	3- 594 I	CA2 SI O4 * H2 O
12	1.4	7	0	8	16	110	29	28- 442 CI	(AU4 ZR)
13	1.4	5	1	0	4	-90	44	23-1360 *I	K2 NA CU F6
14	1.2	5	1	0	5	-50	13	15- 412 II	(V CO GE)
15	1.1	8	2	0	1	-150	100	25- 459 'M	PB2 AG SB3 S7
16	1.0	6	1	2	24	-120	41	17- 316 *I	CR F3 *3 H2 O
17	1 0	5	2	0	4	-160	47	12- 776 DI	AL2 O3.4 1 / 2 SI O2.3 1 / 2 H2 *? H2 O
18	0.9	6	1	2	4	-70	14	23-1120 I	FE S
19	0.8	4	1	2	22	-70	50	29- 836 *I	(LI TA O3)
20	0.5	5	1	5	28	-220	19	20- 960 *I	K2 ZR3 O7
21	0.2	8	3	10	8	-190	26	19- 265 I	CA Y2 O4
22	0.2	5	1	6	22	-210	42	16- 270 'I	NA2 O2
23	0.2	5	1	2	1	-180	81	18-1115 I	RB GA (S O4)2
24	0.2	5	2	1	8	-330	13	15- 248 IM	CA- SR- BA- TH- LA- P- C- O- H2 O
25	0.1	13	3	11	0	-160	49	25- 786 II	NA2 BE F4

Fig. 2

Beyond merely providing an ordered list of possible candidate
matches and then requiring the experimenter to decide which combination best fits his data, SANDMAN attempts a simultaneous phase identification. Highly probable matches which fit well together are indicated by the + sign in front of the JCPDS number. As seen, in
Fig. 2, SANDMAN identifies precisely those phases, and only those
phases, belonging to the true pattern. Similar identification results are obtained for four of the six round-robins. With the plotting capability of the 'PATTERN' command, the experimenter is rapidly
able to verify and/or modify these phase identifications until his
unknown is solved.

When we apply SANDMAN to real data we distinguish two cases:
1) patterns obtained via the APD-3600 automated data collection facility and 2) manually entered patterns obtained on other instruments.
In the first case, SANDMAN's matching and identification performance
is found to be almost as good as its performance on the round-robin
patterns. (This implies that the random errors used to generate the
round-robin patterns were too optimistic. Unfortunately real data
acquired even under the best of experimental conditions also contains
systematic errors.) The powder sample included in the round-robin
test falls into the first category. In the second category, manual-
ly-entered patterns, SANDMAN's performance degrades somewhat further
although it is still greatly superior to existing programs.

As an interesting example of this latter case, we show in Fig. 3
the broad-line elemental Ni sample given by Dr. G. Johnson in the
2dTS search/match users manual. Apparently, this example has been

```
SANDMAN:  MATCH SCORE TABLE - REV 0.04                 7/28/80   8: 6:12
                                                       GJP27A DI

   10  USER LINES                2.0409 > D >  0.7864        BKGND =21 4
   10  LINES SEARCHED WITH I >  1.0             14 SEC SEARCH,    8 SEC MATCH
25867.  INITIAL PATTERNS                836 INSPECTED        22  SCORED
MATCH SCORE TABLE CONTAINS TOP  11 ENTRIES OF   11 ABOVE SEARCH CUTOFF
       *****  REMAINING PHASES MAY NOT BE IN DATA BASE *****

     MLMS   HT/MS/BK/UM  EQ.D.E  I%   JCPDS# DM    FORMULA
     ----   -- -- -- --  ------  --   ------ --    -------
  1   4.4   8  0  0  0   -270   257 +  4- 850 *M   ( NI )
  2   0.8   3  1  0  1    260   257  15- 806 *I    ( CO )
  3   0.5   3  1  1  2    -10   120  27- 619 II    ( AG2 SE )
```

Fig. 3

difficult to solve using existing computer search/match techniques
and is therefore used as a test case. To find the answer requires
opening the 2dTS d-error windows to about 7 PPT. SANDMAN, however,
rapidly identifies Ni as the principle component. The poor line-
intensity matches resulting from the broadness of the lines does not
substantially affect SANDMAN's performance since the probability-based
scoring technique employed is exceptionally robust under adverse conditions. If we look at the output from the simultaneous line match
(Fig. 4), we see that the pattern is remarkably good with $\Delta d/d$ values
of about 1/1000. The problem with the pattern is not the broadness
of the Ni lines, rather that the pattern displays a specimen displacement error of about -260 microns ! This has the effect of shift-
ing peaks by up to 0.16 deg at low 2-theta and reduces the line measurement accuracy from 1/1000 to about 1/100. It is no wonder that
existing search/match programs have difficulty with this pattern.

    In order to write a program as effective and robust as SANDMAN,
two ingredients, aside from clever software, are essential: 1) systematic errors in the incoming data must be accounted for and 2)
sufficient test data must be available to rapidly age the program.

```
SANDMAN  LINE MATCH TABLE - REV 0.04                   7/28/80   8: 6:16
                                                       GJP27A.DI

   10 USER LINES                 2.0409 > D >  0.7864        BKGND =21 4
   10 LINES SEARCHED      I% (MIN DETECTABLE) = 1 0   EQ.D.E = -260 MICRONS
25867.  INITIAL PATTERNS             3 ABOVE MATCH CUTOFF    1  IDENTIFIED
SEARCH/MATCH = 14/  8 SEC       1 SEC IDENTIFICATION      23 TOTAL RUNTIME

          JCPDS NO      4- 850
                        ------
             MLMS         4.4
          HT/MS/BK     8/ 0/ 0
             I%          257
            SPN          5226
                        ------
     USER PATTERN
   D    I   %USD  DD/D   I
 ----  ----  ----  ----  ---
 2.040 100 0 257    1  257.0
 1.770  89.1 120   -1  107.2
 1.250  60.3  89   -1   53.7
 1.062  70.8  72    2   51 3
 1.019  39.8  46   -1   18.2
 0.881  20 0  51    1   10.2
 0 809  50.1  72   -1   36 3
 0.807  50.1   0
 0.788  50 1  78   -1   38.9
 0.786  50.1   0
        ----
 TOTAL I   580.    ( 69% USED  31% REMAINS)
   2  UNMATCHED USER LINES REMAIN
   0  UNMATCHED REF LINES ABOVE BACKGROUND
```

Fig. 4

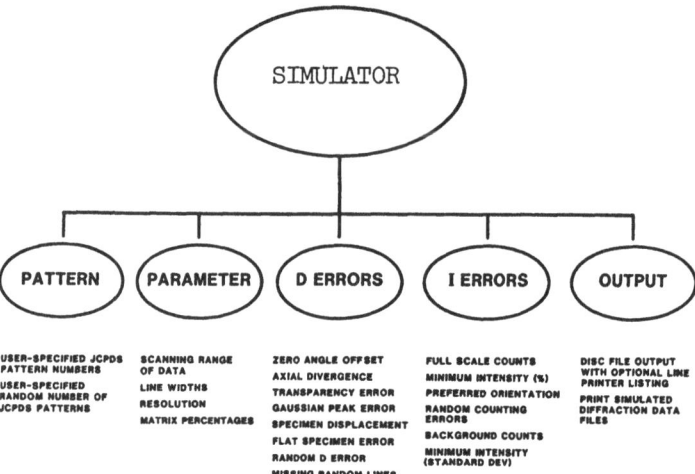

Fig. 5

Experimentally, neither of these can be accomplished in an acceptably
short period of time. Even worse, in certain instances, it may not
be possible to obtain test data with the proper characteristics. For
example, independent control over individual systematic errors is
frequently impossible experimentally. The solution to this problem
is the creation of a simulation program which generates synthetic
diffractograms with systematic errors applied in a controlled fashion.
One is thus able to study the effects of various errors on the search/
match/identify processes, independently, and at will. Further, it
allows rapid generation of enough data to build histograms of critical scoring parameters and optimize them. This technique of modelling has long been used in many areas of science and engineering,
but has seen little, if any, use in the field of X-ray diffractometry.

   SIMULATOR, our diffractogram modelling program, is organized as
an interactive program to control five major areas as seen in Fig. 5.
Data base patterns to be used in generating an artificial diffractogram may be explicitly or randomly chosen. Up to 10 patterns may be
combined to simulate a 10-phase mixture. Their relative proportions
may also be controlled, as well as line widths, line resolution and
scanning limits. Under D-errors, we can independently control five
major systematic errors: Zero angle offset, axial divergence, specimen transparancy, specimen displacement and flat-specimen error.
Under I-errors our present program allows control of such influencing
effects as preferred orientation, random counting errors and background levels. Output capabilities include the ability to write up
to 100 data files at a time and the option of obtaining detailed
listings of the original patterns compared with the final synthetic
diffractogram. The program can be run in batch mode as well as in-

teractive mode and can swap to SANDMAN between runs to permit long overnight and weekend training sessions. Our experience with simulator is that the synthetic patterns it generates can be quite realistic, and it has been used extensively in the development of SANDMAN.

Successful analytic procedures in any field are rooted in a proper understanding of the physical processes involved and in their systematic treatment during the various stages of data analysis. Some specific factors contributing to SANDMAN's success in qualitative X-ray powder diffraction analysis are 1) the global approach taken toward the entire qualitative analysis problem all the way from data collection through the final results, 2) adequate care in packing the reference data base, 3) sensitive peak detection software, 4) use of probability-based scoring techniques borrowed from the field of artificial intelligence, 5) adequate treatment of the important systematic errors, 6) additive instead of subtractive phase isolation techniques, and 7) use of simulation techniques to train the algorithms employed. When these factors are properly incorporated in the software, present-day minicomputers become quite adequate for qualitative analysis.

We would like to thank Dr. J. Ladell for numerous helpful discussions during the course of this work and in the preparation of this paper.

REFERENCES
1) "Control of Systematic Errors In The Computer Controlled Powder diffractometer", R. Jenkins, Y. Hahm, C. Villamizar, W. N. Schreiner, C. Surdukowski, 29th Annual Conference On Applications of X-Ray Analysis, Denver, Colo. (1980).
2) See For Example "Advances Is X-Ray Diffractometry and X-Ray Spectrography", W. Parrish, Ed., Centrex Publ. Co. (1962).
3) "A Second Derivative Algorithm For Identification Of Peaks In Powder Diffraction Patterns", W. N. Schreiner, R. Jenkins, Advances In X-Ray Analysis, 23, p.287 (1979).
4) "A Computer Aided Search/Match System For Qualitative Powder Diffractometry", R. Jenkins, Y. Hahm, S. Pearlman, W. N. Schreiner, Advances In X-Ray Analysis, 23, p.279 (1979).
5) "A Qualitative Analysis Software Package For Use With The Computer Controlled Powder Diffractometer", Norelco Rep. 27, p11(1980).
6) "A Search-Match Method For X-Ray Powder Diffraction Data", R. G. Marquardt, I. Katnelson, G. W. A. Milne, S. R. Heller, G. G. Johnson and R. Jenkins, J. App. Cryst., 12, p.629 (1979)
7) "Preliminary Report On The Design And Results Of The Second Round Robin To Evaluate Search/Match Methods For Qualitative Powder Diffractometry", R. Jenkins and C. Hubbard, Adv. X-Ray Anal., 22, p.133 (1979).
8) "User Guide To Data Base and Search Program (V.18) IBM 360/370", G. Johnson, JCPDS, Swarthmore PA. 19081 (1975).

NBS*AIDS80: A FORTRAN PROGRAM TO EVALUATE CRYSTALLOGRAPHIC DATA

C. R. Hubbard, J. K. Stalick, and A. D. Mighell

Center for Materials Science
National Bureau of Standards
Washington, D. C. 20234

ABSTRACT

Techniques for the computer-assisted evaluation of crystallo-graphic data have been developed to improve the data compilations of the NBS Crystal Data Center and the JCPDS--International Centre for Diffraction Data. The resulting computer program, NBS*AIDS80, can be used for the analysis of unit-cell and powder data by the general scientific community as well. NBS*AIDS80 is written in FORTRAN to permit implementation on a wide variety of computers, and input may be from cards or from a terminal. The research and analysis components that will be of use to the individual scien-tist include the following:
 1) Calculation of the Crystal Data cell, determinative ra-tios, and space group for the comparison and reporting of unit cell parameters in a standard setting and for the identification · of unknowns.
 2) Determination of the reduced cell, reduced form number, and the unit cell with the highest metric symmetry.
 3) Calculation of the molecular weight from the formula using the most recent atomic weights, and comparison of the density calculated by the program with the measured density.
 4) Generation of d-spacings and indices for any input cell and crystal system.
 5) Comparison of input powder data with calculated d-spacings, indexing of lines based on known unit cell parameters, and calculation of figures of merit.

INTRODUCTION

The computer program NBS*AIDS80 is derived from several years
of experience at the National Bureau of Standards in the computer-
assisted evaluation of crystallographic data. The program or its
precursor has been used routinely by the NBS Crystal Data Center
for the analysis of crystallographic data in the preparation of
Crystal Data Determinative Tables[1], and it has recently been
expanded in collaboration with the JCPDS--International Centre for
Diffraction Data for analysis of powder data for the Powder Diffraction File[2]. The combined evaluation of single-crystal and
powder data in a common format enhances the quality of the data in
both data compilations.

NBS*AIDS80 was originally intended for use by the data base
builders in the evaluation of literature data and the creation of
master data files. With such files, research studies can be carried out, indexes can be generated, and publications can be prepared. In addition, the files can be searched in an on-line or
batch mode. However, it has become apparent that the data evaluation routines in NBS*AIDS80 are of value as a research and analysis
tool for the general scientific community as well. In our evaluation of thousands of entries for the Crystal Data and Powder Diffraction Files, we have noted an unexpectedly large number of
errors in the published crystallographic data, especially with
respect to symmetry determination, unit cell transformation,
density calculation, and the indexing of powder data. It is our
hope that by using NBS*AIDS80 research scientists and journal
editors will find and correct such errors prior to publication, and
that additional analysis routines will be suggested for incorporation into the program.

NBS*AIDS80 may also be used as a research aid in conjunction
with routine diffractometry for materials identification and characterization, in addition to the areas mentioned above. For example, in the determination of a crystal structure the program can be
used once unit-cell dimensions have been determined to establish
the probable symmetry of the lattice, the transformation matrix for
the standard setting of the cell and space group, the calculated
density for a check on unit cell contents, and the determinative
ratios needed to find out if the same or a similar compound
previously has been solved. The program should be used again at
the end of the structure determination to give final values for
publication. NBS*AIDS80 is also valuable in powder diffraction
analyses. The program can generate d-spacings for any symmetry and
input cell, and input powder data can be indexed based on a known
unit cell. Once the pattern has been indexed, the program compares
the observed and calculated d-spacings, flags systematic absences
if present, and calculates the figures of merit F(N) and M(20). In
addition, a table of powder data for publication can be

prepared. The research and analysis applications of NBS*AIDS80 are summarized in the following sections.

CELL AND SPACE GROUP TRANSFORMATIONS

In order to compare unit-cell and space-group data from separate experimental determinations, it is essential that all data be reported in a standard orientation. NBS*AIDS80 transforms the input cell and space group to the Crystal Data setting, and the space group number is assigned. The Crystal Data cell provides a standard setting for the publication and comparison of unit cell parameters, and the determinative ratios can be used for direct entry into Crystal Data Determinative Tables for identification of unknowns and location of isostructural materials.

Errors are commonly made in the transformation of unit cell axes and space groups, especially when unusual centerings are encountered. For example, SbSn and NiO crystallize in a rhombohedrally-distorted sodium chloride structure. These cells are frequently reported using F-centered rhombohedral axes; the corresponding hexagonal axes are calculated ignoring the face-centering, leading to incorrect axes and indexing of powder data. A more common use of NBS*AIDS80 would be the calculation of hexagonal unit-cell axes from primitive rhombohedral cell parameters.

Perhaps more importantly, it is essential that a given lattice be characterized by the same cell before comparisons are made. De Camp[3] has given an excellent example involving a lattice that has two different unit cells which appear to be based on the same three lattice translations. Crystals from two samples of condelphine hydroiodide gave apparently the same triclinic unit-cell dimensions when the cells were reduced using two separate reduction algorithms, with only the angle gamma differing significantly (Figure 1). The other differences were assumed to arise from experimental errors. The two cells are shown to define the same lattice when they are transformed using the NBS*AIDS80 algorithm.

	Cell 1	Cell 2	Cell 1'
a	9.34Å	9.32Å	9.34Å
b	17.39	17.45	17.39
c	9.10	9.09	9.10
α	94.85°	94.84	94.85°
β	119.15	118.83	119.15
γ	88.57	86.50	86.71

Figure 1. Unit cell dimensions for condelphine hydroiodide. Cell 1 is transformed to cell 1' by the matrix ($\bar{1}$01,010,001).

DETERMINATION OF METRIC SYMMETRY

The reduced cell[4] and the reduced form number are calculated by NBS*AIDS80 to give the metric symmetry of the lattice[5]. The metric symmetry is the highest symmetry possible for the lattice based solely on geometric considerations. A systematic analysis of the Crystal Data File has shown that the metric symmetry is almost always the same as the true symmetry of the lattice[6]. The metric symmetry is determined from the Niggli reduced cell by the values of the following vector products[5] known as the reduced form:

$$
\begin{pmatrix}
\mathbf{a} \cdot \mathbf{a} & \mathbf{b} \cdot \mathbf{b} & \mathbf{c} \cdot \mathbf{c} \\
\mathbf{b} \cdot \mathbf{c} & \mathbf{a} \cdot \mathbf{c} & \mathbf{a} \cdot \mathbf{b}
\end{pmatrix}
$$

The reduced form can thus be used to indicate the probable symmetry of the lattice. If the reduced form indicates the possibility of higher symmetry than the input lattice, the program will print the cell with the highest possible lattice symmetry. It is well known that cell reduction provides a practical way to prove that a lattice must be triclinic, as the crystal symmetry can never be higher than the symmetry indicated by the reduced form. The use of the reduced form to indicate probable lattice symmetry is particularly important in light of the current trend towards determining the unit cell of a compound solely from diffractometer data. An example of the use of this procedure is given in Figure 2.

	Initial Cell	Reduced Cell	Metric Cell
a	8.095Å	8.096Å	11.698Å
b	8.096	8.095	25.990
c	30.62	25.990	11.194
α	88.67°	90.00°	
β	58.08	90.00	
γ	87.48	92.52	

Reduced Form

$$
\begin{pmatrix}
65.5 & 65.5 & 675.5 \\
0.0 & 0.0 & -2.9
\end{pmatrix}
=
\begin{pmatrix}
\mathbf{a} \cdot \mathbf{a} & \mathbf{a} \cdot \mathbf{a} & \mathbf{c} \cdot \mathbf{c} \\
0 & 0 & -|\mathbf{a} \cdot \mathbf{b}|
\end{pmatrix}
$$

Figure 2.   Determination of the metric cell for an unknown antimony tartrate. The metric cell was found by the program to be orthorhombic B-centered, reduced form number 13.

	Cell 1	Cell 2	Cell 3	Common Cell
a	9.407Å	9.407Å	9.416Å	9.407Å
b	13.773	13.800	13.768	9.407
c	12.829	12.801	12.822	9.416
α	90.06°	90.05°	90.00°	94.06°
β	96.15	95.99	96.14	94.30
γ	89.84	90.01	89.84	94.07

Figure 3.   Alternate A-centered cells for chabazite, all leading
to a common cell.

It should be emphasized that the reduced form as given by
NBS*AIDS80 must be carefully analyzed in order to determine the
true metric symmetry, as the assignment of symmetry is often
dependent on the assessment of experimental errors. A new routine
for symmetry determination is being developed and will be added to
the program. This routine will give all possible symmetries and
pseudosymmetries. In addition, we plan to add evaluation routines
to handle specialized reduced forms that can result from, for
example, the mis-indexing of a powder pattern, the determination of
the subcell of a lattice, or twinning.

One example of metric pseudosymmetry involves the mineral
chabazite, $Ca_2Al_4Si_8O_{24} \cdot 12H_2O$. Chabazite has been reported as
having rhombohedral, monoclinic, or triclinic symmetry. Although
the metric symmetry is nearly rhombohedral, analysis in our laboratory reveals the symmetry to be monoclinic or lower. Figure 3
gives three similar but distinct A-centered cells that can be
related to a common pseudo-rhombohedral cell[7].

Much of the effort involved in the solution of crystal structures can be avoided if the metric symmetry is determined prior to
data collection. In a systematic analysis of organic compounds, we
noted a number of cases in which rhombohedral symmetry was incorrectly reported as monoclinic or even triclinic symmetry. While
the molecular geometries determined were correct, many more structural parameters than necessary were determined and refined leading
to a lower overall accuracy of the structure parameters. Figure 4
gives one such example.

IDENTIFICATION OF UNKNOWNS

The standard Crystal Data cell and determinative ratios calculated by NBS*AIDS80 can be used for the identification of unknown
compounds. For example, the unknown antimony tartrate mentioned
above was identified using Volume 3 of Crystal Data Determinative
Tables. The B-centered orthorhombic cell (a = 11.694(5), b =

	Input Cell	Rhombohedral Cell (Hexagonal Axes)	Reduced Cell
a	19.900Å	11.489Å	11.489Å
b	11.489		11.489
c	21.258	60.590	21.258
α			74.32°
β	108.18°		74.32
γ			60.00

Reduced Form

$$\begin{pmatrix} 132.0 & 132.0 & 451.9 \\ 66.0 & 66.0 & 66.0 \end{pmatrix} = \begin{pmatrix} a \cdot a & a \cdot a & c \cdot c \\ \dfrac{a \cdot a}{2} & \dfrac{a \cdot a}{2} & \dfrac{a \cdot a}{2} \end{pmatrix}$$

Figure 4.   Example of a rhombohedral cell reported as a C-centered
            monoclinic cell. The reduced form (number 9) is clearly
            rhombohedral.

25.961(9), c = 11.192(5), a/b = 0.4504, c/b = 0.4311) was readily
identified as potassium antimony tartrate trihydrate using the portion of the Tables given in Figure 5.

Even though the compound is known, it is always advisable to
obtain the standard cell and check for identification prior to data
collection. The one good crystal in the synthesis frequently does
not correspond to the analysis of the bulk material. For example,
data on orthorhombic sulfur has mistakenly been collected many
times.

An alternative identification scheme will soon be available
using the reduced cells of all materials given in Crystal Data.
This Identification File currently exists at NBS, and it will soon
be available as part of the NIH-EPA Chemical Information System
(CIS)[8]. It will also be issued in a printed format, in which the
reduced cell as given by NBS*AIDS80 can be used for identification.

a/b	c/b	a	b	c	Formula
0.4502	0.3664	8.76	19.43	7.13	$C_{15}H_{10}Cl_2$
0.4506	0.3361	9.251	20.53	6.901	$C_{15}H_{23}Br$
→ 0.4510	0.4316	11.696	25.932	11.192	$K_2Sb_2C_8H_4O_{12} \cdot 3H_2O$
0.4511	0.4473	8.746	19.388	8.672	$CaC_6H_{12}O_6 \cdot 3H_2O$

Figure 5.   Identification of an unknown antimony tartrate using
            Crystal Data Determinative Tables.

Formula	Measured Density	Calculated Density	NBS*AIDS80 Density
$Cd_2P_2O_7$	4.90	5.04	4.886
Ba_3P_2	4.1	4.52	3.371
$GeLi_2Zn$	4.035	4.15	4.353
$Mn_3(PO_4)_2$	4.1	3.92	3.798

Figure 6. Comparison of measured and calculated densities reported
          in the literature with densities calculated by NBS*AIDS80.

CHECKING THE DATA FOR CONSISTENCY

     The program performs a check of the consistency of the unit
cell, Z, and formula through a density calculation. The molecular
weight is computed based on the input formula, and the cell volume
is calculated. Although these computations may seem trivial, our
recent experience as editors reveals that the density calculation
is incorrectly reported in 15% of the cases. The examples given in
Figure 6 are taken from reports of quantitative structure determinations, so there should be no questions about the unit cell
contents. The discrepancies often arise owing to calculation of
the density using preliminary cell dimensions or unit cell contents.
Other sources of error include incorrect computation of molecular
weight, cell volume, and Z or an approximate (or missing) value for
Avogadro's number.

POWDER PATTERN ANALYSIS

     The program can be used to generate 2θ and d-spacings based on
input cell and space group information for any crystal system.
This feature was derived from the program "Lattice Parameter Least
Squares Refinement"[9] with minor modifications. If reflection data
are input (2θ, d, sinθ/λ, or Q) without hkℓ's assigned to any or
all of the reflections, NBS*AIDS80 will assign the indices of the
nearest reflection provided that $|\Delta(2\theta)|$ <0.5 deg. A small segment
of the computer printout from the hkℓ/d-generation and indexing
segment is shown in Figure 7. The 'C' following each observed
d-spacing (D OBS) indicates that the computer code has assigned the
indices. If proper hkℓ's were given as input this column would be
blank. However, if a given hkℓ is not allowed in the input space
group the column would contain an 'N'. Unindexable lines are
marked with a 'U'. The maximum 2θ value for hkℓ/d-generation is an
input parameter or, if left blank, determined by the range of input
lines. The extinction conditions are generated automatically from
the aspect symbol, space group symbol, or cell centering symbol.
They may optionally be entered as input data as well.

MESSAGEPOWDER PATTERN GENERATION/ANALYSIS FOR SYSTEM TET

2-THETA(MAX) = 100.23   D(MIN) = 1.003868

2 CONDITIONS FOR NON-EXTINCTION CALLED FOR
                CLASS CONDITION(S)
                HKL   H+K+L = 2N
                HHL   2H+L = 4N

N	D CALC	D OBS	INT	AUTHORS H K L	PROGRAM H K L	OBS 2-THETA	CALC 2-THETA	DIFF 2-THETA
1	5.4508				1 0 1		16.248	
**	5.4508	5.4502C	80	1 0 1		16.250	16.248	-.002
2	3.8489				2 0 0		23.090	
**	3.8489	3.8472C	65	2 0 0		23.100	23.090	-.010
3	3.1485				1 1 2		28.323	
4	3.1441				2 1 1		28.364	
**	3.1441	3.1434C	100	2 1 1		28.370	28.364	-.006
5	2.7254				2 0 2		32.836	
**	2.7254	2.7250C	15	2 0 2		32.840	32.836	-.004
6	2.7216				2 2 0		32.883	
**	2.7216	2.7194C	20	2 2 0		32.910	32.883	-.027

Figure 7.   A segment of the hkℓ/d-generation output of NBS*AIDS80
            for $NH_4H_2AsO_4$.

REFLECTION SUMMARY FOR ENTIRE PATTERN:
ESTIMATED RESOLUTION = .075 DEG. 2-THETA

            THEORETICAL LINES TOTAL = 75
            THEORETICAL RESOLVABLE = 47
            UNIQUE OBSERVED LINES   = 39

            TOTAL LINES INPUT = 39
            NUMBER INDEXED   = 39
            NUMBER UNINDEXED = 0

*************FOR INDEXED LINES*************

            AVERAGE 2-THETA DIFFERENCE = .012
            MAXIMUM 2-THETA DIFFERENCE = .030
            # WITH DIFF>0.05 (STAR LIMIT) = 0
            # WITH DIFF>0.20 (I LIMIT)   = 0
            M(20) = 59.49 (DLIMIT = 1.3589, # POSSIBLE = 33)
            F(30) = 47.69 (DELTA 2 THETA = .0123, # POSSIBLE = 51)

        Figure 8.   Pattern summary for $NH_4H_2AsO_4$

Following the hkℓ, d and 2θ table, a pattern summary is printed
as shown in Figure 8. Included are the number of lines possible
out to 2θ max, the number of lines resolvable based on an estimated
resolution factor, and the number of lines observed. This summary
expresses the coverage of the experimental measurements. Next is a
tabulation of the number of lines input, number indexed and number
remaining unindexed. For the <u>indexed</u> lines the average and maximum
magnitude of Δ(2θ) are given. An average |Δ(2θ)| of <0.02° should
be expected for high quality data. Data of this quality, or better,
have been reported for nearly two decades primarily due to the use
of internal 2θ standards. Another assessment of the fit between
the observed and calculated 2θ's is the number of lines with |Δ(2θ)|
greater than some arbitrary limit. The JCPDS--International Centre
for Diffraction Data currently uses a limit of 0.05° for "☆"
quality patterns and a limit of 0.20° for "I" quality patterns.
(However, for a PDF card to be assigned either of these quality
marks, chemical and intensity data criteria must also be satisfied.)
Two other assessments of the fit of the input cell and observed
pattern are calculated by the program. M(20) is an indication of
the correctness of the unit cell. According to de Wolff[10] a value
of M(20) >10 coupled with two or fewer unindexed lines indicates
that the cell is substantially correct. A value of M(20) <5
and/or several unindexed lines strongly suggests that the cell is
not appropriate. The Smith-Snyder figure-of-merit F(N)[11] assesses
the overall fit of the experimental data and cell as well as the
coverage or completeness of the observations. The coverage factor
is the ratio of the number of observed and indexed lines, N (up to
30), divided by the theoretical number of lines out to the Nth
observed line. The other term in F(N) is the inverse of the average magnitude of Δ(2θ). For every one of the 105 patterns reported
in Monograph 25, Section 16 and Section 17[12], M(20) exceeded 15 and
F(30) exceeded 13. However, in evaluating data in the literature,
we often find indexed lines with |Δ(2θ)| >0.5°, and patterns with
M(20) <5 and F(30) <5. Often the reasons for the low quality of
fit cannot be explained. However sometimes large discrepancies are
typographical in origin. NBS*AIDS80 has a d/I/hkℓ/2θ table generation subroutine to help eliminate these errors (Figure 9). The
subroutine determines the appropriate number of significant digits
for each d-value based on the average magnitude of error in 2θ for
the entire pattern. This keeps the round-off errors due to format-
ting of d-values less than the average magnitude of the error in
2θ.

PROGRAM DETAILS

NBS*AIDS80 is written in FORTRAN and is currently running on a
UNIVAC 1100 series computer. As overlayed, it occupies about 21K
words of memory and requires from one second for cell analysis to
five seconds for cell and powder data analysis. To improve port-

5.450	80	1	0	1	16.25
3.847	65	2	0	0	23.10
3.143	100	2	1	1	28.37
2.725	15	2	0	2	32.84
2.719	20	2	2	0	32.91
2.438	4	1	0	3	36.83
2.433	4	3	1	0	36.92
2.058	45	3	2	1	43.96
1.9291	1	0	0	4	47.07
1.9241	3	4	0	0	47.20
1.8166	9	3	0	3	50.18

Figure 9: d/I/hkℓ/2θ Table Generation for $NH_4H_2AsO_4$

ability of the program, we have grouped all ENCODE and DECODE
statements together in a single subroutine. Laboratories with
compilers that do not handle ENCODE and DECODE statements may have
to write assembly language subroutines. The program makes use of
the full 128 character ASCII character set. However, input may be
restricted to only upper case for formulas and space group symbols.
The program will be provided free of charge provided that a new,
unlabeled 9 track magnetic tape, certified to 1600 bpi, accompanies
the request. The supplied tape will be written unblocked in ASCII
at 1600 bpi.

ACKNOWLEDGEMENTS

    Work on the program was sponsored by the Office of Standard
Reference Data (NBS), by the Ceramics, Glass, and Solid State
Science Division (NBS), and by the JCPDS--International Centre for
Diffraction Data. The following scientists made invaluable contri-
butions to the design and development of NBS*AIDS80: R. J. Boreni
of the Crystal Data Center, L. Calvert of the Metals Data Center
(Ottawa, Canada), E. Evans of the JCPDS Associateship at NBS, M.
Holmany of the JCPDS, F. McClune of the JCPDS, H. McMurdie JCPDS--
Ceramics Editor, J. R. Rodgers of the Cambridge Crystallographic
Data Centre, A. Santoro of the Reactor Division (NBS), R. Snyder of
Alfred University, and D. Watson of the Cambridge Crystallographic
Data Centre.

REFERENCES

1.   "Crystal Data Determinative Tables." 3rd ed. US Department of
     Commerce, National Bureau of Standards and the JCPDS--Inter-
     national Centre for Diffraction Data, Swarthmore, PA, USA
     (1972, 1973, 1978); Vols. 1-4.

2. "Powder Diffraction File", JCPDS--International Centre for
    Diffraction Data, 1601 Park Lane, Swarthmore, PA 19081.
3. W. H. De Camp, The Existence of Metrically Similar Unit Cells
    Based on the Same Lattice: A Precautionary Note, Acta Cryst.
    B32: 2257 (1976).
4. A. Santoro and A. D. Mighell, Determination of Reduced Cells,
    Acta Cryst. A26: 124 (1970).
5. "International Tables for X-ray Crystallography." Kynoch
    Press, Birmingham, Vol. 1, 3rd ed: 530 (1969).
6. A. D. Mighell and J. R. Rodgers, Lattice Symmetry Determina-
    tion, Acta Cryst. A36: 321 (1980).
7. V. Himes, Data in Figures 2 and 3 were taken from a disserta-
    tion to be submitted to the Graduate School, the Catholic
    University of America, Washington, D. C. 20064, in partial
    fulfillment of the requirements for the Ph.D. Degree in
    Chemistry.
8. CIS Project, Information Sciences Corp.,2135 Wisconsin Ave.
    N.W., Washington, D. C. 20007.
9. H. T. Evans, Jr., D. E. Appleman, and D. S. Handwerker, Report
    #PB 216188, U.S. Department of Commerce, National Technical
    Information Center, 5285 Port Royal Rd., Springfield, VA
    22151.
10. P. M. de Wolff, A Simplified Criterion for the Reliability of
    a Powder Pattern Indexing, J. Appl. Cryst., 1: 108 (1968).
11. G. S. Smith and R. L. Snyder, F(N): A Criterion for Rating
    Powder Diffraction Patterns and Evaluating the Reliability of
    Powder-Pattern Indexing, J. Appl. Cryst., 12: 60 (1979).
12. M. C. Morris, H. F. McMurdie, E. H. Evans, B. Paretzkin, J. H.
    deGroot, C. R. Hubbard, and S. J. Carmel, Standard X-ray
    Diffraction Powder Patterns, NBS Monograph Section 16 and Sec-
    tion 17, National Bureau of Standards, Washington, D. C.
    20234.

REPRODUCIBILITY AND PRECISION OF MEASUREMENTS OF

GUINIER POWDER PATTERNS USING POWDERED SILICON CALIBRANT

A. Brown
Studsvik Energiteknik AB
Nyköping, Sweden   S-611 82

J. W. Edmonds
The Dow Chemical Co.
Midland, Michigan   48640

C. M. Foris
E. I. du Pont de Nemours & Co.
Wilmington, Delaware   19898

ABSTRACT

Calibrated powder patterns of a number of high symmetry crystalline materials were recorded in three different Guinier-type focusing cameras. The films were then measured by three different techniques, one visual and two instrumental. Cell parameters were calculated by three different data reduction procedures. The aim of the work was to establish the level of reproducibility for cell parameters obtained in routine work with these instruments.

1. INTRODUCTION

Guinier powder patterns are noted for their high degree of line resolution. This is largely because the $K\alpha_2$ component of the characteristic radiation is eliminated by the focusing monochromator. For a film cassette located in the subtraction position, line widths are further reduced by the minimization of chromatic dispersion effects (1). Thus, for well-crystallized materials, typical values of the full-width at half-maximum are 0.06° (2θ) and may be as low as 0.04° in the angular range 60-70° (2θ) (2). Conditions are, accordingly, favorable for the precise location of diffraction lines, expressed either as peak maxima or as centroids, even in line-rich powder patterns.

The non-linear relationship between distance, S, measured along the recorded Guinier pattern and the true diffraction angle, θ, is well recognized. Since this relationship varies slightly between exposures, an internal standard should be included in the

x-ray specimen. Calibration of the powder pattern is then possible on the basis of the known Bragg angles for the standard reflections. Measurement of all the reflections in the pattern of a
given crystalline compound, followed by angular correction, indexing and least squares reduction of the data, will frequently yield
cell parameters with standard deviations at the 0.005% level.
However, whether these low σ values correspond to genuinely precise measurements or whether systematic errors are introduced to a
varying extent as a result of different laboratory procedures and
individual operator routines remains to be seen.

To determine the level of reliability of Guinier powder data,
a program of interlaboratory comparisons was begun. The immediate
aim was to specify a set of operating procedures whereby reliable
2θ data can be obtained. The quality of the 2θ data should be such
that, within the limits of the σ values, derived cell parameters
can be reproduced routinely and independently of laboratory methods.
From such a study, hopefully, areas of uncertainty in the overall
procedure can be identified and their influence as sources of systematic error recognized. Extension of this program to a comparison of Guinier and powder diffractometer data for the same materials
is also projected to establish the relative precision for these two
powder techniques.

2. EXPERIMENTAL

2.1 Comparison Materials Compounds chosen for the first stage
of the study were CeO_2, ThO_2, PtP_2, $Pb(NO_3)_2$, Mo_3Sb_7 and As_2O_3.
All are heavy scattering agents for CuKα radiation and all are cubic
with cell parameters that increase progressively through the series
from 5.411 to 11.077 Å. Comparison in this instance indicates whether reproducibility is influenced by the number of measured reflections and their distribution throughout the pattern relative to the
standard reflections.

The x-ray specimens of each compound were taken from a common source of limited size to avoid differences that might arise
from variable traces of impurity in solid solution. Particle size
was such that no further crushing, which might lead to loss of line
definition, would be necessary in the separate laboratories. For
this reason, the lead nitrate was reprecipitated with ethanol from
a saturated aqueous solution, a procedure which seems to have influenced the cell parameter and possibly the stability of the material
in the x-ray beam. The CeO_2 and ThO_2 were of nuclear grade and
had been heated in air to 1200°C for 48 h in platinum crucibles.
The compounds PtP_2 and Mo_3Sb_7 were synthesized by direct reaction
of the elements in evacuated and sealed tubes of vitreous silica.
The As_2O_3 was supplied as a commercial grade material with a conveniently small particle size. All of the samples were well crystallized.

2.2 Internal Standard A small amount of silicon powder from the same source was included in each x-ray specimen. This material is SRM 640, supplied by the National Bureau of Standards. Relevant data used in film calibration are listed in Table 1 (3).

Table 1

DIFFRACTION ANGLES FOR SRM 640 (NBS) SILICON

i	hkℓ	2θ
1	111	28.443
2	220	47.303
3	311	56.122
4	400	69.131
5	331	76.377
6	422	88.031
7	511 333	94.954

$a/\lambda = 3.525176$ at 25°C, $\lambda = 1.54051$Å, $K_i = \theta_i/S_i$

2.3 Specimen Preparation At the DuPont and Studsvik laboratories the mixture of sample and internal standard is dusted onto 3M Magic Transparent Tape. To preserve planarity during exposure this tape is stretched over a metal diaphragm located in the specimen holder (4). At the Dow laboratories a Mylar® film serves as a planar support. The surface is smeared thinly with Vaseline® and the mixture dusted onto the adhesive surface thus provided (5).

2.4 Guinier Photography and Film Reading All three cameras have subtractive geometry with regard to the location of the film cassette in relation to the x-ray beam. The radiation is strictly monochromatic CuKα_1 in each case. The instruments used are of different manufacture and/or cassette radius and different methods are used to measure the films. The situation at the three laboratories can be summarized as follows:

Studsvik (B) Camera type IPT model V-FXC; cassette radius 114.6 mm. Film overprinted with 0.1 mm graduated scale prior to development, and read once visually at 40X magnification to ±0.01 mm (4).

Dow (E) Camera type Huber; cassette radius 114.6 mm. Film scanned by automated microdensitometer in intervals of 0.014 mm at a precision of ±0.001 mm. The reflections are measured as intensity profiles whereby data pairs of intensity/film distance are recorded on a computer mass-storage device. Profile fitting is then employed to establish the peak centroid as a basis for deriving the 2θ value for each reflection.

DuPont (F) Camera type IPT; cassette radius 80 mm with a cutoff imposed at 80° (2θ) yielding five standard reflections. The film is measured on a calibrated, precision screw instrument to ±0.005 mm. Reflections are selected at 6X magnification with a visual system that is parfocalized with the photoelectric system. The photoelectric image is split optically and scanned by a photomultiplier device to produce mirror images of the density profile trace on an oscilloscope screen. The coincidence of the mirror image peaks of the profile trace is the criterion for line position

(maximum light absorption in the film). Line positions (measured
distances) are digitized and density readings are taken manually
from the oscilloscope (normalized to the most intense peak). Data
are recorded for subsequent numerical treatment.

The angular sensitivity, $\Delta 2\theta$, corresponding to a linear interval, ΔS, for the different measuring systems and cassette
radii, is listed in Table 2.

Table 2

ANGULAR SENSITIVITY, $\Delta 2\theta$, FOR INDICATED MEASURING SYSTEMS
        AND CASSETTE RADII

| | $\Delta 2\theta$ Cassette Radius, R | |
ΔS	80 mm	114.6 mm
0.01 mm (Studsvik)	0.0072°	0.0050°
0.005 mm (DuPont)	0.0036°	0.0025°
0.001 mm (Dow)	0.0007°	0.0005°

For Guinier films $\Delta 2\theta = \Delta S(180)/\pi R$
Studsvik (Brown): visual estimation (pre-recorded scale)
DuPont (Foris): microphotometer (split-image comparator)
Dow (Edmonds): microphotometer (computer automated)

2.5 Calibration Procedure and Least Squares Data Reduction
Each laboratory/author employs a data reduction procedure that
evaluates the camera constant or radius at a point corresponding
to each measured reflection. The individual procedures differ
somewhat, however, and can be summarized as follows:

Studsvik (B) All measurements of film distance, S, are made
with reference to the trace on the film produced by the primary
beam. The camera constant, K_i, is derived for each silicon reflection from the measured S_i value and the θ_i angle given in Table 1.
The calibration points are used, three at a time, as the basis for
a second-order Lagrange interpolation. The camera constant is derived for each measurement, S_{hkl}, of a manually indexed reflection.
Values of θ_{hkl} are converted to $\sin^2\theta/\lambda^2$ for use in an iterative
least squares computation of cell parameters and standard deviations.

DuPont (F) Procedure is similar to that used at Studsvik. All
film measurements are referred to the primary beam trace and appropriate camera constants are derived by interpolation. Initially
the program tries to fit the calibration points to a first order
curve. Failing a satisfactory fit, however, a second order curve
is adopted and values of 2θ are derived. The least squares program
does not require an indexed input but indexes the individual reflections and refines the cell parameters simultaneously in an iterative procedure.

Dow (E) The primary beam is not recorded on the film in the
Huber cassette. All film measurements are referenced to the lowest d-value calibrant reflection. The diameter of the cassette is
assumed to be precisely 114.6 mm at this point. Theoretical positions are computed for the remaining reference lines; this permits

the observed reference lines to be identified and their actual positions used. Linear interpolation of the camera radius is then
performed between appropriate pairs of calibration points. The
reflections are manually indexed for input to the least squares
program.

    2.6 Comparison of Data and Results All three sets of films
from the different x-ray instruments were measured on the split
image comparator (F) and on the microdensitometer (E). In the
absence of pre-recorded scales on the Dow and DuPont films, only
the Studsvik set was measured by visual estimation of the line
position (B). The raw measurements of film distance were exchanged between the three authors for treatment by the individual
data reduction procedures.

3. RESULTS

    Exchange of films and measurements, as described in the pre-
vious section, permits comparisons to be made at three distinct
stages of the overall procedure. These stages are:

  - Specimen preparation and recording of powder pattern.

  - Film measurement to produce a set of S values.

  - Application of calibration data to convert S values to
    corresponding values of 2θ, followed by cell parameter
    refinement.

    The results of the comparisons are presented in Tables 3 to 7.
In these tables, the three significant stages are identified at
the left-hand side by the initial of the author responsible. Thus,
in Table 3, the sequence BBB refers to the film obtained at Studsvik, measured at Studsvik, with data reduction performed with the
Studsvik program, under the supervision of Brown. Similarly, in
Table 4, the sequence EFB indicates data obtained from an Edmonds
film, as measured by Foris with final data reduction performed by
Brown. Cell parameters are given in Angström units and the value
$\lambda = 1.54051$ Å is used for $CuK\alpha_1$ radiation. The number in paren-
thesis following each cell parameter is $\sigma \times 10^4$ Å.

4. DISCUSSION

    The data in Table 3 indicate the extent of the agreement
among the three authors. The principal discrepancies are between
EEE and the other two authors for $Pb(NO_3)_2$, Mo_3Sb_7 and As_2O_3.
Smaller discrepancies occur between BBB and the other two authors
for CeO_2 and PtP_2. The slightly higher σ values for the FFF set
are attributed to the relatively low cut off angle of 80° as
opposed to 100° (2θ) used by B and E.

    The data presented in Table 4 were obtained by measuring all
of the films on the DuPont instrument and reducing the data at
Studsvik. The B and F films are in close agreement for all the compounds except Mo_3Sb_7. The greatest changes, relative to Table 3,

Table 3

CELL PARAMETERS, Å: COMPARISON OF INDIVIDUAL RESULTS

Compound	CeO_2	ThO_2	PtP_2	$Pb(NO_3)_2$	Mo_3Sb_7	As_2O_3
Number of reflections	10	10	18	30	32	32
BBB (Studsvik)	5.4117(2)	5.5972(2)	5.6955(1)	7.8574(2)	9.5687(3)	11.0770(2)
EEE (Dow)	5.4122(4)	5.5974(2)	5.6958(2)	7.8560(3)	9.5769(4)	11.0744(3)
FFF (DuPont)	5.4125(4)	5.5974(4)	5.6962(4)	7.8574(4)	9.5685(3)	11.0770(4)

Table 4

CELL PARAMETERS, Å: COMPARISON OF SPECIMEN/CAMERA COMBINATIONS

Compound	CeO_2	ThO_2	PtP_2	$Pb(NO_3)_2$	Mo_3Sb_7	As_2O_3
BFB IPT, 114.6 mm	5.4122(1)	5.5973(3)	5.6959(1)	7.8573(2)	9.5683(1)	11.0768(2)
EFB Huber, 114.6 mm	5.4119(1)	5.5971(2)	5.6956(2)	7.8560(2)	9.5690(1)	11.0752(2)
FFB IPT, 80 mm	5.4120(5)	5.5974(3)	5.6964(3)	7.8574(5)	9.5694(3)	11.0773(6)

are obtained for the B films of CeO_2, PtP_2 and Mo_3Sb_7; similar
changes are obtained for the E and F films of Mo_3Sb_7. This may be
related to sample preparation and the number of silicon calibrant
reflections detected. The E films for $Pb(NO_3)_2$ and As_2O_3 retain
their discrepancies, however, and further investigation is required to ascertain the possible influence of specimen preparation
on these results.

The first three lines of Table 5 show data obtained by measuring the B films by the three different techniques and reducing
all the data at Studsvik (B). The main discrepancies are found to
be in the BBB results for CeO_2 and PtP_2; the cell parameters appear to conform better with those given by E and F when the B films
are measured on their instruments. In general, the data indicate
that correction for anomalous film shrinkage, made by overprinting
a linear scale on the film before development, does not significantly improve the results. On the other hand, visual estimation,
by reading such a scale, may be responsible for the evident discrepancies observed in the BBB results for CeO_2 and PtP_2. The
data in the last two lines of Table 5 confirm the general agreement obtained for instrumental measurement. In this instance the
E films were scanned on the separate E and F instruments.

The data presented in Table 6 indicate that only slight differences, less than 2σ, arise between calibration curves referred
to the primary beam trace or to the silicon reflection with the
lowest d-value observable. The principal discrepancies are for
PtP_2 and Mo_3Sb_7, and these may arise out of the distribution of
lines with respect to the calibrant reflections.

The cell parameters listed in Table 7 were calculated with different least squares programs. The 2θ input data were obtained with
the Dow data reduction procedure. The principal change is in the
parameter for Mo_3Sb_7, as already noted in Table 4. In addition, all
σ values are much smaller for parameters calculated by the Studsvik
program. Comparison of the $\Delta2\theta$ values printed out at the completion
of refinement generally indicates that the Dow program in use for
these studies did not fully converge to give the best possible fit
to the observed 2θ data.

5. GENERAL COMMENTS

The data in Table 4 show that, with the exceptions already
referred to, measurements of Guinier films yield cell parameters
which are reproducible within the limits of the calculated standard
deviations. Silicon is a satisfactory internal standard (at least
for patterns in which all the silicon lines are clearly resolved)
when sufficient material has been added to enable detection of the
weaker silicon lines. A reproducible method of converting film
readings to diffraction angles is provided by interpolation, either
on the basis of a second order curve fit to the lines of the
silicon pattern, or using a linear approximation to this curve

Table 5

CELL PARAMETERS, $\overset{\circ}{A}$: COMPARISON OF MEASURING DEVICES

Compound	CeO_2	ThO_2	PtP_2	$Pb(NO_3)_2$	Mo_3Sb_7	As_2O_3
BBB 0.01 mm	5.4117(2)	5.5972(2)	5.6955(1)	7.8574(2)	9.5687(3)	11.0770(2)
BFB 0.005 mm	5.4122(1)	5.5973(3)	5.6959(1)	7.8573(2)	9.5683(1)	11.0768(2)
BEB 0.001 mm	5.4123(2)	5.5977(3)	5.6961(1)	7.8574(1)	9.5692(2)	11.0772(2)
EFB 0.005 mm	5.4119(1)	5.5971(1)	5.6956(2)	7.8560(1)	9.5690(1)	11.0752(2)
EEB 0.001 mm	5.4119(2)	5.5975(2)	5.6959(1)	7.8563(1)	9.5692(2)	11.0751(3)

Table 6

CELL PARAMETERS, Å: COMPARISON OF CALIBRATION TECHNIQUES

Compound	CeO_2	ThO_2	PtP_2	$Pb(NO_3)_2$	Mo_3Sb_7	As_2O_3
BEB (Dow)	5.4123(2)	5.5977(3)	5.6958(1)	7.8574(2)	9.5688(2)	11.0770(4)
BEB (Studsvik)	5.4123(2)	5.5977(3)	5.6961(1)	7.8574(1)	9.5692(2)	11.0772(2)

Dow (Edmonds) referenced to lowest d-value silicon line observed
Studsvik (Brown) referenced to primary beam trace

Table 7

CELL PARAMETERS, Å: COMPARISON OF LEAST SQUARES PROGRAMS

Compound	CeO_2	ThO_2	PtP_2	$Pb(NO_3)_2$	Mo_3Sb_7	As_2O_3
EEE (Dow Program)	5.4122(4)	5.5974(2)	5.6958(2)	7.8560(3)	9.5769(3)	11.0744(3)
EEE (Studsvik Program)	5.4118(1)	5.5974(1)	5.6957(1)	7.8562(1)	9.5695(1)	11.0749(2)

2θ values used as input data obtained with Dow data reduction program

given by adjacent pairs of silicon lines. A first order curve is less reliable, as is the linear approximation method should any one of the calibrant lines be omitted. Instrumental reading of the film is less open to subjective error than visual reading, while non-uniform film shrinkage appears to be negligible at this level of precision. Least squares lattice refinement programs may induce systematic errors in data sets of the highest quality; programs should be verified for convergence with appropriate test data. A series of tests is being compiled by the Methods and Practices Subcommittee of the JCPDS-International Centre for Diffraction Data.

6. REFERENCES

    1. E.-G. Hofmann and H. Jagodzinski, "Eine neue, hochauflösende Röntgenfeinstruktur-Anlage mit verbessertem, fokussierendem Monochromator and Feinfokusröhre", Z. Metallkd. 46 601 (1955

    2. A. Brown and J. W. Edmonds, "The Fitting of Powder Diffraction Profiles to an Analytical Expression and the Influence of Line Broadening Factors", Adv. X-ray Anal. 23 361 (Edited by J. R. Rhodes, Plenum Press, 1980).

    3. C. R. Hubbard, H. E. Swanson and F. A. Mauer, "A Silicon Powder Diffraction Standard Reference Material", J. Appl. Crystallogr. 8 45 (1975).

    4. A. Brown, "Optimal Calibration Curves for Guinier-Type Focusing Cameras", Adv. X-ray Anal. 21 289 (Edited by C. S. Barret, Plenum Press, 1978).

    5. J. W. Edmonds and W. W. Henslee, "Applications of Guinier Camera, Microcomputer Controlled Film Densitometry to Rapid Routine X-ray Powder Diffraction Analysis", Adv. X-ray Anal. 22 143 (Edited by G. J. McCarthy, Plenum Press, 1979).

AN APPROACH TO THE AUTOMATION OF A MULTIFUNCTION

X-RAY DIFFRACTION LABORATORY

T. L. Nunes

IBM Corporation

Hopwell Junction, New York

A variety of x-ray diffractometers supporting stress measurements, routine phase identification, precision lattice parameter determinations, crystallite size estimates, and high temperature studies have been interfaced to a central minicomputer. The unique attributes of each instrument have been retained while the central minicomputer provides a common control format, data storage area, and analysis program library. Two movable desk top computers provide limited backup for the minicomputer, and serve as test beds for technique development.

The problem of interfacing new automatic equipment from a variety of suppliers to the same minicomputer is simplified by requiring each new instrument to be equipped with an asynchronous ASCII RS-232C I/O port together with the software to support data and command transfers in both directions. We have evolved a hardware interface/controller[1] for older and "home built" instruments.

Two Siemens F-type powder diffractometers and a Rigaku microdiffractometer[2] to which stepping motors have been added are directly controlled by the minicomputer through our own interface. The minicomputer also indirectly controls two Siemens D/500 diffractometers equipped with intelligent controllers. Although the two different types of controllers have different command and response formats, the control format used by an operator at the minicomputer is the same for both.

We have built a variation of our interface/controller to control the two motion drives of a Rigaku Theta – Theta goniometer and the furnace of the hot stage.[3] This experiment is being run through a dedicated small computer to allow rapid adaptation to different patterns of use. Some data processing has been done on the small computer, but most of the data are now transmitted to the mini-computer for display and analysis. Shortly, the control will be transferred to the minicomputer for use elsewhere.

Our approach to laboratory automation has yielded several benefits: First, control via the minicomputer is similar for all instruments, contributing to ease of use. Second, our own program development is confined to the minicomputer, avoiding the need to learn new systems or languages. Third, the commonality of the RS-232C interface allows quick isolation of malfunctions to computer, cabling, or controller and supports easy switching to a backup com-puter. Fourth, data from all instruments are displayed in the same format, clarifying reports and presentations using data from two or more instruments.

A benefit which we hope to realize with the acquisition of new instruments with extensive data analysis packages is to make these analyses available to any instrument in the laboratory by transferring data through the minicomputer to the selected intel-ligent diffractometer for analysis. Thus, program duplication is avoided while the entire laboratory's data analysis capability is upgraded.

REFERENCES

1.   T. L. Nunes, Mini and Micro Computer, Vol. 5, 1980.
2.   G. A. Walker and C. C. Goldsmith, Proceedings of the 1978
     Reliability Physics Symposium, San Diego, California,
     April 18-20, 1978, pp. 56-58.
3.   Rigaku, U.S.A., Inc., Danvers, Massachusetts.

THE USE AND ACCURACY OF CONTINUOUSLY SCANNING POSITION-

SENSITIVE DETECTOR DATA IN X-RAY POWDER DIFFRACTION

Herbert E. Göbel

Forschungslaboratorien der SIEMENS AG

D 8000 München 83, West Germany

ABSTRACT

By collecting the diffracted X-rays in a focussing powder diffractometer with a linear position-sensitive detector (PSD), the
data accumulation time can be widely reduced. Due to the focussing
properties of the geometrical arrangement good resolution can be
simultaneously achieved over a few degrees of the diffraction
angle 2 Theta. The full pattern is collected by scanning the PSD
along the entire 2 Theta arc.[1] Scanning speeds of several hundred
degrees per minute are possible and still yield well plotted diagrams.

This work demonstrates the performance of the system in different tasks of X-ray powder diffraction such as identification of
unknown materials, quantitative analysis, determination of lattice
constants and microcrystalline properties. The results, evaluated
by parts of the program system DIFFRAC 11,[10] show that the accuracy
and resolution of the continuously scanning PSD technique match
these tasks without difficulty and reach the precision limits of
powder diffractometry itself. It is demonstrated that the data
collection velocity, which matches well the data evaluation times,
is not the only advantage compared to conventional diffractometers.

INTRODUCTION

Position-Sensitive Proportional Detectors (PSD) have been used
with good success in some fields of X-ray scattering and diffraction (small-angle scattering, stress analysis and time-resolved
diffraction). Since the angular range of interest can be

investigated simultaneously without loss of resolution, the measuring time can be reduced by up to about 2 orders of magnitude.
A similar gain can be achieved for general powder diffraction in
a combined diffractometer, where a PSD collects data in a reduced
range (about 5°) of the 2 theta arc while scanning the entire range
at high speed.[1] This technique can be used in different focussing
geometrical arrangements (Bragg-Brentano,[1] Guinier[2] or Seemann-
Bohlin[3] and allows the collection of full powder diffraction patterns without serious loss of quality in minutes instead of hours,
unlike conventional systems, where the angular resolution is
achieved by a tiny slit in front of the detector. The function
of the Continuously scanning PSD diffractometer (CPSD) has been
described.[1] Results were presented as qualitative proofs of the
method. This paper reports on CPSD application to different tasks
of XRPD such as

- identification of unknowns

- indexing and lattice constants determination[4]

- quantitative analysis of mixtures

- line profile analysis for determination of crystallite size
  and strain.[5]

It discusses the limits of resolution and absolute accuracy
and points out experimental problems (fluorescence radiation,
maximum counting rate).

EXPERIMENTAL

The experimental equipment (see Fig. 1) consists of a step-
motor-driven diffractometer (Siemens D 500), a 1000 W fine-focus
X-ray tube (XRT), a position-sensitive proportional counter (PSD),[6]
an electronic pulse handling system including the detector bias
supply (HVS), a compact PSD-amplifier (dual AMP/TSCA), a tunable
time-to-digital converter, a combined motor-step counter and binary
adder module (MSC/ADD) and a 4096 multichannel counter (MC). The
system is connected to a PDP 11/34 computer (CMP) with 32 k-words
of memory equipped with RL01 hard discs and a VT 55 semigraphical
scope terminal.

The function of the system has been extensively described[1] and
is only briefly repeated here. The address (2 theta) under which
a diffracted X-ray quantum is stored in the multichannel counter
memory is a combination of a mechanical address (A) representing
the diffractometer angle and an electronic address (a) representing
the incidence position in the PSD (Fig. 2). By adding these two

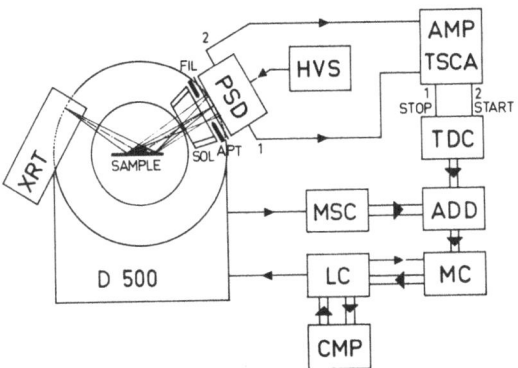

FIL   = Kβ-filter
APT = detector window
SOL = Soller slits
PSD = position sensitive detector
AMP/TSCA = compact PSD-
                    electronics
HVS = high voltage supply
TDC = time-to-digital converter
MSC = motor step counter/divider
ADD = binary adder
MC = multichannel counter
LC = logic controller interface
CMP= computer

Fig. 1.   Block diagram of the electronic equipment

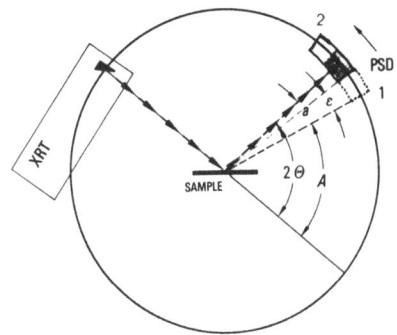

Fig. 2.   Addition principle
of the CPSD-method

addresses the motion of the detector (variation ε) is eliminated and a diffraction line will be collected into the same channels (2 theta) throughout the time the detector scans across it.   In the theta/2 theta mode of a Bragg-Brentano diffractometer the accumulated intensity will be assembled from all crystallites that subsequently come into reflection position by the continuous theta rotation.   The condition that has to be obeyed is to digitize A and a in the same scale (for example by increments of 0.01°, 0.02°, 0.05 or 0.1° of 2 theta).   In this way a quasi-continuous scan of the PSD is possible by which entire diffraction patterns can be collected as in conventional angle dispersive diffraction, however with typical scanning speeds ranging from about 10 to several hundred degrees per minute.

An example is shown in Fig. 3.   The interesting pattern of La_2O_2S was collected from the same specimen shown in (1) and was selected to demonstrate the progress that has been achieved in the latest years with respect to

- extending the angular range of 2 theta (2-3 degrees to 160 degrees, maximum angle for a fixed PSD about 167°)

- reducing the background caused by air scattering, particularly for low angles

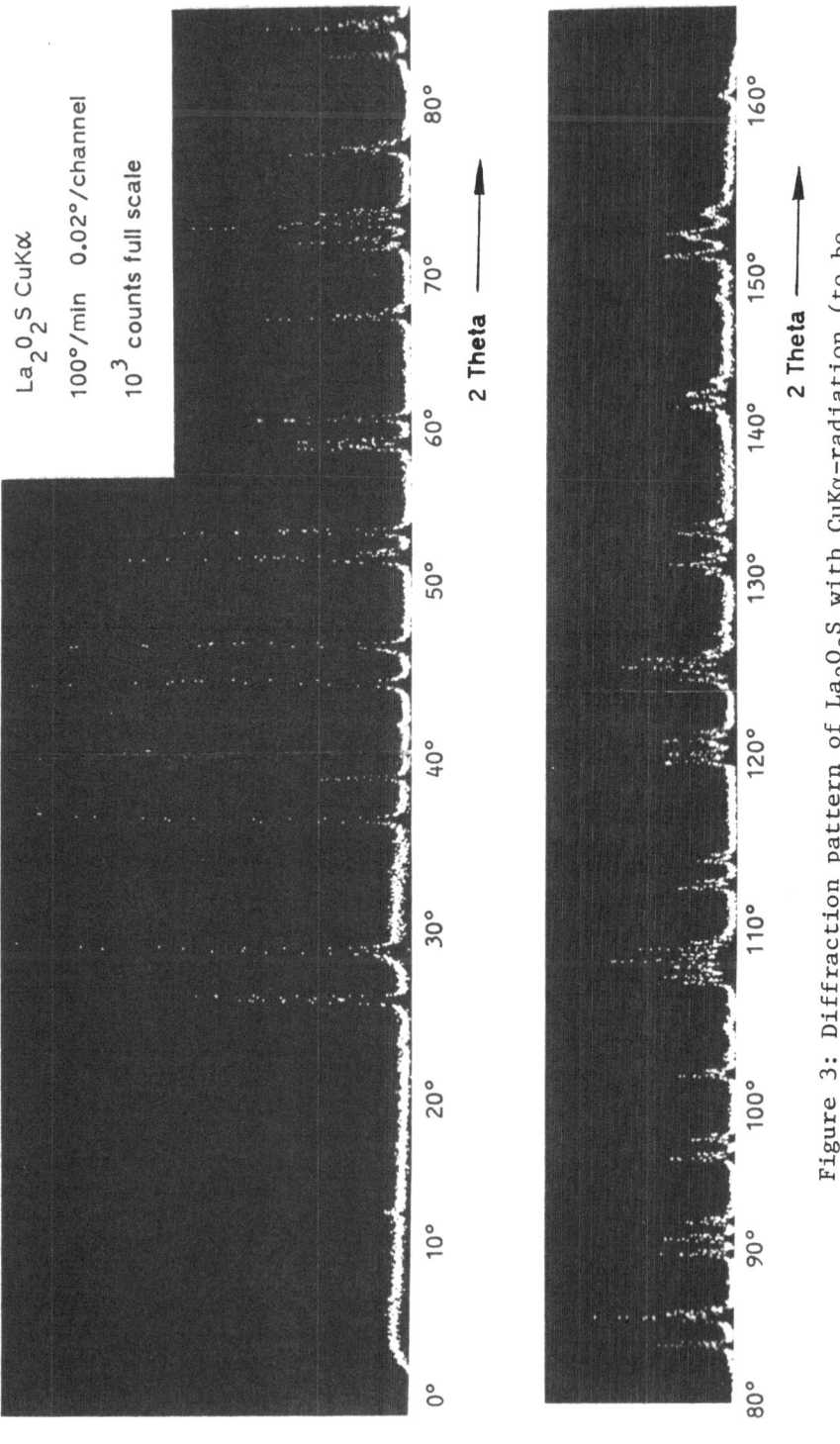

Figure 3: Diffraction pattern of La_2O_2S with $CuK\alpha$-radiation (to be compared with Fig. 5 of $(\bar{1})^2$)

- improving the resolution by higher digitization rates (here 0.02° per channel) and a variable detector window in the low angle region.

For the system alignment two experimental parameters have to be adjusted manually:

1) The detector pulse height for the radiation used has to be adjusted with the detector bias to pass the characteristic quantum energy through a single channel discriminator. Due to long term variations of the gas pressure[7] and the amplifiers the proper setting of this parameter, which influences the peak-to-background ratio of the pattern, has to be checked from time to time or is most favourably monitored on a pulse spectroscope.

2) The clock frequency of the time-to-digital converter (continuously adjustable between 400 and 750 MHz) has to be adapted to the graduation of the 2-theta arc. In RC-encoded linear detectors this adjustment depends on the resistivity of the detector wire, a high-resistive carbon coated quartz-glass fiber. Since this resistivity increases slowly with the exposure time, this adjustment has to be rechecked about every 100 specimens. The lifetime of the wire is restricted to about $5-10 \times 10^9$ counts corresponding to about 200 to 400 hours of operation during which several thousands of specimens can be investigated. Recent improvements of RC-line or delay-line encoded PSPC's using metal wires have increased the stability and lifetime considerably.

The absolute angular accuracy of the system is calibrated by a program CALIB on the basis of an external standard. The program CALIB corrects for zero shifts (linear approach) and nonlinearities from the aberration of measured and calculated line positions. As standard materials, highly pure silicon powder (linear approach) or LaB_6 (calibration curve) were used. A typical correction curve for our vertically mounted D 500 diffractometer is shown in Fig. 4.

For identification of unknown materials, only the linear approach is applied. In the figure the error window of 0.1 (in terms of 1000/d ($\AA^{-1}$) as defined in the Johnson-Vand Search-and-Match program[7] envelopes all measured line positions. For "starred" JCPDS data an error window of 1 is recommended, which is ten times larger. So, nonlinearities can be neglected. Since Si-powder has a too small number of reflections to produce a reliable correction curve, we used LaB_6 as calibration standard, a cubic primitive material with the lattice constant 4.15507 ± 0.00003Å. This material presents for CuKα-radiation 24 sharp and intense diffraction lines almost regularly every 5° of 2 theta from 20° to 150° (Fig. 5).

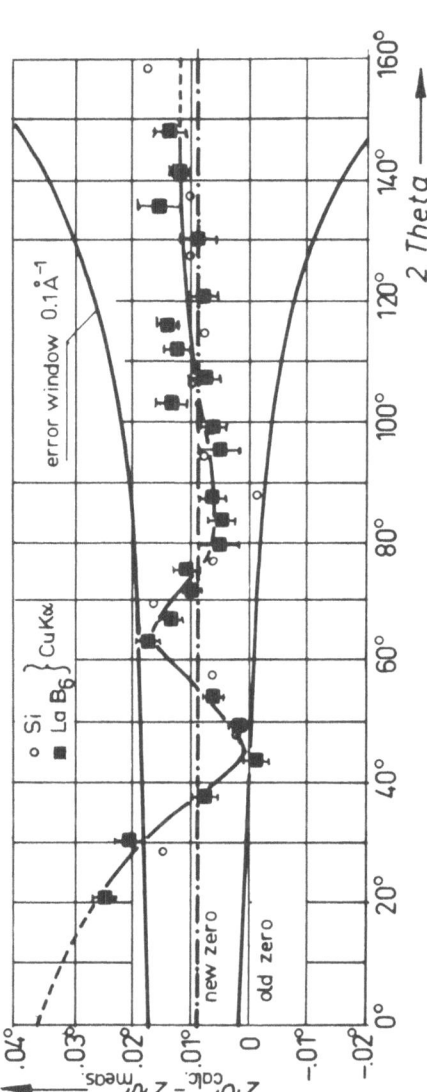

Figure 4: External standard calibration curve of a D 500-diffracto-
meter

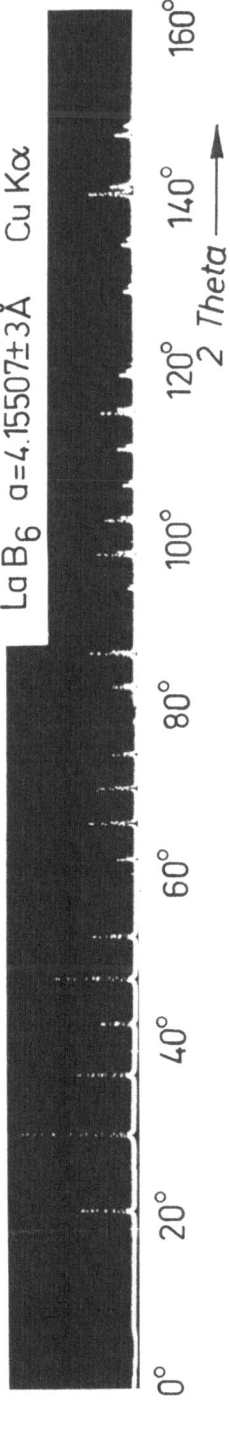

Figure 5. LaB$_6$, an ideal linearity calibration standard for
diffractometers

There are two <u>problems</u> that have to be taken into account
when working with a CPSD system:

1) Geometrical problems leading to a line broadening are due to
nonvertical incidence and to defocussing effects of the off-
center X-rays. Defocussing becomes remarkable at low angles
($2\theta < 20°$), so a relatively narrow detector entrance window
should be used in this region. The geometrical problem was
already mentioned in Ref. 1 and was extensively discussed by
James and Cohen, Ref. 9. As a very useful solution of the defocussing problem we introduced a variable entrance window in
front of the detector[8] with a smaller opening at low angles.
This also reduces the total pulse rate in this region, which
otherwise may overcharge the detector and lead to high dead-time
losses. However, the measured intensities have to be corrected
if compared to JCPDS data.

2) More serious is the problem caused by fluorescence radiation.
This isotropic radiation can be reduced before entering the detector only by using absorption filters, which have a poor
energy selectivity and are effective only for the $K\beta$-lines and
the higher energy part of the fluorescence background. The
more intensive lower energy fluorescence background can only be
discriminated electronically after having entered the detector.
The energy resolution of the PSD of about 20% (for $CuK\alpha$-radiation)
is, however, not good enough to distinguish strictly between
neighbouring elements. Obviously, using $CuK\alpha$-radiation, Mn and
Cr are the closest elements that allow a complete suppression
of their fluorescence background. Fluorescence radiation has
two effects: First, as in conventional systems, it causes a
high background that complicates the detection of small peaks in
the pattern and secondly it charges the PSD over its whole area
and will overfeed the total counting rate capacity of the detector system (about 10^5 counts per second). This leads to severe
dead-time losses. Additionally it will reduce the lifetime of
the wire.

An extreme example is shown in Fig. 6, where a ferrite specimen is analyzed with Cu $K\alpha$-radiation (upper part). The shaded area
represents the integral intensity which is charged to the detector,
and which is mainly fluorescence radiation. The useful peak area
is only a tiny part of the total detector charge. In the lower
part of the figure the same sample is analyzed by using Cr-$K\alpha$-
radiation. While the total charge in the upper part is about 10^5
counts per second, the same respective area in the lower part will
only be $5 \cdot 10^3$ counts per second which go mainly into the peak
intensity. It can also be seen from the picture that the high
total detector charge in the upper case leads to line broadenings
due to pulse pile-up.

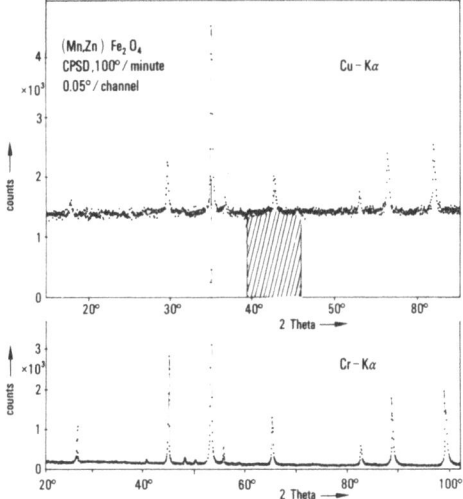

Fig. 6.   Influence of fluorescence radiation

SOFTWARE

For the evaluation of the diffraction data a part of the program system DIFFRAC 11[10] was used.   The whole system is outlined
in Fig. 7 with the parts used in this paper marked by bold frames.
The programs are written in FORTRAN and run on a PDP 11/34 computer
equipped with 64k byte of memory, dual RL 01 hard disc and a VT 55
semigraphical scope terminal under RT 11 (version 3b) operation
system.   For comfortable handling all programs are equipped with an
input dialogue and preset default values for the often returning
data.   Graphical aids are provided to enable the operator to judge
the results and interact if necessary.

Since the results characterizing the accuracy and reliability
of the CPSD data cannot be estimated without reference to the software, a brief description of some of the programs has to be given
here.

The peak-search subroutine used in the "mini-file" based
search and match program "IDENT" distinguishes between peaks and
background from the slope defined by a number of subsequent raw
data points.   If a critical raise is exceeded, a peak maximum is
expected that must at least be three times larger than the statistical error of the background and must form a peak with a minimum
half-width of 0.1° 2 theta.   The list of peaks found can be additionally reduced by setting a threshold relative to the strongest
peak.   The peaks found are then evaluated further and the center
of gravity and maximum location (parabolic approach) are calculated
for the peak position and the integral and peak intensities as well

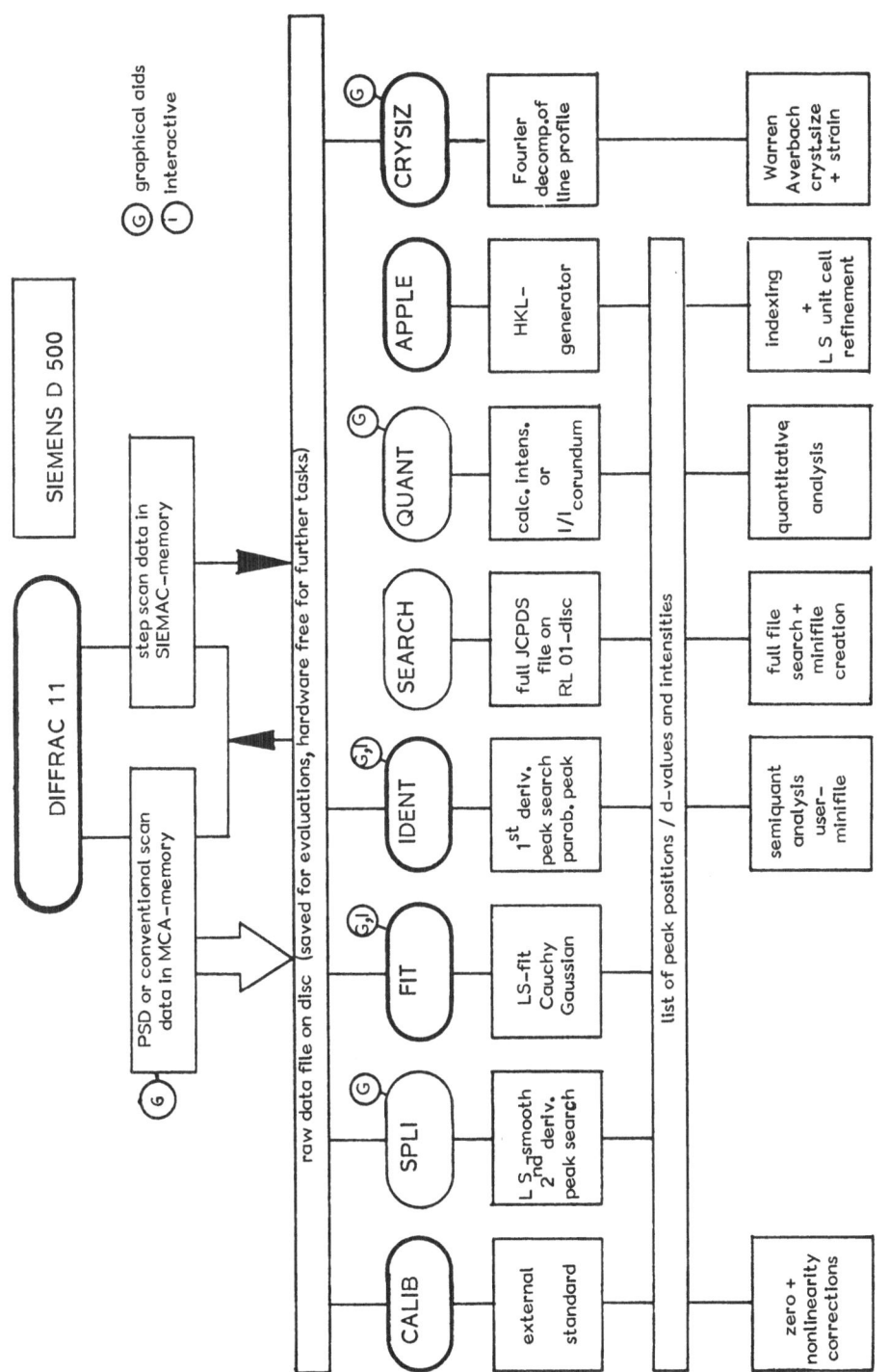

Figure 7: The DIFFRAC 11 Software System

as the half widths of the diffraction lines are evaluated. The
routine was developed as a fast (about 1000 raw data points per
second) and reliable peak detection program in full patterns,
exact enough for problems, with high overdetermination as in identifications on a JCPDS-data base, and least squares unit cell
dimensions refinements using for example Appleman's program
"APPLE".[4]

More precise positions, intensities and widths of single
peaks or (overlapping) groups of peaks can be determined by a
least-squares profile fitting program "FIT". It is based on the
Marquardt algorithm[11] for fast convergence. As theoretical profiles Gaussian and Cauchy curves are provided whereby one doublet
$K\alpha_1$ and $K\alpha_2$ can be described by one set of 3 parameters (position,
height, width of the $K\alpha_1$-line) that are adapted simultaneously.
Every parameter can be fixed if the proper result can be predicted.

For peak profile analysi   order to determine microcrystalline properties like crystallite sizes or crystal defect induced
lattice distortions, a program "CRYSIZ" is applied using the
Warren-Averbach approach.[5] To distinguish between the two line
broadening effects, for non-cubic systems two orders of reflections from the same lattice plane have to be selected out of the
pattern.

RESULTS

The accuracy of the data is demonstrated in a very well known
sample: α-quartz. It is applied to the problems of identification, unit-cell dimensions refinement, profile fitting and quantitative analysis. For the first two applications, the patterns were
collected at a scanning speed of 400 degrees per minute, the fast-
est gear of the D 500 diffractometer, in order to show the performance under extreme conditions. In usual analyses, where peaks
of one percent of the strongest line have to be reliably detected,
the scanning speeds range between 10 and 100 degrees per minute.

The data shown in the example for identification (Fig. 8) and
unit-cell dimensions refinement (Table 1) of α-quartz were evaluated using a linear system calibration (see Fig. 4). They are not
selected data but may be called typical. In terms of $1000/d$ $(\mathring{A}^{-1})$
the average deviation between measured and standard 5-490 d-values
is 0.23 $(\mathring{A}^{-1})$, which is far below the "error window" of 1 $(\mathring{A}^{-1})$ in
the Johnson-Vand search and match program.[7] In fact, values in
this region correspond to the truncation errors of starred JCPDS
cards. This can be seen from Table 1 where indexings are being
shown of both the starred card 5-490 and the CPSD data from the
upper example applying Appleman's program.[4] The standard data

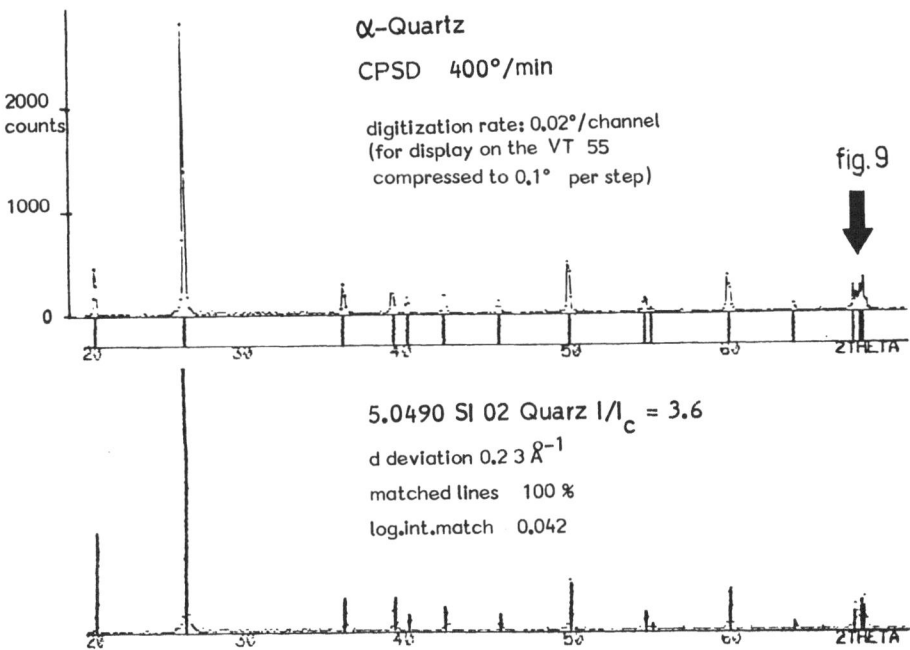

Figure 8: *Identification of a specimen at 400°/min
Upper part: raw data with markers indicating the detected
peak positions (threshold 3%)
Lower part: JCPDS-card 5-0490 superimposed to raw data.

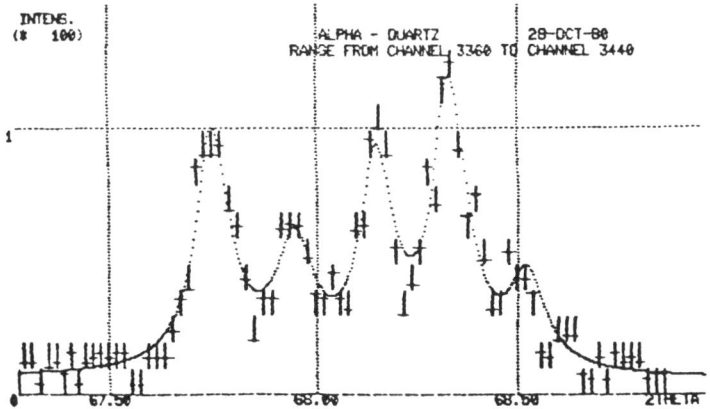

Figure 9: *Least-squares fit of Cauchy functions to the detail
indicated by the arrow in Fig. 8

*Hard Copies of the VT 55 Terminal

TABLE 1. Unit cell refinement of α-quartz from measured data run
with 400°/min scanning speed (upper part) and starred
JCPDS-file data (lower part)

SIO2   31 OCT-80

DIRECT CELL:
A = 4.71300 B = 4.91300 C = 5.40500 ALPHA = 90.000 BETA = 90.000 GAMMA = 120.000 VOL = 112.905

USED WAVELENGTH IS 1.540562
REFLECTIONS ARE CALCULATED TILL 37.0 1-THETA: FOR THE FIRST 2 CYCLES ONLY TILL 20.0
MINIMAL ALLOWED TOLERANCE IS 0.010, MAXIMAL TOLERANCE IS 0.050
CALCULATED FOR HEXAG SYSTEM

PRIMITIVE BRAVAIS-LATTICE

CONDITIONS FOR NON-EXTINCTION
              00L: L=3N

	A	B	C	ALPHA	BETA	GAMMA	VOLUME	
DIRECT CELL	4.91317	4.91317	5.40513	90.000	90.000	120.000	112.9993	CPSD
CORRECTIONS	0.00000	0.00000	0.00000	0.000	0.000	0.000	0.0000	measurement

LARGEST RESIDUAL REDUCED TO UNIT WEIGHT-0.00753   OBS 10   400°/min
STANDARD ERROR UNIT WT OBS 0.00412   DEGREES OF FREEDOM 14
STANDRD ERRS 2.92E-04 2.92E-04 3.96E-04 0.00E-01 0.00E-01 0.00E-01 1.24E-02

HKL LISTING   SIO2   31-OCT-80   HEXAG SYSTEM   (### REFERS TO FIXED, R TO REJECTED)

N	H	K	L	D CALC	D OBS	1-THETA CALC	1-THETA OBS	1-THETA DIFF	WEIGHT
1	1	0	0	4.25493	4.25691	10.430	10.425	0.005	1.0
2	1	0	1	3.34336	3.34339	13.320	13.320	0.000	1.0
3	1	1	0	2.45659	2.45668	18.274	18.273	0.001	1.0
4	1	0	2	2.28135	2.28083	19.733	19.738	-0.005	1.0
5	1	1	1	2.23645	2.23618	20.146	20.149	-0.003	1.0
6	2	0	0	2.12747	2.12738	21.227	21.228	-0.001	1.0
7	2	0	1	1.97965	1.97912	22.898	22.905	-0.007	1.0
8	1	1	2	1.81785	1.81754	25.070	25.075	-0.005	1.0
9	0	0	3	1.80177		25.310			
10	2	0	2	1.67168	1.67160	27.438	27.439	-0.001	1.0
11	1	0	3	1.65915	1.65956	27.663	27.655	0.008	1.0
12	2	1	0	1.60821		28.618			
13	2	1	1	1.54144	1.54168	29.981	29.976	0.005	1.0
14	1	1	3	1.45288	1.45297	32.017	32.015	0.002	1.0
15	3	0	0	1.41831		32.895			
16	2	1	2	1.38204	1.38214	33.873	33.870	0.003	1.0
17	2	0	3	1.37494	1.37500	34.072	34.070	0.002	1.0
18	3	0	1	1.37187	1.37181	34.158	34.160	-0.002	1.0
19	1	0	4	1.28794	1.28782	36.732	36.736	-0.004	1.0

	A	B	C	ALPHA	BETA	GAMMA	VOLUME	
DIRECT CELL	4.91325	4.91325	5.40505	90.000	90.000	120.000	112.9974	standard
CORRECTIONS	0.00000	0.00000	0.00000	0.000	0.000	0.000	0.0000	5-490

LARGEST RESIDUAL REDUCED TO UNIT WEIGHT-0.01237   OBS 1
STANDARD ERROR UNIT WT OBS 0.00733   DEGREES OF FREEDOM 17
STANDRD ERRS 4.21E-04 4.21E-04 6.19E-04 0.00E-01 0.00E-01 0.00E-01 1.94E-02

HKL LISTING   SIO2 JCPDS   31-OCT-80   HEXAG SYSTEM   (### REFERS TO FIXED, R TO REJECTED)

N	H	K	L	D CALC	D OBS	1-THETA CALC	1-THETA OBS	1-THETA DIFF	WEIGHT
1	1	0	0	4.25500	4.26000	10.430	10.417	0.012 /	1.0
2	1	0	1	3.34333	3.34300	13.320	13.322	-0.001 /	1.0
3	1	1	0	2.45663	2.45600	18.273	18.263	0.011 /	1.0
4	1	0	2	2.28128	2.28200	19.734	19.727	0.006	1.0
5	1	1	1	2.23646	2.23700	20.146	20.141	0.005	1.0
6	2	0	0	2.12750	2.12800	21.227	21.221	0.005	1.0
7	2	0	1	1.97966	1.98000	22.898	22.894	0.004	1.0
8	1	1	2	1.81783	1.81700	25.071	25.083	-0.012 /	1.0
9	0	0	3	1.80168	1.80100	25.311	25.321	-0.010	1.0
10	2	0	2	1.67156	1.67200	27.438	27.432	0.006	1.0
11	1	0	3	1.65908	1.65900	27.664	27.665	-0.001	1.0
12	2	1	0	1.60824	1.60800	28.617	28.622	-0.005	1.0
13	2	1	1	1.54145	1.54100	29.981	29.991	-0.010	1.0
14	1	1	3	1.45284	1.45300	32.018	32.014	0.004	1.0
15	3	0	0	1.41813	1.41800	32.894	32.903	-0.009	1.0
16	2	1	2	1.38204	1.38200	33.873	33.874	-0.001	1.0
17	2	0	3	1.37491	1.37500	34.073	34.070	0.003	1.0
18	3	0	1	1.37189	1.37200	34.155	34.152	0.003	1.0
19	1	0	4	1.28788	1.28800	36.734	36.730	0.004	1.0

have larger deviations from the calculated positions of an optimum
unit cell than the here evaluated CPSD data. "Error window" = 1,
which is usually recommended for reliable full file search runs,
seems to be routinely achieveable.

For a precise determination of unit-cell dimensions or for
indexing, an absolute accuracy better than 0.01 - 0.02 degrees of
2 theta is required to produce reliable data. This is about the
accuracy shown in the example of Table 1. To be on the safe side,

however, the least squares profile fitting routine "FIT" should be
used to localize the diffraction peaks. With this method accuracies of a few thousandths of a degree can be achieved even for
overlapped groups of peaks. An example for FIT is shown in Fig. 9
with the "quintriplet" of α-quartz. This group of diffraction
peaks is often used to demonstrate the resolution of diffractometers. The standard errors for the peak positions were calculated to be between 0.003° and 0.005°.

In quantitative analysis of mixtures preferred orientation and
crystallite statistics are important sources of errors. While the
first of these problems can only be avoided by using complicated
preparation techniques, the second problem is usually solved by
spinning and oscillating the specimen in the diffractometer, which
will not only require special specimen holder constructions and
often also special substrates, but will become impracticable if
high measuring velocities are necessary. The CPSD technique has
the sample oscillation inherent in its principle: In the coupled
theta/2 theta mode of the diffractometer the specimen is rotated
in the primary beam by half the angle of the detector window, for
example, 5 degrees if the detector is 10 degrees open. So, subsequently, all crystallites come into reflection position that
are oriented within this angular range. In conventional systems
without the above-mentioned auxiliary attachments this range would
be only a few hundredths of a degree, and for a minor phase only
very few crystallites will then have the right orientation to
contribute to the diffracted intensity. An example is shown in
Fig. 10, where the quartz content in alumina ores was to be measured. While a conventional system produced a strong scattering
of the plotted points, the same specimens in a CPSD-system showed
a smooth curve of the measured intensities versus the quartz
content. In both cases the integral intensity of the (101)-line

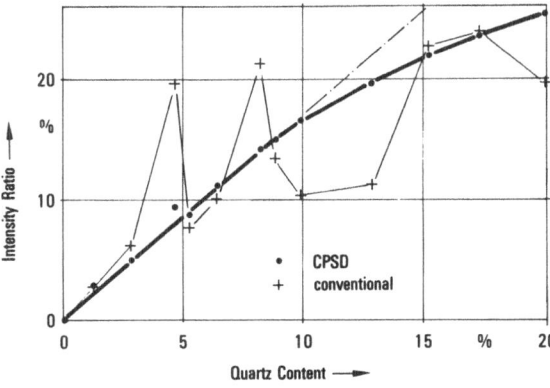

is related to the inten-
sity of the strongest
peak of $Al(OH)_3$, the
major component, corrected
for additional impurities.

One of the most critical
analyses of X-ray powder
diffraction is the deter-
mination of microcrystal-
line properties such as
strain, induced by crystal
defects, and crystallite
size and obtained from

Fig. 10.   Quantitative analysis: Quartz
in alumina

the Fourier decomposition of line profiles. The apparatus function is fixed by a calibration specimen, for example highly perfect
silicon powder. As application the crystallite sizes of differently treated TiO_2(rutile) powder samples were determined. The
powders, produced by hydrothermal decomposition of a water solution of $TiCl_4$ under the influence of ammonia, were annealed at
different temperatures in air to form a suitable crystallite size
for ceramic materials production. The profiles of the (110) and
(220) reflections were used to distinguish between crystallite
size- and strain-broadenings applying the Warren-Averbach approach.
It turned out that the strain contributions were negligible and
the broadenings were clearly due to the crystallite size. Examples
for the observed line profiles are shown in Fig. 11. The average
crystallite size of a series of specimens is plotted against the
annealing temperature in Fig. 12. In the logarithmic scale the
crystallite size forms an expected straight line. Additional
points, measured on a conventional diffractometer, fit well into
this curve. This is, of course, not surprising since the results
have to be independent of the apparatus function.

     As in quantitative analysis, here also the improved statis-
tics is the main advantage compared to conventional diffractometer
systems, where single larger crystallites may produce sharper peaks
than actually correspond to the average grain size.

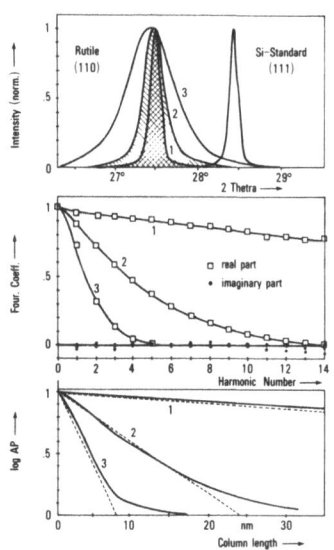

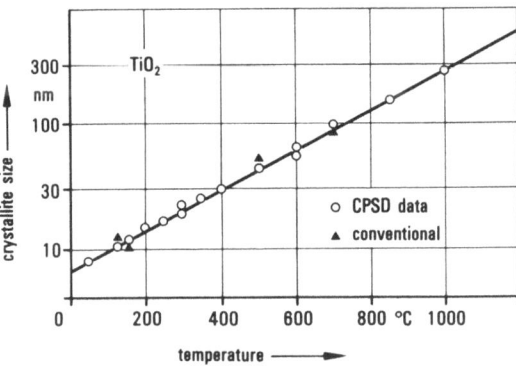

Fig. 11. Fourier analy-
sis of line profiles for
crystallite size deter-
mination in TiO_2 annealed
24 hrs at 1000°C,[1] 350°C,[2]
and as precipitated.[3]

Fig. 12. Crystallite size of rutile
powder annealed 24 hours at different
temperatures

CONCLUSIONS

The CPSD method has been developed not only as a very fast, but also a very accurate diffraction technique suitable for even the most ambitious tasks of XRPD. Since the system is a combination of mechanical and analogue or digital electronic quantities, the alignment of the system is more critical than in purely mechanical diffractometers. Above all, the zero-shift has to be rechecked from time to time. After improvements in the beam path the only remaining background problem is fluorescence radiation. The CPSD method puts also new requirements on diffractometer automation and data evaluation. The control can be simplified and reduced to only a few functions. On the other hand the evaluation systems have to be fast to keep up with the measuring speed, and reliable to reduce human control of the results. These have to be informative enough (graphical aids) to enable the responsible analyst with one fast glance to judge the sense or nonsense of the findings.

ACKNOWLEDGEMENTS

This publication was not the work of a single person. It resulted from the cooperation of preparative chemists (Dr. L. Vité, Mira Dobner, Hannelore Mews), experienced electronic and detector experts (N. Halder, M. Henne) and busy software specialists (Barbara A. Jobst, Dr. M. Peterat). Prof. Dr. G. G. Johnson was very helpful in establishing the APPLE program and in discussions of the IDENT program. The author is indebted to all who contributed to this study, which was supported by the Siemens X-ray Analytical Systems Division.

REFERENCES

1. H. E. Göbel, A New Method for Fast XRPD Using a Position Sensitive Detector, in: "Advances in X-Ray Analysis, Vol. 22," G. J. McCarthy, C. S. Barrett, D. E. Leyden, J. B. Newkirk, and C. O. Ruud, eds., Plenum Publishing Corp., New York (1979)
2. H. E. Göbel, ACA Winter Meeting (March 1979), Honolulu, Hawaii
3. H. E. Göbel, Ger. Offen. 2,637,945 (March 1979)
4. D. Appleman, ACA Meeting 1963, Cambridge, Mass. Program modified for PDP 11 by G. G. Johnson, G. Power, S. E. Stern and Barbara A. Jobst (1979).
5. B. E. Warren, R. L. Averbach, J. Appl. Phys. 21:595 (1950) and J. Appl. Phys. 23:497 (1952).
6. N. Broll, M. Henne, W. Kreutz, Analytical Application Note 271 (1979), Order No. E 689/2021 from Siemens AG, Karlsruhe, W. Germany.

7. G. G. Johnson and V. Vand, Ind. Eng. Chem. 59:19–31 (1967).

8. H. E. Göbel, Ger. Offen. P 28 33 078.1 (1979).

9. M. R. James and J. B. Cohen, J. Appl. Cryst. 12:339–345 (1979).

10. DIFFRAC 11, Software System for XRPD (1979/80) developed at Alfred University (Alfred, NY), Siemens Research Laboratories (Munich, W–Germany) and Siemens Corporation (Cherry Hill, N.J.) in contact with the JCPDS.

11. D. W. Marquardt, J. SIAM 11:431–441 (1963).

A NEW MODEL OF X-RAY POSITION SENSITIVE DETECTOR

DEVELOPED IN FRANCE

Louis Castex
Laboratoire de Mécanique-Physique, ERA CNRS n° 769
F-33405 Talence

Jean Lou Lebrun
Département Matériaux, Ecole Nationale Supérieure d'Arts
et Métiers, F-75640 Paris

Serge Bras
Société Instruments SA, F-91160 Longjumeau

Since 1973 X-ray position sensitive detectors (P.S.D.) have been employed for stress analysis. The first models,[1-3] based on the gas flow of Ar-CH4, were composed of a quartz fiber anode (20 μm diameter), coated by a thin carbon layer. From 1975 the gas filling has become permanent. Hence the use of fiber P.S.D. (F-P.S.D.) was made easier; nevertheless, two important inconveniences remained. On the one hand, an intensive X-ray beam could destroy rapidly and locally the carbon film, which makes the detector malfunction; on the other hand, the pulse height analysis (P.H.A.), carried out in 1976, has proved to be ineffectual[4] because of the design of the F-P.S.D. itself: its fiber impedance does not remain constant. Thus the appearnace, in 1979, of a new P.S.D. with a metallic wire (W-P.S.D.) presented a significant advantage (see Fig. 1). The anode--made of tungsten or molybdenum gilded--can resist intensive X-ray beams perfectly. Hence the life of the detector is only limited by a possible ionization of the filling gas, which approximately correspands to 10^{12} counts. A gas refilling--which will be provided in the standard equipment--allows us to use the detector again. Moreover, the P.H.A. reveals a good efficiency, as was observed when irradiating the (222) plane of type 1010 steel by Kα radiation of copper (Fig. 1). Similar results have been obtained with the (213) plane of titanium alloy TA 6V and with the (331) plane of Inconel 718, with which one could not previously get any reliable results in the area of stress analysis.

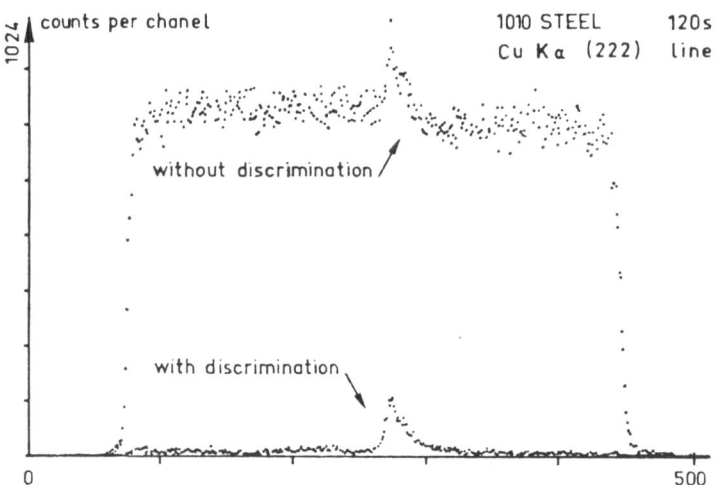

Figure 1.   Example of the W-P.S.D. associated P.H.A. efficiency

On account of all these qualities, the new W-P.S.D. (Fig. 2)--
the features of which are listed and compared with those of the
F-P.S.D. in Table 1--is now working in some laboratories in France
concerned with stress analysis, quantitative phase analysis, diffraction profile analysis and small angle scattering.

For our purpose, we successfully used the W-P.S.D. in stress
analysis, instead of the F-P.S.D., which was considered too fragile
and too limited owing to the unrelible P.H.A.. It is worthwhile to
notice, however, that the problems related to the geometry of P.S.D.
(5) remain of course the same in the case of the W-P.S.D..

REFERENCES

(1) Borkowski, Kopp "new type of P.S.D. of ionizing using risetime
    measurement." Review of Scientific Instruments, 1968, volume 39,
    n° 10, 1515-1522.
(2) Gabriel, Dupont "position sensitive proportional detector for
    X-ray cristallography." Review of Scientific Instruments, 1972,
    volume 43, n° 11, 1600-1602.
(3) Allemand   "détecteur à localisation de particules."
    Brevet n° 73-46051 du 21 décembre 1973.
(4) Groupement Français pour l'Analyse des Contraintes par Diffracto-
    métrie de Rayons X.
    - réunion du 15 novembre 1978, Grenoble, Miège.
    - réunion du 16 mars 1979, Nancy, Bras.
(5) James, Cohen   "geometrical problems with a P.S.D. employed on a
    diffractometer, including its use in the measurement of stress."
    Journal of Applied Cristallography, 1979, 12, 339-345.

Table 1

Main Features of F-P.S.D. and W-P.S.D.

model		fiber P.S.D.	wire P.S.D.
origin		(1) and (2)	(3)
anode		quartz fiber coated with carbon	wire of tungsten or of molybdenum gilded
cathode		aluminium plate	backgammon game
filling gas	90%	Ar or Xe	Xe
	10%	CH4	CH4
pressure	MPa	.1	.1 to .14
energy resolution (8.05 KeV)		20%	13%
F.W.H.M.	μm	180	130
location linearity		2%	1%
counting rate limit (counts per second)		10^4	10^4
active length mm		55	50
active height mm		8	12^+
volume space mm		110x68x44	140x46x78
life duraion		600 h^+	10^{12} counts^{+++}

+   6 mm for the best resolution,
++   mean value observed in 7 years,
+++ ionization limit of the filling gas.

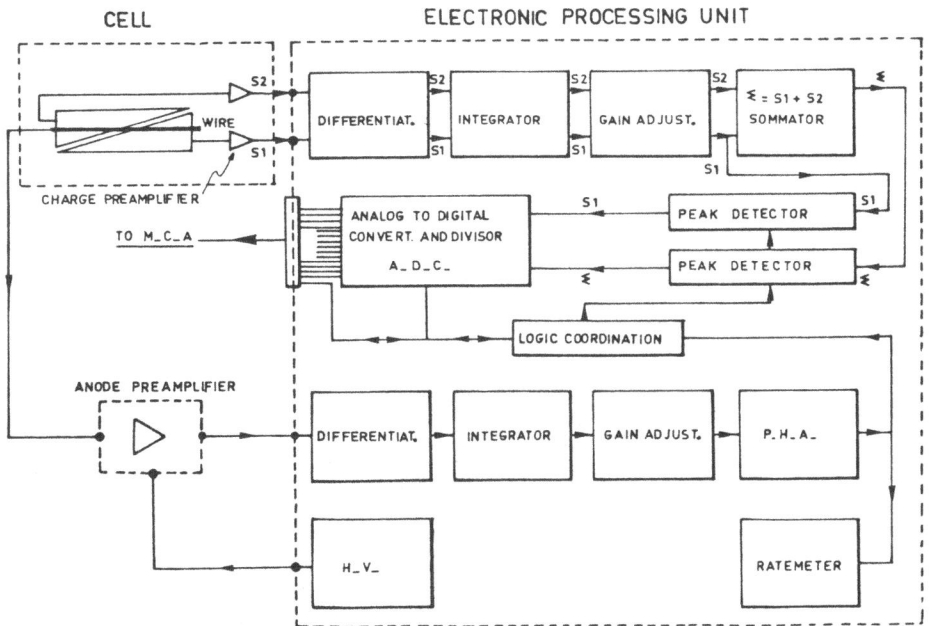

Fig. 2. Block-diagram of the W-P.S.D.--X-photon position = $S1/\Sigma$‾‾

USE OF A POSITION SENSITIVE DETECTOR

Macrostress automatic measurements, Quantitative phase analysis,
Microstress analysis.

J. L. Lebrun, J. M. Sprauel and G. Maeder

Dept. Materiaux, E.N.S.A.M.

151, Bd de l'Hôpital, 75640 PARIS Cedex 13, FRANCE

EQUIPMENT

Our goniometer used is a classical one with step scanning motors both on θ and 2θ movements. The theoretical radius is 250 mm
but it can be changed easily.

Two kinds of PSD have been used,[1] the older one of a resistive
type with a carbon coated quartz fiber and, for this year, a capacitive one with a gold-plated molybdenum wire anode, which is more
reliable, with a very good pulse height analyser. It is connected
to a 48K calculator (IN 90) with a floppy disk unit. A special
interface has been made which is able to drive separately the 2θ and
θ movements of the goniometer, using the output signals of the calculator. After the different settings of the goniometer and the
initialization of the positions in θ and 2θ, every movement can be
ordered either by a very simple software from the keyboard or inside a more complex program.

MACROSTRESS AUTOMATIC MEASUREMENTS

Simultaneously with the equipment described above, we have
developed a complete software, making the experiment fully automatic.
A simplified flow chart is given on figure 1. This program is very
flexible and works with all kinds of samples and materials. Mainly,
we can choose the number of ψ positions (either positive or/and negative), the value of the sample oscillations if necessary, the statistical precision, L.P.A. correction or not ($\theta = \theta_0 + \psi$).

The previous peak is treated during the following acquisition.
The peak position is taken as the maximum of a 3rd degree polyno-

mial fitted in between the inflexion points. The treatment enables
us to draw the integrated intensity (in c/s units), the background
and peak level and the integral breadth in reciprocal lattice units.
After the last peak has been treated, the slope of the line $2\theta = f(\sin^2\psi)$ is calculated with an 80% confidence interval and then
the stress in MPa units.

These results let us control the reliability of the measurement.
Indeed, should a bad diffraction occur for one or a few angles, we
would be able to notice it on either intensity or breadth parameters
Moreover, the profile breadth provides interesting information
about the microstructural state of the metal.

The period of measurement is at present limited by the length

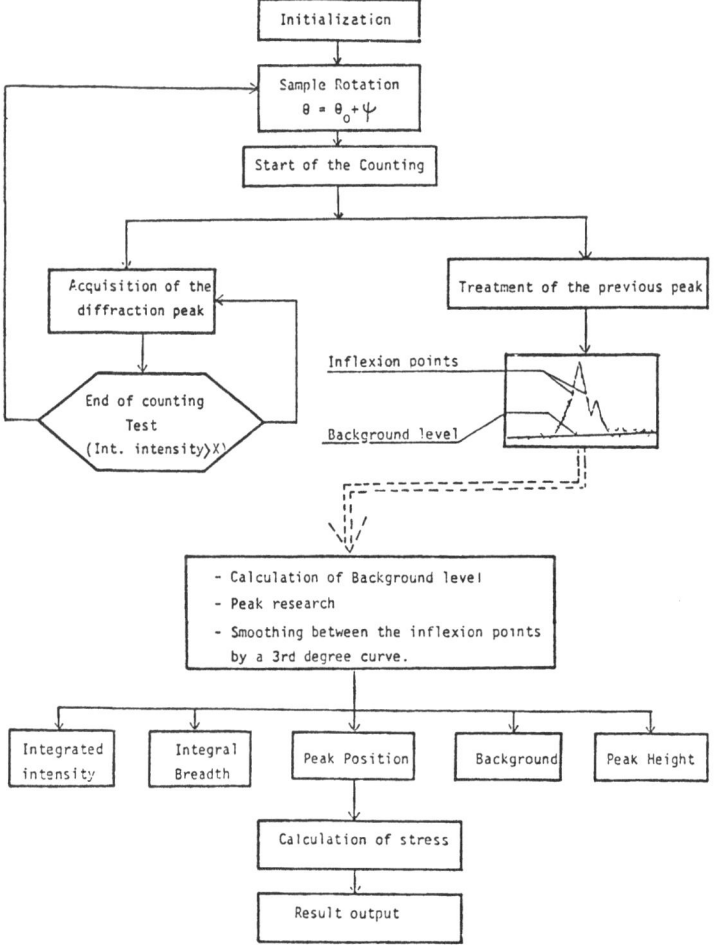

Fig. 1 : Flow chart for Macrostress Automatic Measurements

of the peak treatment (about 80s). With very easy measurement
(heat treated steel for example) the total time required for a 3
to 5 ψ measurement is 5 to 7 minutes. But a much greater advantage
is obtained in more difficult cases such as measurement on titanium
or nickel base alloys, particularly when they are heavily cold worked, by machining for example, where the PSD plays all its role.
A 5 ψ measurement on a milled titanium part takes about 20 minutes.

A special advantage of the PSD is to detect immediately the
need and then the effects of sample oscillations on the shape of
the peak diffraction as shown on figure 2.

QUANTITATIVE PHASE ANALYSIS (Residual Austenite)

The X Ray diffraction technique is commonly used, and we shall
not describe the theoretical principles of the method[2]. However,
it is important to note that the main source of errors lies in the
presence of a crystallographic texture. The use of pole figures[3]
though being a precise method is much too long and complicated for
common phase analysis. However, it is interesting to use a few
peaks of each phase α and γ to lower the texture problems.

With Mo radiation, the $(111)\gamma + (110)\alpha$, $(200)\gamma$ and $(200)\alpha$
peaks are between 18.5 and 29.5θ and the $(220)\gamma$, $(211)\alpha$, $(311)\gamma$
and $(222)\gamma + (220)\alpha$ peaks between 31 and 42°. So, with the PSD,
two acquisitions are sufficient to get these five isolated and
four overlapping peaks (Fig. 3). Therefore, the program pilots
the goniometer to the desired positions, records the peaks

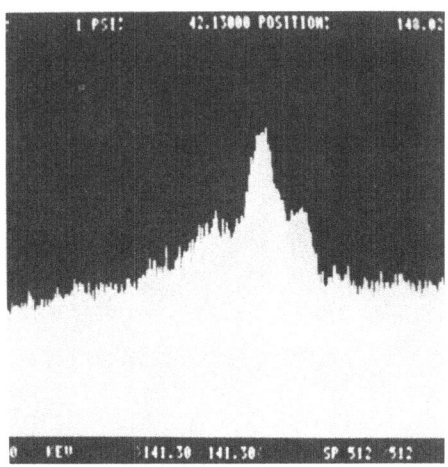 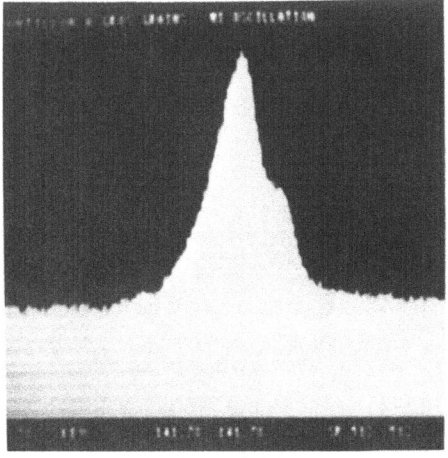

a) beginning of acquisition   b) end of acquisition
   without oscillation   with oscillations

Fig. 2: Visualization of the need of sample oscillations
         during peak recording (Aluminum Weldment)

and measures the integrated intensities after background has been
eliminated. The volume of Austenite can then be calculated by different methods : a quick one uses the sum of the intensity of α
and γ isolated peaks ; another one uses the different pairs of these
isolated α-γ peaks - we can thus note the importance of the texture
and the last one which takes also the 4 overlapping peaks in an
iterative calculation from GIAMEI and FREISE.[4]

A complete measurement lasts about 5 minutes for a specimen
with more than 10% of austenite. For specimens with a smaller
amount, the counting statistic must be better; the length of the
acquisitions is then greater.

The advantage of the PSD for this quantitative phase analysis
is the simultaneous record of a few peaks of different phases.
With the automatic calculation, the analysis can be extended to analysis of 3 or more phases [2].

MICROSTRESS ANALYSIS

This kind of analysis is very interesting for determination
of the microstructure of materials, for example after machining[5] or
plastic deformation[6]. However, it is not yet very developed, mainly
because of the length of the experiments. Indeed, the analysed peaks
must be recorded with a great statistical precision ; this is very
long with classical step-scanning installations.

With the PSD, the entire peaks are recorded at once. Figure 4
shows the broadened peak (422) of a machined aluminium part and,

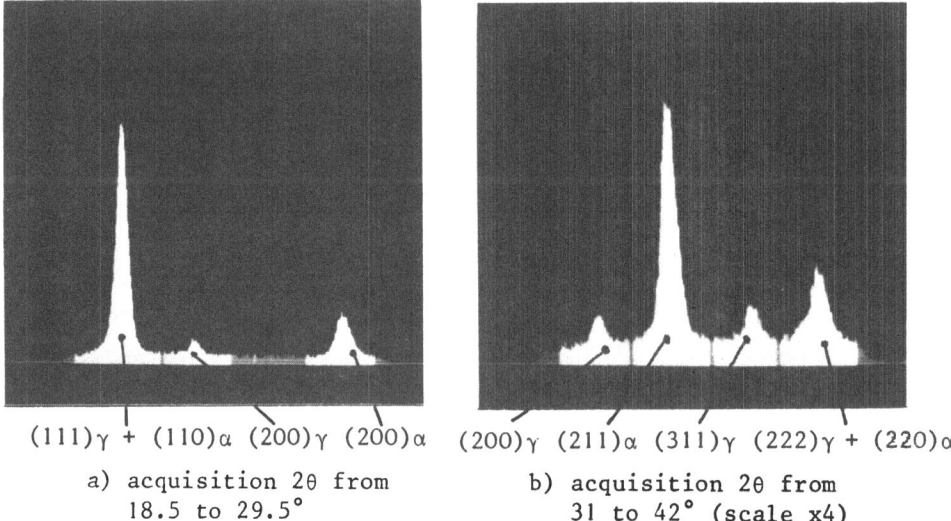

(111)γ + (110)α (200)γ (200)α   (200)γ (211)α (311)γ (222)γ + (220)α

       a) acquisition 2θ from   b) acquisition 2θ from
          18.5 to 29.5°   31 to 42° (scale x4)

          Fig. 3 : Residual Austenite automatic measurement
                   (16 NCD 13 carburized steel)

superposed the instrument peak (here a well recrystallized aluminium sheet). After smoothing, theoretical corrections and background substraction of the peaks, a Fast Fourier Transform is performed. The calculator then eliminates the instrumental broadening by the STOKES correction. The results can then be visualized on the screen as shown on figure 5 (real and imaginary parts).

The separation of particle size and strain broadening is done on this single analysis with a method first proposed by GANGULEE[7] and then MIGNOT and RONDOT[8]. The mean square strain is modelized into an hyperbolic form $\langle \varepsilon_L^2 \rangle = C_1 + C_2/L$ and the particle size broadening in a linear form $A_L^p = (1 - L/D)$. The real part of the Fourier transform is thus fitted to a 3rd degree polynomial whose coefficients are used for the calculation of the searched parameters. In the case of Figures 4 and 5 the results are D = 136 A and $\langle \varepsilon_L^2 \rangle = (1.16 + 95/L).10^{-6}$.

Like for the other uses of the PSD, everything is done automatically in a 20 minute minimum period since the peak must be very well recorded. This peak can be the same as in the case of the macrostress analysis. It brings 2 advantages: precision in the analysis is better, thanks to the large Bragg angle, and we can follow the evolution the microstructural parameters D and $\langle \varepsilon_L^2 \rangle$ together with the macrostresses. This is very important for the understanding of the mechanical behaviour of the materials like, for example, shot peened layers, machined surfaces of fatigued parts. This microstructural analysis used to be done, but only roughly, by the evolution of the breadth of the diffraction peaks[9].

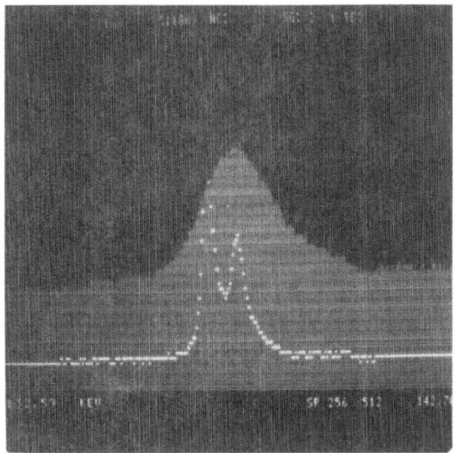

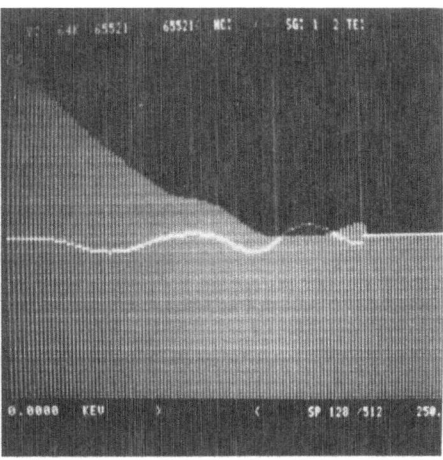

Fig. 4 (422) Aluminium peaks bars- Broadened (machined Al) dots – Instrumental (recrystallized Al)

Fig. 5 Fourier coeff. after STOKES correction. Real (bars) and imaginary (dots) parts.

With the PSD, it can be completed with the most sophisticated
methods in a very reasonable period of time even with difficult
materials such as titanium alloys.

CONCLUSION

This very quick and schematic description of macrostress automatic measurements, quantitative phase analysis, and microstress
analysis, shows all the advantages we have drawn from the use of the
PSD with its calculator and the automatization of the goniometer.

   - a great reduction in the length of experiments,
   - a better accuracy in the measurements,
   - a complete control of the parameters that give the relia-
     bility in the results,
   - the possibility of getting together much more information
     on the macro and microstructural characteristics of the
     material.

REFERENCES

1. L. CASTEX, J.L. LEBRUN, and S. BRAS, "A New Model of X Ray
   Position Sensitive Detector Developed in France", this conference.

2. "Quantitativen Rontgenographischen Phasenanalyse", Harterei-
   Techn. 27 : 229-278, (1972)

3. C.R. HOUSKA, and V. RAO, "Determination of Volume Fraction in
   Multiphase Systems Using Incomplete Pole Figures", Met.
   Trans., 9A : 1483, (1978)

4. A.F. GIAMEI, and E.J. FREISE, "Optimization of X Ray Diffraction
   Quantitative Analysis", Trans. TMS-AIME, 239 : 1676 (1967)

5. S. MASSON, "Application de la méthode de Warren-Averbach à l'é-
   tude de l'écrouissage superficiel par usinage d'un acier austé-
   nitique", Mem. Scie et Tech. de l'Armement, 46 : 1015, (1972)

6. J.L. LEBRUN, G. MAEDER, and P. PARNIERE, "Influence of the Cold
   Rolling Reduction on the stored Energy in a Low Carbon Steel
   Sheet", 5th I.C.O.T.O.M., Aachen, Ed. by G. Gottstein and K.
   Lucke, Springer-Verlag, 513, (1978)

7. A. GANGULEE, "Separation of the Particle Size and Microstrain
   Components in the Fourier Coefficients of a Single Diffraction
   Profile", J. Appl.Cryst., 7 : 434, (1974)

8. J. MIGNOT and D. RONDOT, "Méthode de Séparation des Dimensions
   de Domaine et des Microdéformations à partir des coefficients
   de Fourier d'un seul Profil de Raie de Diffraction X", Act.Met.,
   23 : 1321, (1975)

9. T. GOTO, "Application of X Ray Stress Measurement to Pre-Service
   inspection and In-Service inspection", I.C.M. 2, Boston, 1614,
   (1976).

A VERSATILE X-RAY STRESS ANALYZER USING A

POSITION SENSITIVE DETECTOR

Yasuo Yoshioka
Musashi Institute of Technology
1 Tamazutsumi, Setagaya, Tokyo 158, Japan

Ken-ichi Hasegawa and Koh-ichi Mochiki
Faculty of Engineering, University of Tokyo
7 Hongoh, Bunkyo, Tokyo 113, Japan

INTRODUCTION

Instrumentation for X-ray stress analysis has been advanced rapidly in the last few years. Especially, the time required for data accumulation has been remarkably reduced without motion of the detector by using a new X-ray detector called a position sensitive detector (PSD). Applications of PSD to the field of X-ray stress analysis were carried out by James and Cohen,[1] Ruud and Barrett,[2] and the authors.[3,4] In our laboratory, several position sensitive proportional counters (PSPCs) were designed and manufactured for residual stress measurements. Results show that the PSPC method is a powerful alternative to the conventional counter method or the film method.

This paper reports a design of a versatile PSPC X-ray stress analyzer for use in industry and the laboratory.

DESIGN

This analyzer is designed in an attempt to measure the stress in ferritic steel using Cr-Kα X-rays. The basic configuration is shown in Fig. 1. The angle between an X-ray beam from a tube with Cr target and the bisector of a PSPC is set to be 24° and the intersection coincides with the specimen surface, which is also the center of rotation for ψ angle setting, for the diffraction angle from the αFe(211) line of about 156° in 2θ.

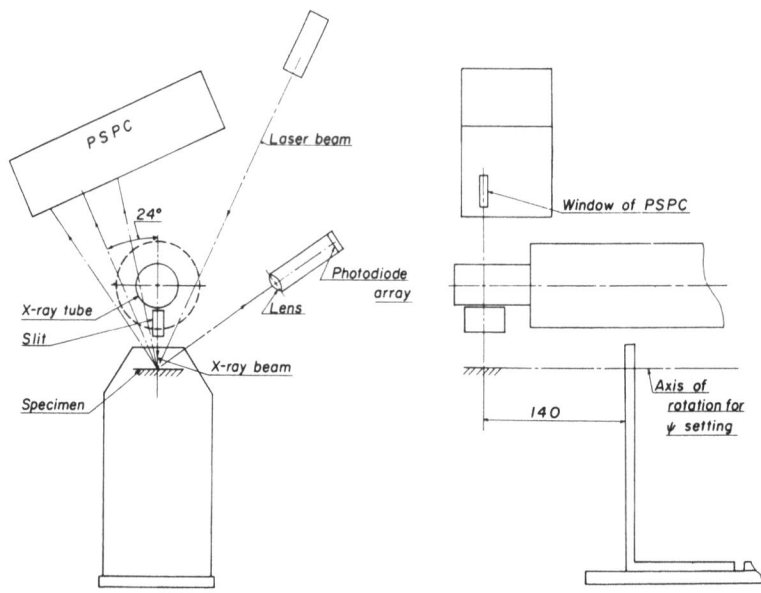

Fig. 1.   Schematic construction of X-ray stress analyzer.

A water-cool type X-ray tube, manufactured by Toshiba Co., Tokyo, Japan, is chosen. It enables the recording of the profile at high diffraction angle with high intensity since the tube head has a dimension of only 40 mm effective diameter with its shielding case. The distance between the tube window and the specimen surface is 65 mm. The X-ray beam passes through a soller slit or a pin-hole and irradiates the specimen. The beam size can be easily varied within a range from 2 x 2 mm^2 to 2 x 10 mm^2 by exchange of a sector mask in front of the soller slit, and a 2 mm or 1 mm diameter spot is obtained by use of pin-hole slit.

Both the PSPC and the tube manually or electrically rotate around the beam position on the specimen over a ψ range of -20° to 50°. A stepping motor is used in the case of an electrical drive.

The PSPC, which had to be behind the tube, is a gas-flow type with a uniform angular resolution described in Ref. 3. It is capable of recording the diffraction profile over a range of about 24° in 2θ when a distance between the PSPC and the specimen is set to be 200 mm. It was connected to a microcomputer via two charge sensitive preamplifiers, a mixer, two stretchers, a divider and a 512 channel pulse height analyzer. A profile at each ψ angle is collected simultaneously in the PHA, the peak of profile is determined by the half-value breadth method, and a stress value is calculated from peaks measured at several ψ angles by the $\sin^2\psi$ method

The positioning of the specimen is very critical in the PSPC method, as well as the film method because a receiving slit can't be used in front of the PSPC window. In this case, since the specimen should be held to the determined position within a range of 0.5 mm, the specimen is precisely positioned with the help of a laser beam and a photodiode array. As shown in the figure, a laser beam is directed on the position for stress measurement. The light-spot is seen through a lens with an incident angle of 30° to the laser beam and it images on a linear photodiode array which has an aperture length of 6.5 mm. This image is recorded on a CRT screen as a peaky profile, and it moves horizontally according to the specimen displacement. In this system, a displacement of 50 μm is detectable with high sensitivity. Fig. 2 shows a photograph of the apparatus under measurement in a welded specimen.

EXAMPLES

Profile Observations

Fig. 3a shows an example of a profile diffracted from an annealed low carbon steel specimen. The X-ray tube is operated at 30 KVP and 10 mA, and a beam size is about 2 x 10 mm^2 on the specimen. When a peak count was preset to be 2048, the time for forming a profile was only 14 seconds. The profile in Fig. 3b is of a

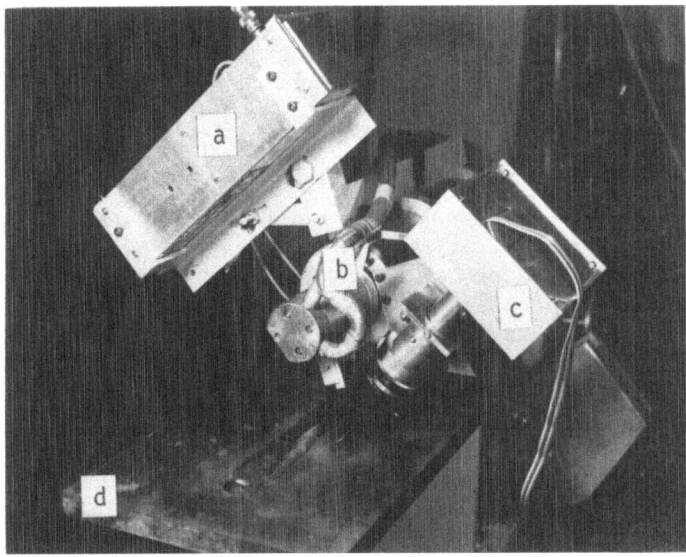

Fig. 2. A photograph of X-ray stress analyzer and specimen. a) PSPC; b) X-ray tube; c) position sensor system; d) specimen.

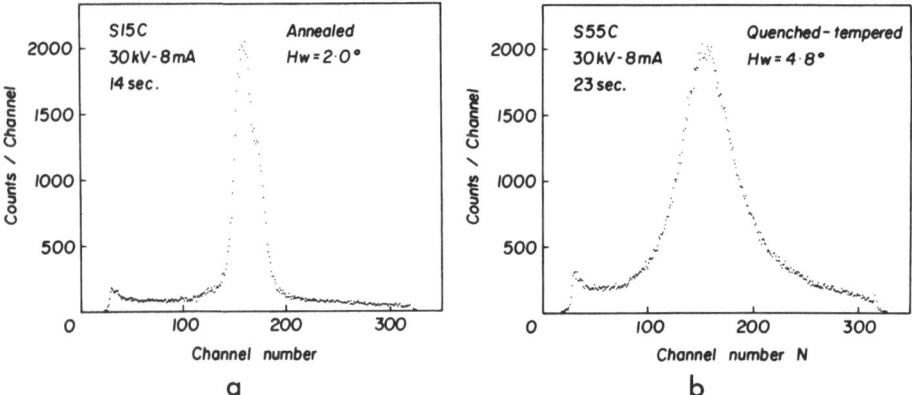

Fig. 3. Diffraction profiles from annealed low carbon steel specimen
        (a) and quenched-tempered steel specimen (b).

quenched and tempered steel. The count time was 23 seconds. Each
half-value breadth in 2θ is 2.0° and 4.8° respectively. If such
profile is recorded by a standard diffractometer using a scanning
speed of 4° per minute, it would have required about 5 minutes.
The time is reduced by factors of 13 to 21.

The precision of peak position determination was examined by
replicate measurements. As the breadth of the profile increases
the peak position becomes more sensitive to small fluctuations in
the data. However, the standard deviation of peak position was
less than 0.05° in 2θ even for the specimen that has a breadth of
4.8° if the peak count is preset to more than 2048. This value is
sufficiently good for stress measurement.

Stress Measurements

Several bending stresses were applied to a quenched and tempered steel specimen with a rectangular shape, and X-ray stress at
each load was measured by the $\sin^2\psi$ method. Fig. 4 shows an example of the $\sin^2\psi$ diagram at zero applied load. It indicates a
nearly linear relationship. As shown in Fig. 5, the relation between applied stress and stress by X-rays indicates a nearly
straight line of slope almost equal to 1.

CONCLUSIONS

A versatile X-ray stress analyzer has been developed using a
position sensitive proportional counter for residual stress measurement in large specimens. This analyzer provides an excellent
advantage of reduction in the time to measure stress in comparison

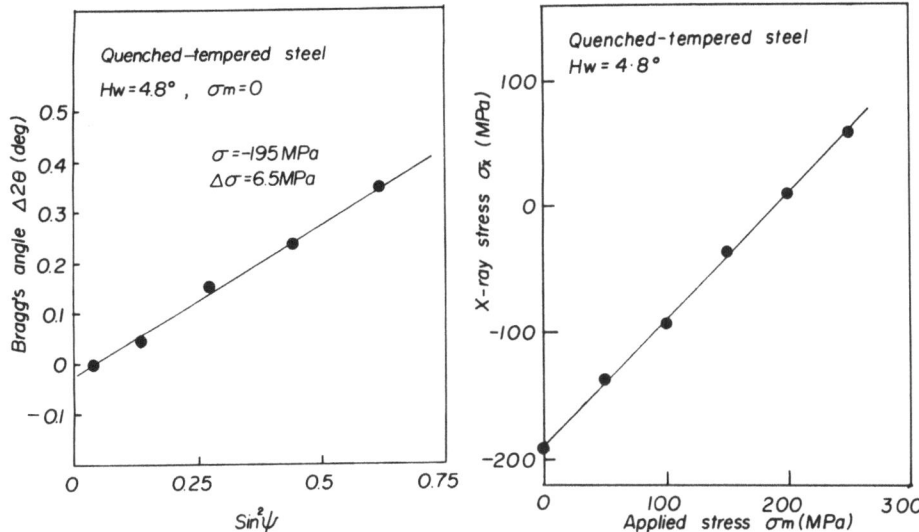

Fig. 4. An example of $Sin^2\psi$ Fig. 5. Relation between applied
        diagram.   and X-ray stress.

to the standard diffractometer method, and it will be useful in field and shop stress measurements. For example, nonlinear $sin^2\psi$ relations, which are becoming a problem, require experimental data of 2θ at many ψ angles for accurate analysis. This analyzer is most suitable for this purpose because the data can be accurately and rapidly accumulated.

REFERENCES

1. M. R. James and J. B. Cohen, "The Application of a Position Sensitive Detector to the Measurement of Residual Stresses," Advances in X-Ray Analysis, 19:695 (1975).
2. C. O. Ruud and C. S. Barrett, "Use of Cr K-Beta X-Rays and Position Sensitive Detector for Residual Stress Measurement in Stainless Steel Pipe," Advances in X-Ray Analysis, 22:247 (1979).
3. Y. Yoshioka, K. Hasegawa and K. Mochiki, "Study on X-Ray Stress Analysis Using a New Position-Sensitive Proportional Counter," Advances in X-Ray Analysis, 22:233 (1979).
4. Y. Yoshioka, K. Hasegawa and K. Mochiki, "A Position-Sensitive Proportional Counter for Residual Stress Measurement by Means of Microbeam X-Rays," Advances in X-Ray Analysis, 23:325 (1980).

INTEGRAL TYPE, POSITION-SENSITIVE PROPORTIONAL CHAMBER WITH

MULTIPLEXER READOUT SYSTEM FOR X-RAY DIFFRACTION EXPERIMENTS

Koh-ichi Mochiki, Ken-ichi Hasegawa and Akira Sekiguchi

Dept. of Nuclear Engg., University of Tokyo
7-3-1 Hongo, Bunkyo-ku, Tokyo 113, Japan

Yasuo Yoshioka

Musashi Institute of Technology
1, Tamazutsumi, Setagaya-ku, Tokyo 158, Japan

INTRODUCTION

Recently intense X-ray sources have been used in diffraction experiments on stress analysis, dynamical structure analysis, and so on. Since in these experiments one-dimensional position-sensitive X-ray detectors are very useful to reduce the measuring time, various kinds of X-ray detectors and processing systems have been developed. These detectors may be divided into two groups, the pulse type and the integral type. The pulse type employing either the charge division method, or the delay line method, processes each signal produced by an incident X-ray photon. To achieve a precise measurement, a sufficient gas gain and a relatively long processing time are needed. Therefore, the maximum counting rate depends on both detector characteristics and a signal processing time. On the other hand, the integral type, such as self-scanning photodiode array detectors (1, 2) periodically processes the charges produced by X-rays in each pixel. The total maximum counting rate of the detector is the sum of the maximum counting rate of each pixel and is higher than that of the pulse type. Therefore, for high flux experiments the integral type would be more suitable than the former. From this point of view, we propose a new integral type position-sensing system which consists of descrete electrodes, capacitors for charge accumulations and an analog readout electronics (3).

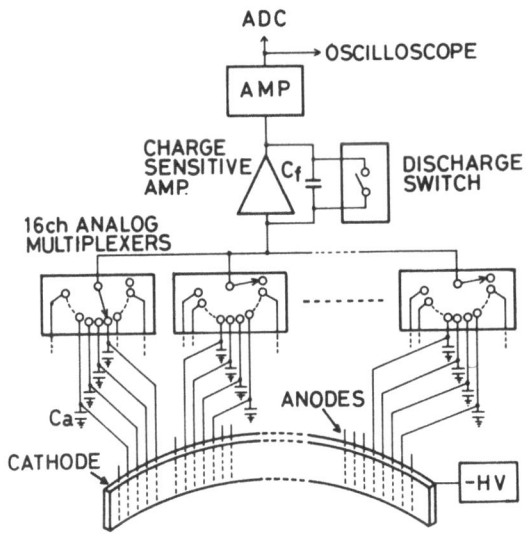

Fig.1. Schematic representation of the system

PRINCIPLE OF THE DETECTING SYSTEM

In order to represent the principle of the detecting system, the schematic diagram is shown in Fig.1. The one-dimensional detector consists of a multianode proportional chamber and capacitors, Ca's, connected to each anode. Each anode is regularly connected to each input terminal of CMOS analog multiplexers. The charges induced by incident X-rays are accumulated in the capacitors. Therefore, each capacitor operates as an integrator. After discharging the feedback capacitor, Cf, of the charge-sensitive amplifier by an analog switch, the accumulated charge in Ca is fed to an IC charge-sensitive amplifier by addressing the associated channel of the analog multiplexer. The output amplitude of the charge-sensitive amplifier is proportional to the amount of charge produced by a primary ionizing event, a gas gain, an X-ray incident flux and an accumulating time, and inversely proportional to the feedback capacitance, Cf. After transference of the charge from the accumulating capacitor to the feedback capacitor, the channel of the analog multiplexer is driven to the off-state and the charge-sensitive amplifier holds onto the output voltage so as to be processed by an analog-to-digital converter. In this manner all channels of the analog multiplexers are sequentially driven by addressing signals, and a train of the voltage pulses is obtained from the charge-sensitive amplifier. As the incident position of the X-ray photons is identified by the anode number, i.e. the channel number, the position resolution is limited by the anode interval.

Table 1. Parameters of the position-sensitive detector.

Number of anodes	128 (64)
Intervals of anodes	1 mm (2 mm)
Anode-cathode gap	2.5 mm
Anode wire	Ni-Cr, 20 μmϕ
Active anode length	20 mm
Radius of anode wire surface	286.5 mm
Active region	$\pm$12.8° (128 mm)
Gas	Ar + 10%CO_2, gas flow
Capacitance of the accumulating capacitor Ca	$\sim$ 300 pF

DETECTOR

Based on this principle, we have constructed a curved multi-anode detector. The main parameters of the detector are given in Table 1. An aluminized mylar window and an aluminum plate are used as cathodes. A negative high voltage is applied to them.

READOUT SYSTEM

As shown in Fig.1, the analog part of the readout electronics basically consists of 8 multiplexers with 16ch input terminals (HARRIS HI-506A), a charge-sensitive amplifier (HARRIS HA-5115) and an analog switch (HARRIS HI-201) for discharging the feedback capacitor. Since the on-resistance of the multiplexer is about 2 kΩ and the accumulating capacitance is about 330 pF, the time required to transfer the accumulated charge to the feedback capacitor must be longer than 5 μs for 10 bits analog-to-digital converter. In this system the processing time required for one channel is adjusted to 100 μs; initial discharge 6 μs, charge transfer 40 μs and output hold 54 μs. To obtain the distribution of an incident X-ray flux by the successive readout method, therefore, it takes for the accumulating time plus total processing time 12.8 ms.

TEST RESULTS

Dynamic Range, Gas Gain, Accumulating Time, Diffraction Experiment

The characteristics of the readout system were tested with a current source which supplied a known charge for one of the accumulating capacitors. We obtained the results that the maximum output voltage of the charge amplifier is 10 volts and the dynamic range of 62dB, i.e. about 3 decades has been obtained, which is sufficiently good for the general use of the detecting system.

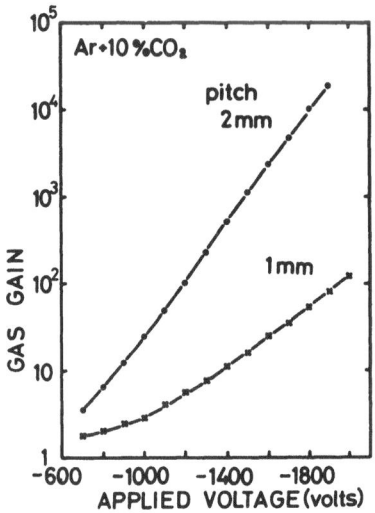

Fig.2.   Gas gain-applied voltage characteristics

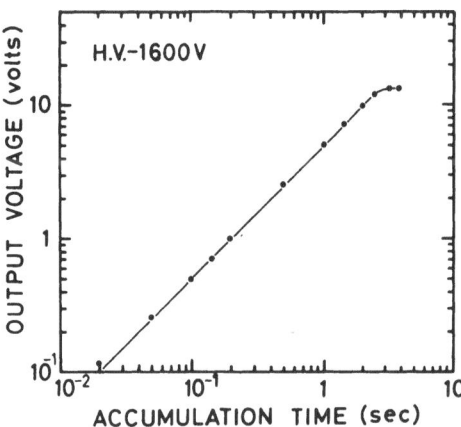

Fig.3.   Linearity of output voltage vs. accumulating time

Since in this type of detector the gas gain can be widely selectable, we measured the characteristics in two cases; intervals of anodes 1 mm and 2 mm. From the results shown in Fig.2 it was confirmed that in the former case the gas gain can be selectable over the range of 2 decades and in the later case over the range of about 4 decades. Figure 3 shows the dependenceof the output amplitude on the accumulating time. Good linearity over 2 decades was obtained.

The diffraction pattern shown in Fig.4 was obtained to confirm the applicability of this detecting system.

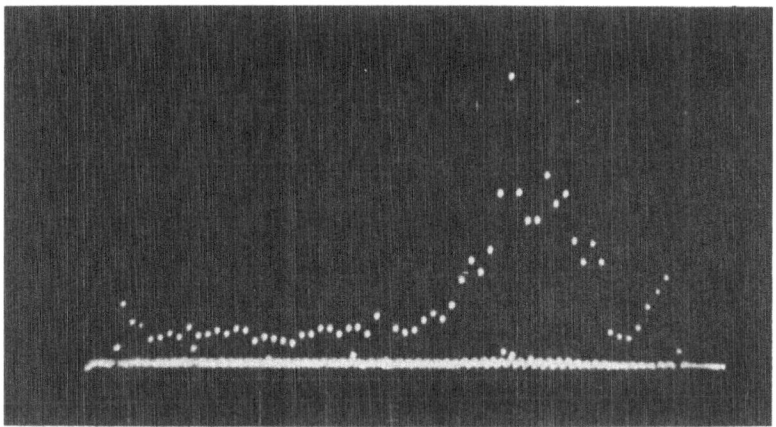

Fig.4. Diffraction pattern of polyethylene; anode pitch 2 mm

CONCLUSION

The following characteristics of the system are found;
(1) the dynamic range of the readout electronics and accumulating time is sufficiently wide, (2) the gas gain is widely selectable, (3) the uniformity of the gas gain of each anode is relatively poor, but the correction of it could be performed with the help of a computer. Though the detector system has not been tested in an extremely high intensity field of $\sim 10^7$ cps per anode, it would be useful if it is operated at low gas gain.

REFERENCES

(1) R.C.Gamble, J.D. Baldeschwieler, and C.E.Giffin, Rev. Sci. Instrum., 50, 1416-1420 (1979).
(2) D.A.Steffen, and C.O.Rund, Adv. in X-ray Analysis, 21, 309 (1978).
(3) K.Hasegawa, K.Mochiki and A.Sekiguchi, "Integral Type, Position-Sensitive Proportional Chamber with Multiplexer Readout System" to be published.

AN AREA-IMAGING PROPORTIONAL

COUNTER FOR X-RAY DIFFRACTION

C. Richard Desper
Polymer Research Division
Organic Materials Laboratory
Army Materials & Mechanics Research Center
Watertown, Massachusetts 02172

and

Ronald Burns
Xentronics Corporation
Cambridge, Massachusetts 02140

INTRODUCTION

Slow data acquisition rates have, in the past, limited the use of X-ray diffraction for characterization of polymeric materials. Photon counting techniques yield quantitative data in digital form for computer analysis. However, a great deal of data acquisition time is required when data is taken sequentially; i.e., when each intensity determination (at a particular goniometer setting) requires a separate time interval, during which intensity at other angle values is ignored. The problem is particularly acute for oriented polymers since two or more diffraction angles are involved: The Bragg angle along with at least one orientation angle. For this reason, an area-imaging (two-dimensional) proportional counter has been developed for use with a four-circle X-ray diffraction system. Although basically a single-crystal unit, this goniometer has been used in this laboratory[1] and others for studies of oriented polymers. The receiving pinhole collimator and aperture have been removed, and the area-imaging counter has been mounted on the detector arm track. The original receiving aperture is still used for alignment, and the area detector is positioned with its center at the receiving aperture center position.

PRINCIPLES OF OPERATION

Counter Design

The counter consists of an aluminum body, a beryllium X-ray window, an X cathode, a Y cathode, and an anode plane. A high purity xenon-methane gas mixture is used for photon detection (see Fig. 1). An 8 Kev X-ray photon enters and is absorbed in the xenon gas creating a cloud of approximately 400 ion pairs. The free electrons are attracted to the 10μm diameter anode wires, biased at 4-5 Kv. As the electrons approach the anode wires amplification of 10^5 occurs. This signal is capacitively coupled to the cathode wire planes, which are each made of a single high resistance wire. The two ends of each cathode drive four pre-amplifiers, which, along with appropriate gas purifiers, are housed within the detector case. The detector is similar in principle to one described by Borkowski and Kopp.[2]

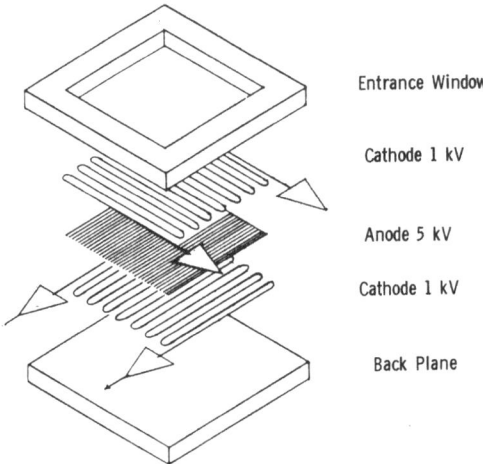

Entrance Window

Cathode 1 kV

Anode 5 kV

Cathode 1 kV

Back Plane

Figure 1. Area imaging proportional counter

Signal Processing

The four cathode signals yield the X and Y position of each photon event from the RC propagation delays of the cathode wires. The associated electronics, shown in Figure 2, incorporate double delay-line shaping, crossover timing, and time-to-amplitude conversion to generate pulses proportional in amplitude to the X and Y positions. The shaping amplifier outputs are also summed in a fifth channel to yield a signal proportional to the photon energy. A logic pulse (the "energy pulse") is generated when the sum amplitude is within prescribed limts.

The circuitry also employs a coincidence gate which is useful
but not essential to the system. This gate receives three input
signals: The energy pulse and single channel analyzer pulses from
the two time-to-amplitude converters, each generated when the X or
Y position are within prescribed limits. A coincidence output
occurs when all three inputs are present, and is used to limit

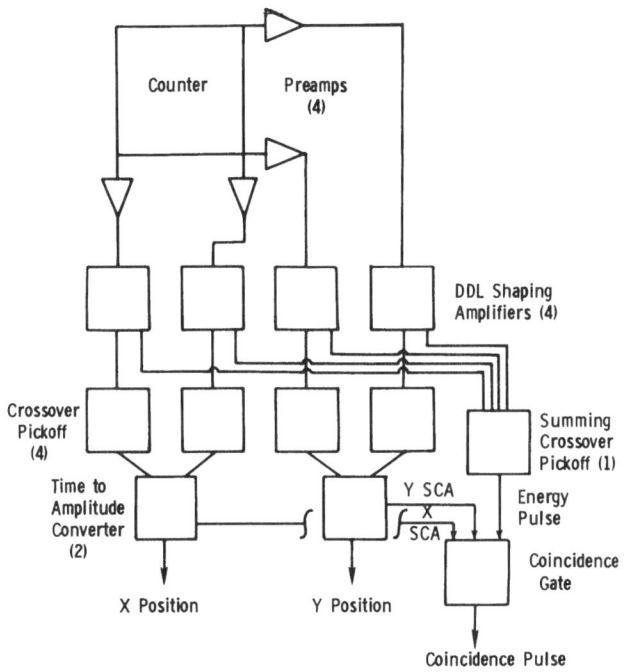

Figure 2.   Block diagram of counter
and amplifier circuits

subsequent readout to events which are in the required energy and
position ranges. Use of the coincidence gate renders detector
noise totally negligible, since, for a coincidence output to appear,
signals must appear on all four cathode lines, satisfy discriminators in five crossover pickoffs, and satisfy discriminators in
two time-to-amplitude converters. However, the coicidence gate
may be omitted, using the energy pulse in place of the coincidence
pulse.

Readout

Readout is obtained both on a storage cathode ray tube (CRT),
as an image of the intensity on the detector face; and in a multiple
channel analyzer (MCA), in digital form. The present MCA has one
analog input, storing data as either counts vs X or counts vs Y.

A new MCA is planned, with dual analog inputs to store data as counts vs (X,Y). The MCA output is diplayed on its own CRT in graphical form.

PERFORMANCE

Basic Detector Performance

    The performance parameters of the detector are summarized in Table 1. Several comments are appropriate:

1. The detector gain has remained unchanged for four months, indicating the effectiveness of the gas purifiers.

2. Energy resolution is measured in the sum channel, where the non-uniformity of gain with X and Y cancels out.

3. Spatial sensitivity is higher for the X axis due to greater capacitive coupling of the X cathode wire with the anode. As a result, the square detector face yields a rectangular image of aspect ratio 1.25.

4. The spatial resolution is determined by the broadening of a finely collimated Kratky camera beam, running at minimum tube power.

5. The image is distorted next to the edges due to reduced spatial sensitivity.

Table 1.   Detector Performance

Active area	100 mm X 100 mm
Fill gas	85% Xe, 15% CH_4, 2 atm
Window	1 mm Beryllium
Window absorption	22% at 8 Kev
Detector bias (optimum)	4.2 Kv at 8 Kev
Energy resolution	20% FWHM
Gain uniformity	15%
Spatial sensitivity	2.4 nsec/mm (X)
	1.9 nsec/mm (Y)
Spatial resolution	1 mm FWHM
Linearity	5% over image area
	(stable to 1%)
Preamp output rise time	200 nsec (90%)
Preamp output decay time	3 μsec (90%)
Deadtime loss at 20 KHz	6%
Deadtime loss at 100 KHz	30%

Applications

Diffraction patterns have been obtained using the goniometer, equipped with a copper tube, a graphite incident beam monochromator, and a 0.5 mm pinhole collimator, at a typical sample-to-detctor distance of 280mm. At this distance the detector spans a range of 20° with a resolution of 0.2°. Factors other than detector resolution also contribute to overall resolution. As a benchmark, a highly oriented polyethylene strand of 0.7 mm diameter was examined. A quantitative equatorial 2θ scan was obtained in 30 seconds; then the (110) orientation distribution (χ scan) was measured in another 30 seconds (see Figure 3). For comparison, approximately one hour was required for each of these scans in previous work[3] with such samples. Thus the experiment has been accelerated by a factor of over 100. A further speedup by a factor of 2.5 is anticipated when the present x-ray tube, rated at 600 watts, is replaced by a more modern tube rated at 1500 watts.

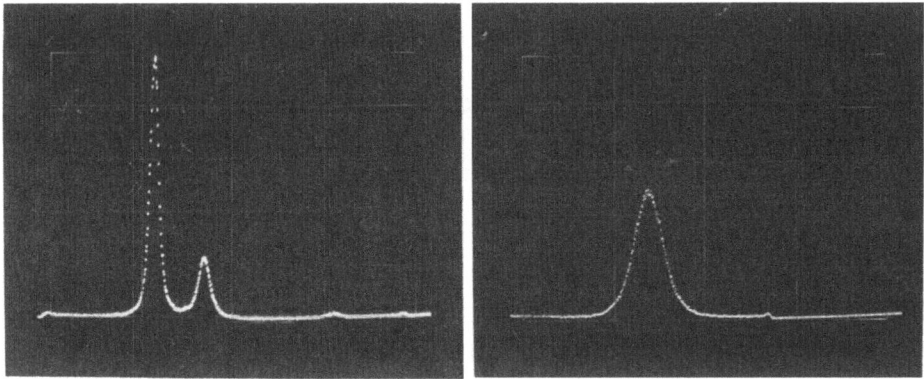

Figure 3.   (a) Equatorial scan of polyethylene strand, showing (110) and (200); (b) Orientation distribution of (110) intensity.   Full scale:5000 counts.

Angular range may be traded off for angular resolution, and vice versa, by changing the sample-to-detector distance. Moving in to a 100 mm distance, the detector spans a 53° range at 0.6° resolution. Figure 4a is a urea rotation photograph showing 19 diffraction spots obtained at 100 mm. The pattern was measured in 90 seconds, limited by the motor drive speed. For higher angular resolution, the distance is increased. Working at 600 mm, the $\alpha_1 - \alpha_2$ splitting is readily resolved for the (531) line of silicon powder (see Figure 4b). Resolution of this doublet is notable, particularly considering that the instrument optics are not designed for high resolution powder work.

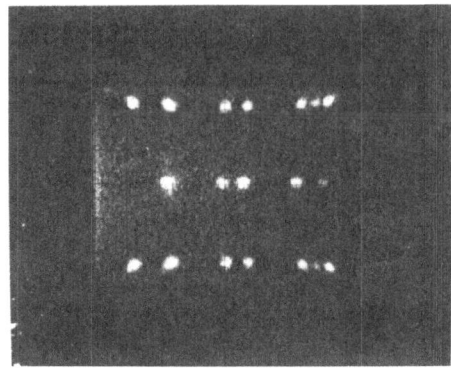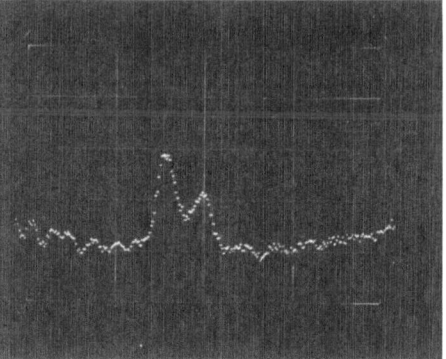

Figure 4.   (a) Urea crystal: rotation photograph
about c axis; (200) at center; (b) Silicon powder: (531)
reflection, 5000 counts full scale, splitting 0.42°.

CONCLUSION

Other materials studied to date include aramid fibers, carbon
fibers, and polymer containing ZnO filler, which cannot be discussed
here in detail. The area-imaging detector is of greatest value
where advantage may be obtained from its two-dimensional nature, as
in the oriented fibers discussed here. Where orientation is absent,
the value of the area detector may lie in capturing all of, or a
large part of, the Bragg diffraction rings. The rapidity of the
measurement also offers the possibility of time-slicing studies of
crystallization or deformation processes.

REFERENCES

1. C. R. Desper, A computer-controlled X-ray diffractometer for
      texture studies of polycrystalline materials, in: "Advances
      in X-ray Analysis, Vol. 12," C. S. Barrett, G. R. Mallet,
      and J. B. Newkirk, eds., Plenum, New York, 1969.
2. C. J. Borkowski and M. K. Kopp, Design and properties of
      position-sensitive proportional counters using resistance-
      capacitance position encoding, Rev. Sci. Instrum. 46:951
      (1975).
3. C. R. Desper, J. H. Southern, R. D. Ulrich, and R. S. Porter,
      Orientation and structure of polyethylene crystallized under
      the orientation and pressure effects of a pressure capillary
      viscometer, J. Appl. Phys. 41:4284 (1970).

STRESS MEASUREMENT IN STAINLESS STEEL BY USE OF MONOCHROMATIC

Cr-Kβ X-RAYS AND A POSITION SENSITIVE DETECTOR

Yasuo Yoshioka
Musashi Institute of Technology
1 Tamazutsumi, Setagaya, Tokyo 158, Japan

Ken-ichi Hasegawa and Koh-ichi Mochiki
Faculty of Engineering, University of Tokyo
7 Hongoh, Bunkyo, Tokyo 113, Japan

INTRODUCTION

X-ray stress analysis in austenitic stainless steel is gene-rally carried out on the γFe(311) diffraction line produced by Cr-Kβ X-rays. However, it is often pointed out that not much reliance can be placed on the precision of the stress because the contrast between a diffraction peak and its background is poor. In addition, to measure the stress is sometimes impossible on a specimen which has martensite structure produced by the strain induced transforma-tion, because the αFe(211) line appears in the neighborhood of the desired γFe(311) line, since Cr-Kβ X-rays accompany Cr-Kα from a conventional X-ray source. If Cr-Kα X-rays can be eliminated and only Cr-Kβ X-rays directed on the specimen, only the γFe(311) line with high contrast will be obtained and one can expect to measure the residual stress with high precision.

Diffraction from a bent single-crystal is commonly used to eliminate undesired wavelength in diffraction work using the para-focussing method. But a flat-crystal monochromator should be used in X-ray stress work because of the advantage of the parallel beam method. However, as the diffraction intensity is very weak after the monochromatization, the time required for a detector scanning becomes much longer than usual with the conventional goniometer method. If a position sensitive proportional counter (PSPC) is applied, the time required for measurement will be remarkably re-duced. In the present study, monochromatic Cr-Kβ X-rays were pro-duced by use of a single crystal monochromator, and stresses in specimens of austenitic stainless steel were measured by use of a PSPC system.

EXPERIMENTAL

Monochromatization of X-Rays and Apparatus

A perfect LiF single crystal with a flat face was used for the monochromatization of X-rays. The general X-ray optical alignment is shown in Fig. 1. Radiation from a fine focus X-ray tube with a Cr target passes through a soller slit. The crystal is placed with an incident angle of 31.2° in 2θ for Cr-Kβ X-rays. The monochromatized Kβ X-ray beam irradiates the specimen. The angle between the incident beam and the bisector of a PSPC, which has an effective length of 80 mm, is set to be 31.5° for the γFe(311) line of 148.5° in 2θ, and the distance from specimen to counter is 180 mm. The measurable range of Bragg's angle is about 148.5 ± 10° in 2θ. A conventional X-ray source, which is non-monochromatized, was also prepared, to compare with the effectiveness of the monochromatized.

A PSPC with a high counting rate, reported by authors,[1] was used. This charge-division type PSPC consists of a high-resistance wire cathode and nine wires of anode with the dimensions of the entrance window 20 mm in height and 100 mm in width. The output pulses from the PSPC are processed by a 512 channel pulse height analyzer and microcomputer. The minimum angular resolution of this system is about 0.2° in 2θ (FWIM) and a good linear response with respect to the path angle of X-ray beam is obtained.

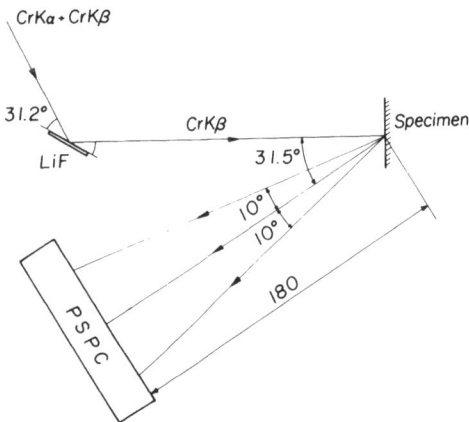

Fig. 1. X-ray optical alignment for stress measurement by means of monochromatic Kβ X-rays.

Specimen Preparation

All specimens used were fabricated from 8 mm thick and 40 mm wide plate of type 304 stainless steel, this material being in the solution-annealed condition. Several kinds of specimens were prepared as follows.

1. For profile observations and residual stress measurements, spcimens were cut to a size 50 mm length from the plate. They vere classified due to surface states as follows:

    Specimen A: As received
             B: Machined by a WA30 grinder
             C: Machined by a sintered carbide tool.

2. For calibration of X-ray stress by mechanically applied stress, a specimen for a bending test was machined in the form of a rectangular beam with dimensions of 100 mm length, 30 mm width and 3 mm thickness. Surfaces were finish ground in the same way as specimen B. This is called specimen D.

Stress Measurements

X-ray stresses were measured by the $\sin^2\psi$ method. The diffraction peak was determined using the half-value breadth method. The measurements were all made at an input voltage of 30 KVP and 20 mA. Calibration of the PSPC was accomplished using the Cr powder (211) line and Fe (211) line simultaneously.

RESULTS

Observation of Diffraction Profile

Fig. 2a shows diffraction profiles from specimen A, the top being obtained by the dichromatic X-rays and the bottom by the monochromatic. The ratio of peak intensity Ip to background intensity Ib is about 6.1 on the dichromatic profile. However, background height is reduced after the monochromatization, Ip/Ib being increased to 23.4. As shown in Fig. 2b and c, the effectiveness of monochromatization is more evident on a specimen which has a large internal strain by machining such as specimens B and C. Subtraction of background is clearly achieved. Although the dichromatic beam gives also the αFc(211) peak from martensite with Cr-Kα X-rays together with the γFe(311) peak, in the top profile of Fig. 2c the αFe(211) peak is completely eliminated using the monochromatic Kβ X-rays. The time required for data accumulation was about 5-10 minutes when the peak count is preset to be 2048.

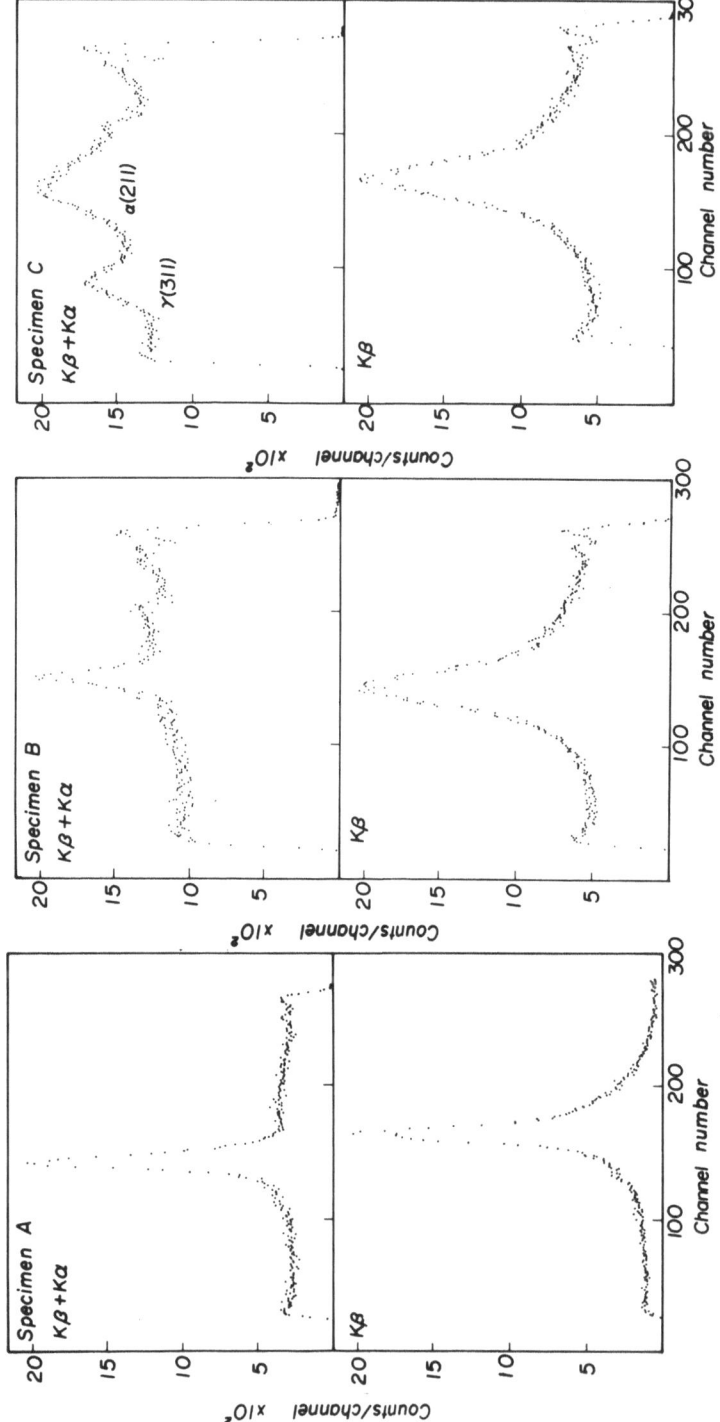

Fig. 2 Comparison of diffraction profiles from various type 304 stainless steel specimens obtained by dichromatic Cr X-rays (top) and by monochromatic Cr-Kβ X-rays (bottom). a) Specimen A: as solution-annealed; b) Specimen B: machined by a WA30 grinder; and c) Specimen C: machined by a sintered carbide tool.

If a conventional goniometer method is employed under the same X-ray source, more than ten times longer would be necessary.

Residual Stress Measurements

The $\Delta 2\theta$ vs. $\sin^2\psi$ relations for specimens B and C are shown in Fig. 3. $\Delta\sigma$ means 95% confidence limit calculated from the least squares of the $\Delta 2\theta - \sin^2\psi$ relation. The bulk elastic properties for type 304 stainless steel, 193 GPa elastic modulus and 0.3 Poisson's ratio, were assumed for the stress calculation. A good linear relation is found to hold on the $\sin^2\psi$ diagram of each specimen, and the stress values are approximately equal to those measured on the $\gamma Fe(220)$ line by Cr-Kα X-rays. On the other hand, oscillations in $\Delta 2\theta - \sin^2\psi$ of specimen B were tried with the di-chromatic beam and the stresses obtained did not agree with those obtained by the monochromatic beam. In addition, it was impossible to measure a residual stress in specimen C by the dichromatic beam because a separation of the $\gamma Fe(311)$ peak from the overlapped $\alpha Fe(211)$ peak was very difficult.

Calibration of X-Ray Stress by Applied Stress

A series of bending loads were applied to specimen D, and the X-ray stress at each load was measured by the $\sin^2\psi$ method. Fig. 4 shows the relation between mechanically applied stresses and those measured by X-rays. All of the points lie on a good straight line using the monochromatic beam as shown by closed circles. The slope of this line by least-squares is calculated at 1.02 which is almost equal to the theoretical value of 1. On the contrary, many data using dichromatic radiation and oscillation show irregularities when X-ray stress is plotted against mechanical stress.

CONCLUSIONS

The· stress measurement in stainless steel was carried out by the use of monochromatic Cr-Kβ X-rays and a PSPC system. Results indicate that the monochromatic X-ray technique is suitable for stress analysis in 304 type stainless steel in combination with the PSPC system. This technique is indispensable in the case of measurement in specimens that contain martensite.

REFERENCE

1. Y. Yoshioka, K. Hasegawa and K. Mochiki, "A Position-Sensitive Proportional Counter for Residual Stress Measurement by Means of Microbeam X-rays," _Advances in X-Ray Analysis_ 23:325 (1980.

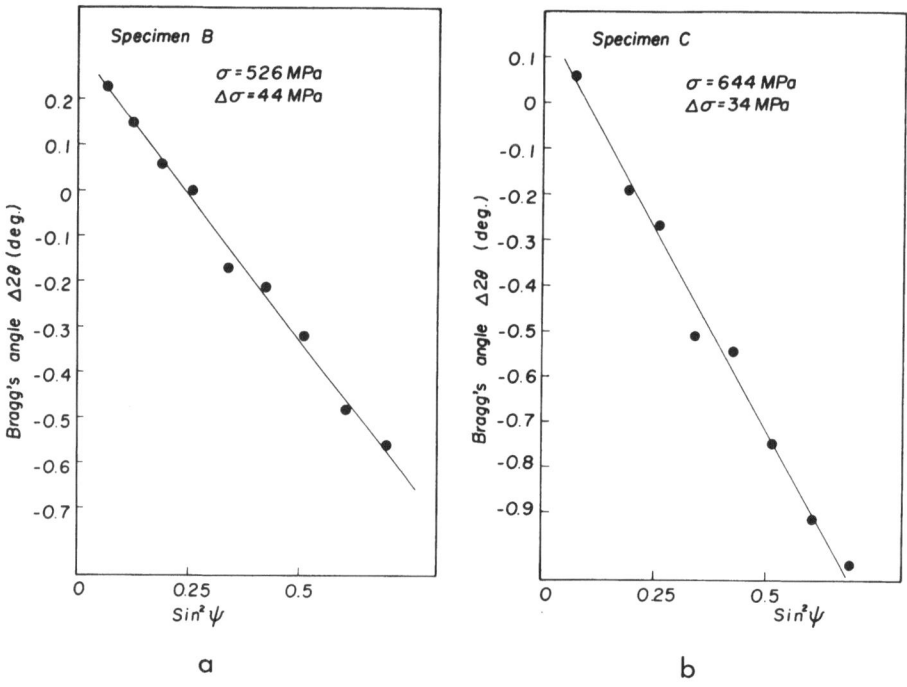

Fig. 3. Examples of $\mathrm{Sin}^2\psi$ diagram of specimen B (a) and specimen C (b).

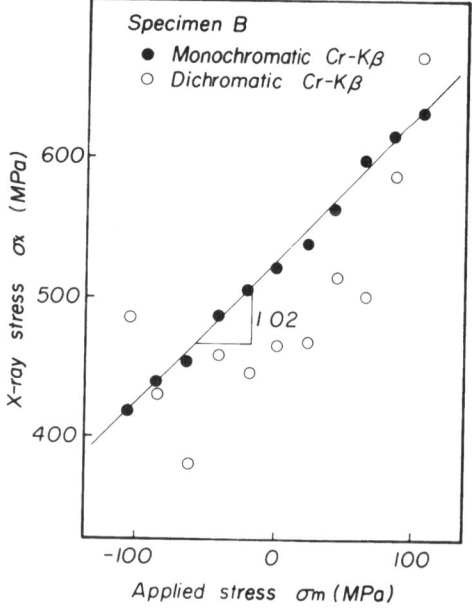

Fig. 4. Relation between applied stress and X-ray stress.

A HIGH-SPEED SIGNAL PROCESSOR USING A DIGITAL DIVIDER FOR POSITION SENSITIVE PROPORTIONAL COUNTERS

S. Kobashi, K. Mochiki, K. Hasegawa, A. Sekiguchi,
H. Hashizume and Y. Yoshioka*

Faculty of Engineering, University of Tokyo,
Hongo, Bunkyo-ku, Tokyo 113, Japan

* Musashi Institute of Technology,
Tamazutsumi, Setagaya-ku, Tokyo 158, Japan

INTRODUCTION

Most position sensitive proportional counters (PSPCs) currently
used in X-ray diffraction experiments have a dead time longer than
5 µs. Though such PSPCs are useful in measuring weak diffraction
diagrams, a faster counter is needed to detect strong X-ray diagrams
produced with synchrotron radiation sources. The long dead time of
PSPCs using a charge division position read-out is due to the slow
analog division circuit plus analog-to-digital converter employed
in the present system. A fast processor can be built utilizing two
high-speed ADCs to digitize voltage signals from the detector,
followed by a digital divider to compute position of detected
photons. The present paper describes the design of such a processor
and some preliminary testings of its performances.

PRINCIPLES OF OPERATION

In the block diagram of Fig. 1, signal pulses from the detector
are first processed by analog circuitry, newly designed to obtain
optimum system performances. The charge sensitive preamplifiers
integrate signal currents at the extremities of a charge splitting
resistive line. This line can be the internal signal pick-up electrode of the detector or externally coupled to the segmented
cathode structure. The output pulses are amplified and shaped to
near-Gaussian waveforms. The shaping amplifiers use eight cascaded
active integrators of 25 ns time constant. The summing amplifier
adds the shaped pulses of the right and left channels to produce

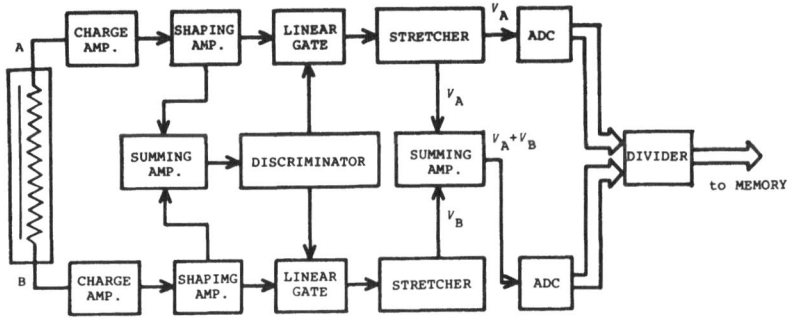

Fig. 1.   System block diagram.

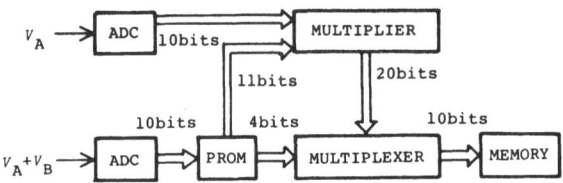

Fig. 2.   Block diagram of the digital divider.

input to the pulse-height discriminator. The amplitude of the
resultant pulse is proportional to total charge deposited in the
detector by an X-ray photon. The linear gates sample the useful
500 ns of the signal pulses. Signal voltages V_A and V_B, defined by
the stretchers, are added with the second summing amplifier to
provide a denominator signal for the digital divider.

     The divider, shown in Fig. 2, first digitizes V_A and $V_A + V_B$
using two high-speed, 10-bit binary successive approximation-type
ADCs (Datel G10B2A). These ADCs can digitize an input signal in
1 μs. We use hereafter symbol * to denote digitized quantities.
The output $(V_A + V_B)$* of the denominator ADC is used to address the
read-only memory, which stores the floating decimal-point values of
$1/(V_A + V_B)$* in 11 bits for all possible 1024 values of $(V_A + V_B)$*.
This memory can be read in 450 ns. The ratio $V_A*/(V_A + V_B)$*,
representing position of a detected photon, is found by calculating
the product of V_A* and $1/(V_A + V_B)$* with an IC multiplier (TRW TDC
1003J) of 170 ns processing time. The result is given in 20 bits,
of which significant 10 bits are sampled by the multiplexers
according to bit-selection information stored in the accessed
channel of the ROM. The position signal thus obtained is used to
address an ordinary data storage memory of 1024 channels.

     The processor described herein can determine position of a
detected photon within 2 μs.

TEST RESULTS

A processor with an analog divider dose not have a good position
resolution over a wide denominator input range. Improving the
performance is one of the purposes as well as achieving high speed
signal processing. The position resolution of the processor has been
examined by measuring the fwhms of simulated position peaks generated
with a pulser. The shaping amplifiers were fed with two series of
pulses having a fixed amplitude ratio $v_A + v_B$ from the same source.
We modulated the pulser output $v_A + v_B$ so that the upper half (-5 ∿
-10 V) of the input voltage range for the denominator ADC was scanned
with a uniform weight. The result shows a position resolution of
4/1024 for more than 80 % of the working range. The observed
integral non-linearity was less than 2/1024.

The differential non-linearity of the ADCs used in this processor is less than 1/2 LSB, which is much greater than that of the
Wilkinson-type ADCs. However, this imposes no important limitation
on the differential linearity of position determination in the
counter system. The detector produces distributed pulse heights
even for X-rays incident at a fixed position because of statistical
fluctuation of gas gain. Consequently, the non-linearities of the
ADCs are averaged over a certain input range, resulting in a good
linearity of the system[1].

The processor has been tested with a detector 7 cm long, 1 cm
high and 1.0 cm thick filled with 90 % argon plus 10 % methane at
760 torr. The cathode was made of a nickel-chromium wire of 28 Ω/cm
wound around an epoxy plate with a pitch of 0.5 mm. The total
cathode resistance was 50 kΩ. Figure 3 shows the spatial response
of the overall system, recorded by exposing the detector to flat
MnKα X-rays. The shallow dips in the center are likely to be
produced by imperfections in the detector.

To investigate the intensity linearity, the detector was
exposed to a beam of CuKα X-rays transmitted through a variable
number of alminium foils. The beam irradiated an area 1 mm wide and
1 cm high on the detector window. Figure 4a shows response curves
observed at four different settings of the discriminating levels.
At high beam intensities a reduced gas gain in the detector produced
an energy loss-peak shifted to a lower energy with a concomitant
broadening of the peak width. This explains non-linear curves in
Fig. 4a as well as their dependences on discriminator level. The
excellent high count rate capability of the processor is demonstrated
by curve M, which extends up to 200,000 cps. This was recorded with
a detector having four anode wires[2], which retains a high gas gain
at this count rate owing to distributed space charge between anode
wires. Reduced signal amplitude from the detector at a high beam
intensity leads to a degraded precision of digital division in the
processor. This results in a lower resolution of calculated

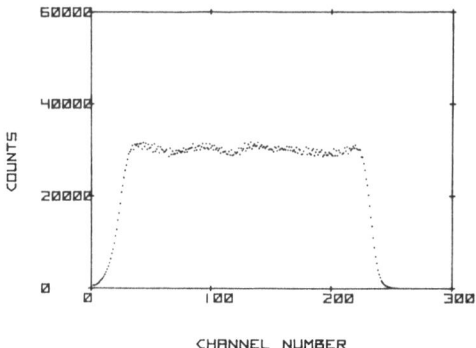

CHANNEL NUMBER

Fig. 3. Spatial response of the counter system.

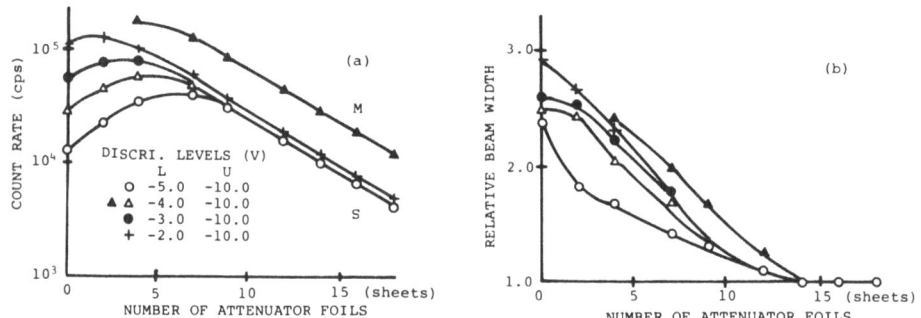

Fig. 4. (a) Intensity response of the counter system. (b) Spatial
resolution vs. X-ray beam intensity. Data points in curves S were
observed with a single-anode-wire detector, while those in curve M
were obtained from a multi-anode-wire detector.

position, as seen from increased widths of the observed beam profiles
illustrated in Fig. 4b.

REFERENCES

1. S. Kobashi, K. Mochiki, H. Hashizume, K. Hasegawa and
    A. Sekiguchi, A fast digital processor for position sensitive
    proportional counters, to be published.
2. Y. Yoshioka, K. Hasegawa snd K. Mochiki, A position sensitive
    proportional counter for residual stress measurement by means
    of microbeam X-rays, Advances in X-ray Analysis 23:325 (1980)

X-RAY SPECTROMETER FOR EXAFS

USING A POSITION SENSITIVE DETECTOR

K. Taniguchi, K. Oka*, N. Yamaki** and S. Ikeda**

Department of Solid-State Electronics
Osaka Electro-Communication University
Hatsumachi, Neyagawa, Osaka 572, Japan
*Department of Research, Union Giken Co., Ltd.
4081-1, Tsuda, Hirakata, Osaka 572, Japan
**Department of Chemistry, Osaka University
Machikaneyama, Toyonaka, Osaka 560, Japan

An active recording x-ray crystal spectrometer for extended x-ray absorption fine structure (EXAFS)[1,2] has been built using a position sensitive detector of the self scanning photodiode array (SSPA) type. The SSPA detector has energy and position sensitivity for x-rays. The spectrometer was applied to the measurement for EXAFS of the several compounds in foil, powder and liquid states. The spectra can be obtained rapidly, and compare very well with other methods. We found that the SSPA detector is very useful for the measurement of EXAFS.

Fig. 1 shows a block diagram of the EXAFS apparatus. The SSPA of Matsushita[3] is used as a one-dimensional x-ray detector. It has better position sensitivity (28 μm) than a position sensitive proportional counter, and has no dead time limitation. The width of the p-type region of each photodiode is 12 μm and the length is 0.5 mm; the total length of the array is 14.4mm. The SSPA consists of 512 photodiodes and MOS switches. The MOS switches are turned on and off from one diode to another. The video signals thus produced are amplified, and peak signals that are stored are converted to 12 bit digital signals and connected to a 64 kB micro-computer. The readout is performed for every pair of photodiodes. Therefore 512 data of signal intensity correspond to x-ray intensities at the corresponding positions of the photodiode pairs, all collected simultaneously.

An x-ray tube with Cu-target is used to check the detector and

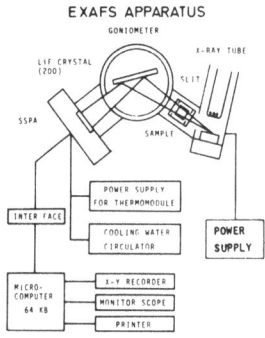

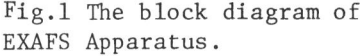

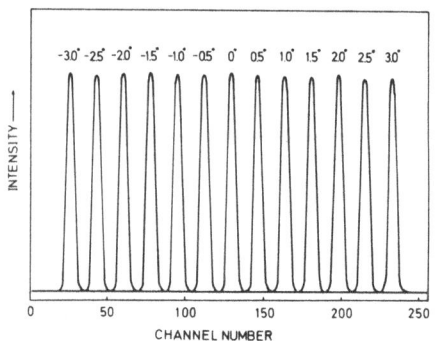

Fig.1 The block diagram of   Fig.2 The signals of direct
EXAFS Apparatus.   beam from Cu-tube.

monochromator system. The focusing of the tube is of the normal
type. Fig. 2 shows signals of the direct beam, after passing
through an absorber. The position of the signal peaks changes with
the changing of angle 2θ. The peak height fluctuates about 2%
depending on position. The precision of position is close to the
precision of 2θ. Cu-K$\alpha_{1,2}$ characteristic x-rays produce well
separated Kα_1 and Kα_2 peaks in both first and second orders. The
line widths are determined by both the width of the target focus
and the dispersion caused by the mosaic structure of the crystal.[4]
The direct x-ray beam from the tube is detected with a slit of 0.5
mm width placed at the center of the goniometer. The detector
(cooled to $\sim$ 0°C) rotates on an arm at angles 2θ while the crystal
rotates at θ. Peak heights of the L spectral lines of W on the
SSPA agreed with the theoretical relative intensities of these:
Lβ_4:Lβ_1:Lβ_3:Lβ_2 = 4:50:6:20.

RESULTS

     Figure 3 shows spectra of x-ray intensity with and without
passing through a sample of Fe foil of 10 μm thickness. The x-ray
tube used is fine focus 2 kW type with W-target. The monochromator
is a crystal of LiF (200). The spectrum without sample is marked
I_o. The roughness of the spectra is caused by the appearance of
several specific x-ray emissions and absorptions. Other causes may
be the mosaic structure and the roughness of the crystal surface.
The intensity marked I with the Fe sample decreases suddenly at a
smaller angle than that at the K absorption edge of Fe. Fig. 4
is a plot of the total absorption coefficient of the sample, μx,
calculated from the spectra of the intensity of x-rays with and
without passing through the Fe sample. The EXAFS appear clearly.

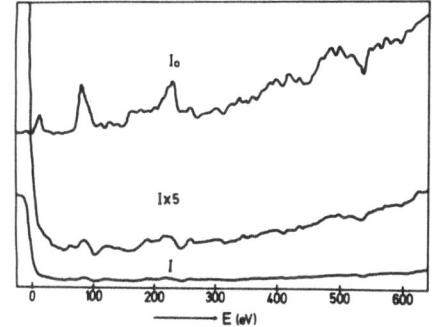

Fig.3 X-ray absorption spectra
of Fe foil sample.

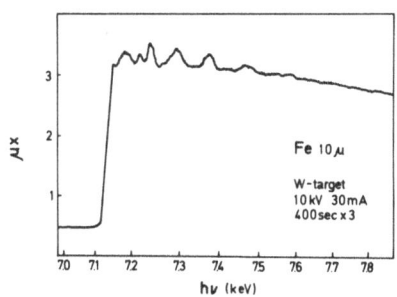

Fig.4 The EXAFS of Fe foil
detected by SSPA.

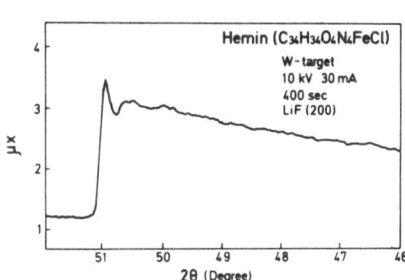

Fig.5 The EXAFS of hemin
detected by SSPA.

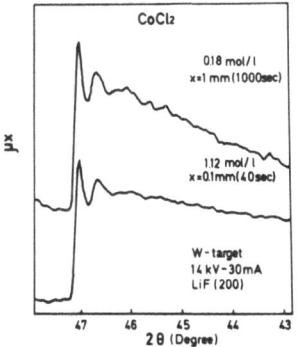

Fig.6 The EXAFS of $CoCl_2H_2O$
solution detected by SSPA.

The EXAFS spectrum obtained with a thin pellet of hemin, made by pressing the powder, is shown in Fig. 5. Some roughness in the curve is ascribed to fluctuations in the pellet thickness. Fig. 6 is the EXAFS spectrum for an aqueous solution of $CoCl_2$. At high concentrations (lower curve) a 40 sec accumulation time was used. For a low concentration a 100 sec accumulation was used (upper curve); water was used as a reference and with the thicker samples the absorption of water overlaps the $CoCl_2$ curve.

In conclusion, the SSPA is found useful and rapid for EXAFS, comparing well with other methods. The Fourier transformation of the Fe data shows good agreement with atomic distance, indicating satisfactory measurement precision. However, the SSPA is subject to radiation damage and may be affected by flourescence or Compton scattering. These are subjects for our future work.

REFERENCES

1. L. V. Azaroff, Rev. Mod. Phys. 35:1012 (1963)
2. D. E. Sayers, F. W. Lytle and E. A. Stern, Advances in X-Ray Analysis 13:248 (1970).
3. T. Takamura et al., National Technical Report 21, No. 6, 692 (1975).
4. H. Compton and S. K. Allison, X-Rays in Theory and Experiment, 2nd ed., Van Nostrand, New York (1935).

X-RAY STRESS MEASUREMENTS IN GROUND SURFACE

OF STEEL BY POSITION SENSITIVE DETECTOR

M. Kawata and M. Morinaga
Toyohashi University of Technology
Toyohashi, Aichi 440, Japan

Y. Yoshioka
Musashi Institute of Technology
1, Tamazutsumi, Setagaya, Tokyo 158, Japan

ABSTRACT

The residual stress produced by surface and cylindrical grinding was measured by a position sensitive proportional counter. In
medium carbon steels the residual stress depends largely on the
grinding conditions. For a gentle cylindrical grinding compressive stress was observed, but for a rough grinding tensile stress
was present in the workpiece.

INTRODUCTION

Grinding causes plastic deformation in the near-surface
region of the workpiece. The importance of thermal, burnishing
and mechanical cutting processes during grinding has been stressed
with respect to residual stress states.[1] Recently, it has been
reported that the stress state is no longer a plane stress state,
but it is a three-dimensional stress state.[2] The usual $\sin^2\psi$
method is not applicable to the analysis of such a stress state.
Dölle and Hauk[2] proposed a new method for this problem, and Dölle
and Cohen[3] applied it to the study of surface grinding. The purpose of the present paper is to show how the grinding parameters
affect the magnitude and the sign of residual stresses in both surface and cylindrical grinding of steels.

EXPERIMENTAL

The specimens used in the present experiment are medium carbon steels. The chemical compositions are (0.48% C, 0.25% Si, 0.75% Mn, 0.012% P and 0.007% S). The specimens were heat-treated in two ways; either by water-quenching from 1123°K and then tempering at 870-920°K (hardness H_V = 270) or air-cooling from 1123°K (H_V = 200). These specimens were ground under the conditions listed in Tables 1 and 2.

The X-ray measurements were carried out on a Rigaku diffractometer, employing the Cr-Kα radiation. Tube voltage and current were 30 kV, 8-20 mA. A position sensitive proportional counter (PSPC) was used, which is a resistive wire cathode type.[4] With this PSPC system the 211 reflection was measured. The time for measurements of one peak was 20-60 sec. The peak position was determined by the half-value breadth method.

RESULTS

Stress analyses were performed following the method proposed by Dölle and Hauk.[2] The geometry and the axes used for the analysis are illustrated in Fig. 1(a) for surface grinding and (b) for cylindrical grinding. The residual strain ε'_{33} was measured

Table 1. Conditions of Surface Grinding

Grinding method	Dry grinding
Wheel	WA46J, O200 x 13 mm
Wheel speed	30 m/sec
Down-feed rate	20 µm/pass x 5 passes
Grinding direction	reciprocal traverse
Feed speed	1.6×10^{-2} m/sec, 3.0×10^{-2} m/sec
Dressing	20 µm/pass x 1 pass

Table 2. Conditions of Cylindrical Grinding

Finishing	Down-feed rate	Feed speed (m/sec)	Work velocity (m/sec)	Dressing
Rough	40µm/pass 10 passes	2.0×10^{-2}	0.37	20µm/pass 2 passes
Medium	20µm/pass 5 passes	1.0×10^{-2}	0.37	10µm/pass 2 passes
Fine	5µm/pass 10 passes	1.7×10^{-3}	0.15	3µm/pass 2 passes

* Wet grinding(W-900H solution), Wheel: WA60K, O240 x 13 mm.

at various ϕ and ψ angles. The strain components ε_{ij} of the sample were determined by the least-squares method.[3] The stress components σ_{ij} were then calculated by using the X-ray elastic constants; $S_1=6.4 \times 10^{-4}$ MPa^{-1} and $1/2S_2=-1.5 \times 10^{-4}$ MPa^{-1} (5). The accuracy of determined stress components was estimated to be about 15% from the statistical analysis.

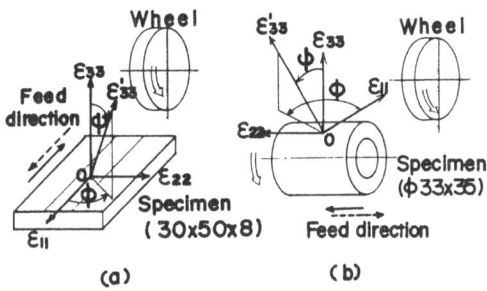

Fig. 1   Geometry for surface grinding (a) and
         cylindrical grinding (b).

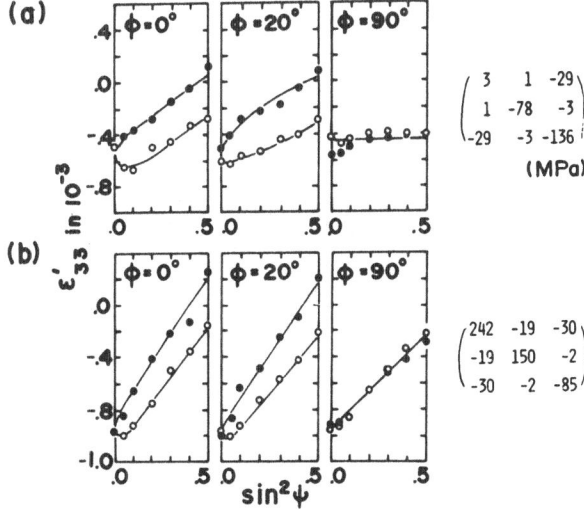

$$\begin{pmatrix} 3 & 1 & -29 \\ 1 & -78 & -3 \\ -29 & -3 & -136 \end{pmatrix}$$
(MPa)

$$\begin{pmatrix} 242 & -19 & -30 \\ -19 & 150 & -2 \\ -30 & -2 & -85 \end{pmatrix}$$

Fig. 2   Plots of ε'_{33} vs $\sin^2\psi$; Feed speed–(a)1.6×10^{-2}
         m/sec and (b)3.0×10^{-2}m/sec.

Surface Grinding

The plots of ε'_{33} vs $\sin^2\psi$ are shown with the values of stress components σ_{ij} in Fig. 2. There is a clear splitting between the curves of $\psi \geq 0$ (symbol o) and $\psi \leq 0$ (symbol •) at $\phi = 0°$ and $20°$. This is the so-called ψ-splitting. This ψ-splitting does not appear at $\phi = 90°$, which is the direction perpendicular to the grinding direction. Thus, the residual stress (and strain) depends on the grinding directions.[3] From Fig. 2 it is found that the feed speed of workpiece affects the magnitude of residual stress the most predominantly of the several parameters of surface grinding. The feed speed of (b) is about twice as large as that of (a). The measured residual stress of (b) was larger than that of (a). It is also noted that at $\phi = 90°$ the slope of ε'_{33} vs $\sin^2\psi$ plots was nearly zero for (a), whereas it was steeper for (b). This difference may be due to the plastic flow on the ground surface. The slower feed speed seems to be appropriate for reducing tensile residual stress.

Cylindrical Grinding

In Fig. 3 the results of three different grinding operations are compared; (a) rough finishing, (b) medium finishing and (c) regular fine finishing. Surface roughness of the workpiece was 2.0 μm for (a), 1.1 μm for (b) and 0.8 μm for (c), respectively. The ψ-splitting appeared in each figure except for the angle $\phi = -90°$. It is evident that there are obvious differences among the results of (a), (b) and (c). In (a) the residual stress was in tension, while in (c) compressive stress was observed. The result of (b) was between (a) and (c), as might be expected. The effect of feed speed on the residual stress was examined in the ranges of 0 to 0.01 m/sec. The results show that with increasing feed speed the residual stress shifted to the tensile side, in accordance with the results of surface grinding (Fig. 2). These results imply that gentle grinding induces a compressive residual stress, but rough grinding induces a tensile stress to the workpiece. The appearance of tensile residual stress may be mainly due to the thermal stress and mechanical cutting action of abrasive grains, but the burnishing action could place the workpiece in a compressive stress state.[1]

Irrespective of surface or cylindrical grinding, there was a similar tendency in the residual stress states between the quenched and tempered specimen and the air-cooled specimen.

From the present experiment the residual stress measured by the X-ray method was found to be sensitive to the grinding parameters. The residual stress tensor is a good representation for the grinding operation. X-ray examination of the workpiece

could be useful for developing and controlling the grinding process.

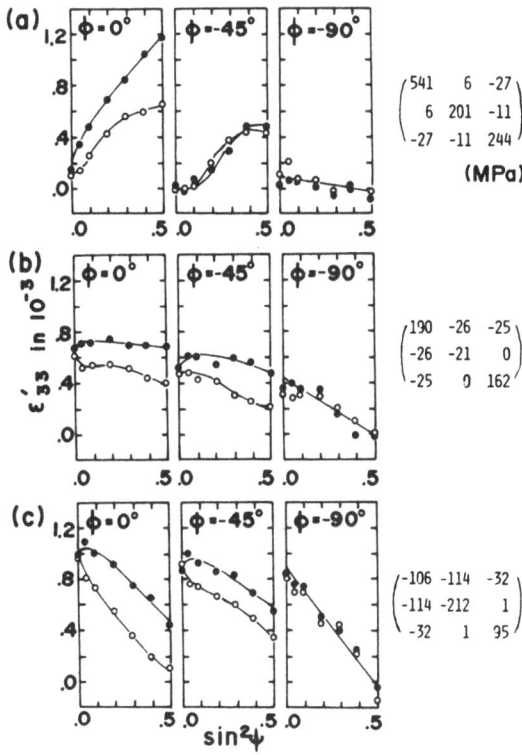

Fig. 3   Plots of ε'_{33} vs $\sin^2\psi$ for (a) rough,
(b) medium and (c) fine finishing.

REFERENCES

1. M. Wakabayashi and M. Nakayama,"Experimental Research on
    Elements Composing Residual Stresses in Surface Grinding",
    Bull. Japan Soc. of Prec. Engg., 13:75 (1979).
2. H. Dolle,"The Influence of Multiaxial Stress States, Stress
    Gradients and Elastic Anisotropy on the Evaluation of (Residual)
    Stresses by X-Rays", J. Appl. Cryst., 12:489 (1979).
3. H. Dölle and J. B. Cohen,"Residual Stresses in Ground Steels",
    Metallurgical Transactions 11A:159 (1980).
4. Y. Yoshioka, K. Hasagawa an- K. Mochiki,"Study on X-Ray Stress
    Analysis Using a New Position-Sensitive Proportional Counter",
    Advances in X-Ray Analysis, 22:233 (1979).
5. E. Macherauch,"X-Ray Stress Analysis", Experimental Mechanics,
    6:140 (1966).

TIME-RESOLVED X-RAY POWDER DIFFRACTOMETRY USING LINEAR

POSITION-SENSITIVE PROPORTIONAL COUNTERS

Herbert E. Göbel

Forschungslaboratorien der SIEMENS AG

D 8000 München 83, West Germany

ABSTRACT

Several examples of time-resolved X-ray powder diffraction studies with linear position-sensitive proportional counters (PSPC) are described. The changing patterns are recorded by spectrum multiplexing in a multichannel counter device. The time resolution for single processes is limited to seconds or tenths of seconds mainly because of the counting statistics of the observed peaks. Repetitive processes, however, where the interesting peaks can be accumulated over a large number of periods, permit much higher time resolution. It is finally determined by the timing of a detected X-ray quantum. Using a continuous beam source like a conventional X-ray tube the incidence of a photon into the PSPC can be timed with an accuracy of 20 ns (fwhm). A pulsed synchrotron radiation source, like a storage ring, will be able to define this resolution to its pulse duration, which is less than 200 ps.

Before approaching those limits this paper presents the high-temperature phase transformations in iron for different steel specimens as an example of single processes. It also discusses domain-wall effects and electrostriction in ferroelectric perovskites as a function of polarization at frequencies of up to 10 Hz, and elastic lattice distortions of a steel sheet vibrating at about 200 Hz. The timing limit of 20 ns for the PSPC detection system was measured at the storage ring DORIS at DESY, Hamburg.

INTRODUCTION

The following of lattice changes in solids due to an external
condition (temperature, pressure, electric or magnetic fields) can
often not be investigated statically by X-ray diffraction by using
a number of equilibrium points, and almost never under normal
conditions by using quenched specimens. X-ray diffraction analysis has to be integrated into the experiment and has to follow
the process dynamically, recording intermediate states similarly
to the snapshots in a movie film.

Position-sensitive proportional counters (PSPC) promise ideal
qualifications for these investigations since they allow the observation of an angular range simultaneously large enough to re-
ceive the full information upon the changes of interest, and they
are single quantum detectors indicating the event of an incident
X-ray immediately. By spectrum multiplexing in a multichannel
counter system a process can be traced dynamically until well
plottable patterns have been collected for every intermediate
state. This sets a limit for the observation velocity of single
processes since these detectors have a maximum pulse rate capability
of about 10^5 counts per second. Since a total number of 10^4–10^5
counts will be necessary for a profile, the observable processes
have to be not much faster than seconds. If faster single processes have to be investigated, high intensity X-ray sources and
integrating detectors have to be used.[1,2] Repetitive processes
do not have this limitation. If any repetition rate is permitted,
the profiles can be accumulated over a large number of periods.
The time resolution will then be only a question of the timing of
the experiment under analysis and the X-ray detection time resolution.

EXPERIMENTAL

The electronic conception for collecting diffraction patterns
of intermediate states during dynamic processes is called spectrum
multiplexing. For this purpose the address range of a multichannel
counter (MCA) of, for example, 4096 channels (12 bit), is sub-
divided into a number (any power of 2) of ranges with a complementary width to form a 12 bit address, for example 32 ranges of 128
channels (Fig. 1). While the 128 channels, representing the least
significant seven bits of the address, contain the position information 2 theta of an incident X-ray on the PSPC, the most significant 5 bits will be coupled to the experiment under analysis. All
12 bits are coincidentally transmitted to the MCA to sort any
incoming X-ray to the corresponding state of the experiment.

For slow processes (< 1 kHz) the coincidence problem can be

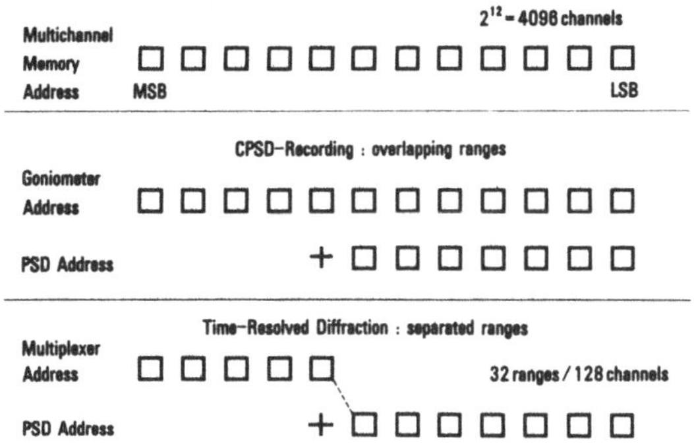

Multichannel
Memory 2^{12} = 4096 channels
Address   □ □ □ □ □ □ □ □ □ □ □ □
                MSB   LSB

CPSD-Recording : overlapping ranges

Goniometer
Address   □ □ □ □ □ □ □ □ □ □ □ □

PSD Address   + □ □ □ □ □ □ □

Time-Resolved Diffraction : separated ranges

Multiplexer
Address   □ □ □ □ □   32 ranges / 128 channels

PSD Address   + □ □ □ □ □ □ □

Fig. 1.   Bit-balance for multiplex recording of PSPC data

ignored and a simple range switching will not obscure the differences of subsequent intermediate states (quasi-static time-slicing).
A block diagram of an experimental arrangement for this case is
shown in Fig. 2.

   Both Fig. 1 and Fig. 2 show the close relationship to the
continuously scanning PSD diffractometer (CPSD[3]) used to record
entire diffraction patterns at high speed.   Both systems use similar modules and may also be combined.   By oscillating the PSPC in
a small angular region (about 1°) using the CPSD-technique, possible small inhomogeneities in the detector system are wiped out.
The profiles show the pure counting statistics and the detector
wire is loaded uniformly, which will prolong its life.

   A typical multiplex recorded peak profile as a function of a
process variable x(t) is shown in Fig. 3.   The possibility of using

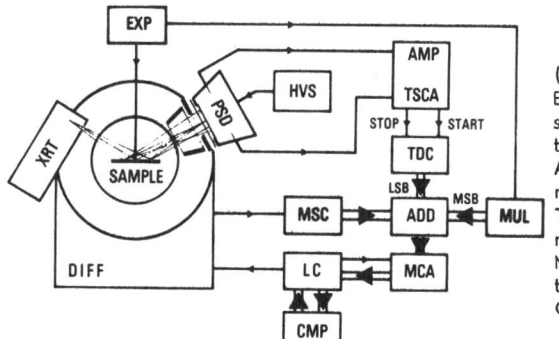

(XRT = X-ray tube, DIFF = diffractometer,
EXP = experimental parameter at the
sample, PSD = position-sensitive proportional counter, HVS = detector bias,
AMP/TSCA = compact pulse processing
module for RC-line encoded PSPC's,
TDC = time-to-digital converter, MSC =
motor step counter, ADD = binary adder,
MUL = multiplexer device, LC = logic controller, MCA = multichannel counter,
CMP = computer)

Fig. 2.   Block diagram of a PSD diffractometer for time-resolved
         experiments at low frequencies (0 - 1 kHz)

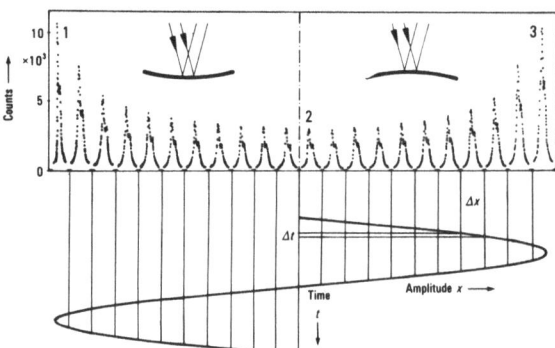

a CPSD system in a multi-plex mode for time-resolved studies was also taken into account in the data evaluation software DIFFRAC 11.[4] In the pro-grams FIT and SPLI comfort-able options for multiplex patterns are installed.

Fig. 3. Multiplex observation of lattice distortions in a vibrating membrane (see example 3, Fig. 9).

The device for the multiplexer depends strongly on the experi-ment. As an all-purpose device we constructed a timer with con-stant intervals mainly used in slow processes to be observed line-arly in time. The time slices can be varied from 0.01 s to 100 s. The second universal device is a fast analog-to-digital converter (conversion time 1.5 μs) used in experiments where the process variable is an analogue electrical quantity.

For fast processes (> 10 kHz) the steps of the process vari-able will be of the same order of magnitude as the processing time of the PSPC pulses (about 2 μs). With these processes, special care has to be taken regarding the coincidence of the multiplexer address and the position address of the incident X-ray. The cir-cuitry is split into a fast branch for the timing of a detected X-ray in its phase relation to the process under investigation and a slow branch to produce the multiplexer address as well as the PSPC address. The fast branch has to open a gate for as long as is required for the data transfer to the multichannel counter ("fast-slow coincidence").

The time resolution that can be achieved with the detection system for a single wire PSPC was tested in a pulsed beam experi-ment at the storage ring DORIS of the German electron synchrotron facility DESY at Hamburg (Fig. 4). The anode signals coming from both ends of the RC-encoded detector wire are added and amplified using a fast timing-filter amplifier with 100 ns shaping time (TFA, Canberra 2110). A constant fraction trigger (CFT, Canberra 1428A) and a time-to-pulse height converter (TPC, Ortec 457) will transform it to an amplitude proportional to its timing relative to the primary X-ray beam pulses.

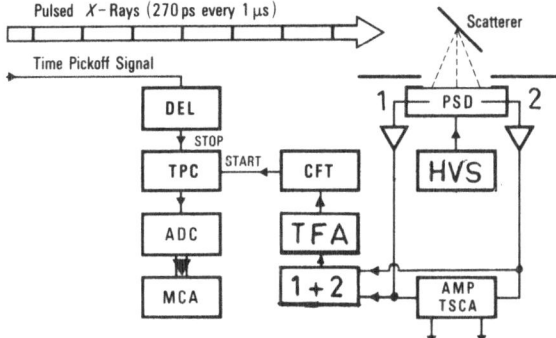

Fig. 4.   Time-resolution experiment at DORIS storage ring

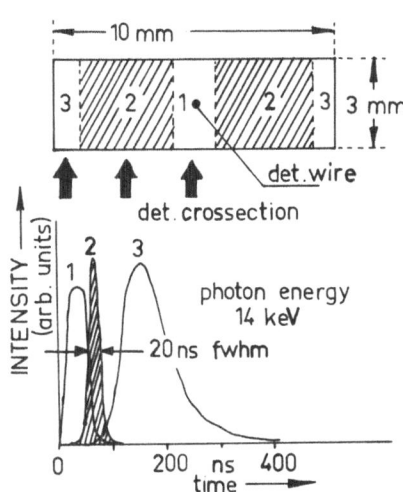

Fig. 5.   Drift-times and time-resolutions depending on the lateral incidence.

The results of this experiment are shown in Fig. 5. Depending on the lateral incidence of the X-rays the best resolution with 20 ns (fwhm) is achieved in the shaded areas of the detector, while the central part and the fringe areas, due to their inhomogeneous electrical field distribution, produce a poorer time resolution (also a poorer energy resolution!). The drift times of the primary ionization electrons from the fringes to the anode wire range at about 100 ns. So if the largest possible detector window is used for the experiment (excluding only the fringe areas), a time resolution of 100 ns can be achieved. When using a pulsed X-ray source like the DORIS storage ring this resolution is sharp enough to define a detected X-ray within one pulse period of 1 μs. The final timing of the photon can then be defined to the pulse duration of 270 ps of the primary beam.

RESULTS

Three examples of time-resolved X-ray diffraction measurements are presented.

1) High temperature <u>phase transformations in steel</u> samples during fast heating (single process).

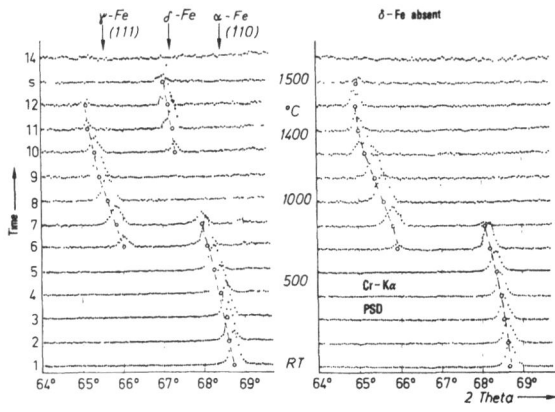

The (110) reflections of α- and δ-Fe and the (111) reflection of γ-Fe are plotted as a function of heating time (in steps of 1 s) till above melting temperature, Fig. 6. It is obvious that in high-carbon steel the δ-phase is absent. Additionally the thermal expansion co-efficients can be deter-mined.

Fig. 6. High-temperature phase transformations in low-carbon (left) and high-carbon steel during fast heating.

2) <u>Domain-wall effects and electrostriction in ferroelectric ceramics</u>

Distorted perovskite structures like the La-modified lead zirconium-titanium oxide PLZT form ferroelectric domains. The polarization vector points into the distortion direction of the pseudocubic cell, i.e. into the c-axis for tetragonal or the (111)-direction (hexagonal c-axis) for rhombohedral distortion. Adjacent domains may be antiparallel (180° wall) or perpendicular (90° wall).

An applied external electrical field will enhance one type of domains by moving the domain walls until all domains have the right orientation with respect to the field direction. X-ray diffraction is able to distinguish between 90° wall domains, but not, however, between 180° wall domains.

In a PLZT 7/65/35 ceramic material with rhombohedral symmetry the domains have statistical distribution in the depolarized state. In an applied external field one type of orientation will be domi-nant, that means that a preferred orientation of the crystallites into the field direction will be visible.

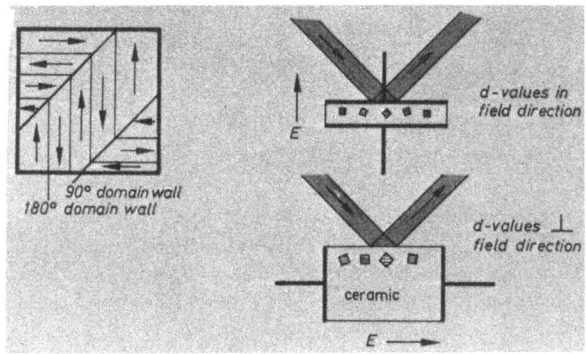

Fig. 7. Domains in ferroelectric
materials and directions of
observation.

In Fig. 7 the directions
of observation are shown,
and Fig. 8 demonstrates
the orientation effects
of the external field as
a function of the polari-
zation P. The intensity
distribution between the
left and the right series
of diffraction profiles
should be complementary.

In addition to the inten-
sity changes also the
line shifts are visible
because of the electro-
strictive effects.

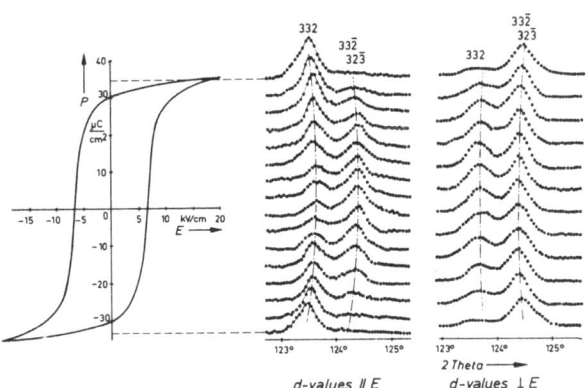

Fig. 8. Hysteresis loop and diffraction lines of a rhombohedral
PLZT 7/65/35 ceramic.

These measurements were
performed at a frequency
of about 10 Hz, since
DC-measurements are com-
plicated by delayed domain
wall motions. An
equilibrium will be
achieved only after very
long times.

3) Lattice distortion in an oscillating steel membrane (<u>dynamical</u>
    <u>stress analysis</u>)

    Peak stresses during transient torques and resonances can
cause fatigue of construction materials. They can normally only
be tested under the dynamical conditions of the working part. X-
ray diffraction is able to measure these quantities without electrical contact.

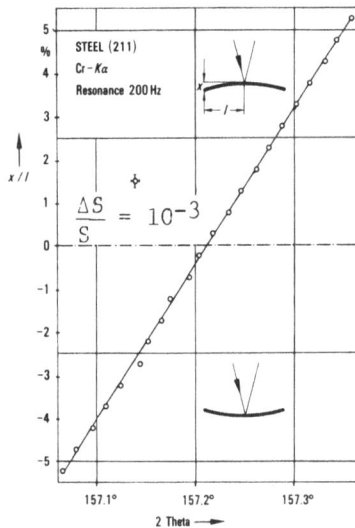

An example is shown in Fig. 9 on a steel sheet oscillating at 200 Hz resonance frequency. The lattice distortion, vertical to the surface represented by the position of the (211)-reflection with Cr-Kα-radiation is plotted against the bending of the membrane. From the linearity of the measured data and the typical error bars the relative accuracy of this method can be deduced. The raw data, shown in Fig. 3, were evaluated by a L.S. profile fitting program "FIT".[4]

Fig. 9.   Lattice distortion at the surface of an oscillating steel sheet.

CONCLUSIONS

Time resolved diffraction opens a new field for X-ray analysis by integrating it into a dynamical experiment. PSPC's seem to have the best qualifications for these studies since they permit the timing of a scattered photon with an accuracy down to 20 ns.

Due to the maximum counting rate of about 10^5 counts per second the minimum total counting time for one time slice has to be between 0.1 and 1 seconds to produce well plottable patterns, thus restricting the time-resolution for single processes or for processes allowing only a reduced number of repetitions, as for example in biological specimens.[5] Future experiments at high time resolution may be imagined, such as pulsed laser induced reactions (phase transformations, recrystallizations) or materials strains under shock waves or ultrasonic vibrations.

ACKNOWLEDGEMENTS

The author is indebted to Dr. L. Pleugel from the Institut für Eisenhüttenkunde der R.W.T.H. Aachen (high-temperature phase transformation in iron) and to Dr. G. Wolfram (domain effects in PLZT) for suggesting these measurements and preparing the specimens. F. Erbakan helped to perform the dynamical stress measurement and Barbara A. Jobst installed the multiplex spectra readout

into the DIFFRAC 11 software system. Special acknowledgements
have to be rendered to Dr. C. Hermes and Dr. F. Parak from M.P.I.
für Biochemie at Martinsried for making the beam facility at DORIS
(DESY, Hamburg) available for the time resolution test.

REFERENCES

1. R. E. Green and J. A. Dantzig, Flash X-Ray Diffraction Systems,
    in "Advances in X-Ray Analysis, Vol. 16," L. S. Birks, Charles
    S. Barrett, John B. Newkirk, and Clayton O. Ruud, eds.,
    Plenum Publishing Corp., New York (1973).

2. Q. C. Johnson, A. C. Mitchell, Paper H 5 at the ACA Winter
    Meeting, March 1979, Honolulu, Hawaii

3. H. E. Göbel, A New Method for Fast XRPD Using a Position Sensi-
    tive Detector, in "Advances in X-Ray Analysis, Vol. 22,"
    G. J. McCarthy, C. S. Barrett, D. E. Leyden, J. B. Newkirk,
    and C. O. Ruud, eds., Plenum Publishing Corp., New York (1979).

4. H. E. Göbel, "The Use and Accuracy of CPSD-data in XRPD (see
    contribution in this volume).

5. H. E. Huxley, A. R. Faruqui, J. Bordas, M. H. J. Koch, J.
    Milch. Nature (1980) in press.

TECHNICAL FEASIBILITY OF A BOREHOLE PROBE FOR IN-SITU X-RAY DIFFRACTION ANALYSIS

G. Borgonovi, D. Epperson, G. Houghton and V. Orphan

Science Applications, Inc.
1200 Prospect Street
La Jolla, California   92038

INTRODUCTION

The mining industry uses coring as a standard method for the evaluation of ore deposits. Since coring is costly and time-consuming, increasing attention has been given to logging techniques capable of providing information about the orebody, thus allowing replacement of at least part of a coring program. An evaluation of the feasibility and potential for application of a variety of borehole logging techniques(1) concluded that X-ray diffraction (XRD) is one of the promising techniques for borehole assay, along with neutron activation analysis, X-ray fluorescence, photoluminescence, and infrared spectroscopy.

This paper represents the results of a study(2) to develop an XRD borehole probe, that is an instrument which allows one to collect and interpret diffraction patterns from the surface of a borehole. The main emphasis of the study was on technical feasibility, rather than on applicability. For design purposes, it was assumed that the instrument would be used for in-situ identification of minerals. There are, however, other potential applications for an XRD borehole probe. For example, after successful demonstration of a prototype, the instrument could be redesigned for the determination of average mineral grain size or for the determination of components of the in-situ stress field in rocks.

REQUIREMENTS AND ALTERNATIVES

The purpose of the XRD borehole probe is to obtain interpretable diffraction patterns in-situ. Under these conditions, there is no control over the microscopic characteristics of the sample. In addition, there is a number of practical constraints dictated by the economics of present day borehole drilling and logging technology. For example, the diameter of the probe should be limited to a few inches.

Table 1 lists the major factors to be considered in the design, the alternative solutions, and the potential problems which arise. Some of the solutions for different design factors are correlated to some extent.

As far as the diffraction scheme is concerned, the possibilities are angular analysis or energy analysis. Energy analysis (energy dispersive diffraction) requires a polychromatic spectrum and a fine grained (powder) sample. Angular analysis not limited to a single scattering plane is more flexible since it provides useful information for both polycrystalline samples and single crystals. In the first case filter monochromatization of the incident beam is required

The second factor is the X-ray source. Isotopic sources, chiefly Fe-55, have been considered, but the intensity obtainable is very low, and self-absorption prevents the use of thicker sources. Small X-ray diffraction tubes as needed for the probe do not exist, however preliminary designs have been obtained from specialized manufacturers. Construction of an adequate tube and of a small high voltage power supply is considered to be feasible.

Table 1.   Main Factors, Solutions, and Problems for XRD Borehole Diffractometry (Solutions Underlined Represent Design Choices).

DESIGN FACTOR	ALTERNATIVE SOLUTIONS	POTENTIAL PROBLEMS
Diffraction Scheme	Monochromatic Beam, Angular Analysis	Not adequate for minerals with large grains
	Polychromatic Beam, Angular Analysis	Not adequate for minerals with small grains
	Polychromatic Beam, Energy Analysis	Not adequate for minerals with large grains
	Polychromatic Beam, Energy and Angular Analysis	Beyond state of the art
X-Ray Source	Isotopic Source (Fe-55)	Not enough intensity
	X-Ray Tube	Needs special development, including power supply
Cooling	Coolant Lines from Surface	Too cumbersome, not practical
	Transfer Heat to Borehole Wall	Poor coupling and thermal conductivity
	Self-Contained Heat Sink	Limited running time
Data Acquisition	Film	Requires withdrawal of probe for data analysis
	Conventional Radiation Detector	Limited space for scanning arrangement
	Electro Optical Position Sensitive Detector	New technology, delicate equipment
Data Analysis	Off-Line	Long delay before data utilization
	On Line	Need for data transmission and on site computer capability

Cooling of the X-ray tube is an important factor. It is not practical to carry coolant lines from the surface and it is not effective to dissipate the heat to the borehole wall. Accordingly, the probe should carry its own heat sink, which implies a limited measurement time.

A possibility for the detector is X-ray sensitive film. This solution could be used as an intermediate step for the evaluation of the probe, but would require withdrawal of the probe for data

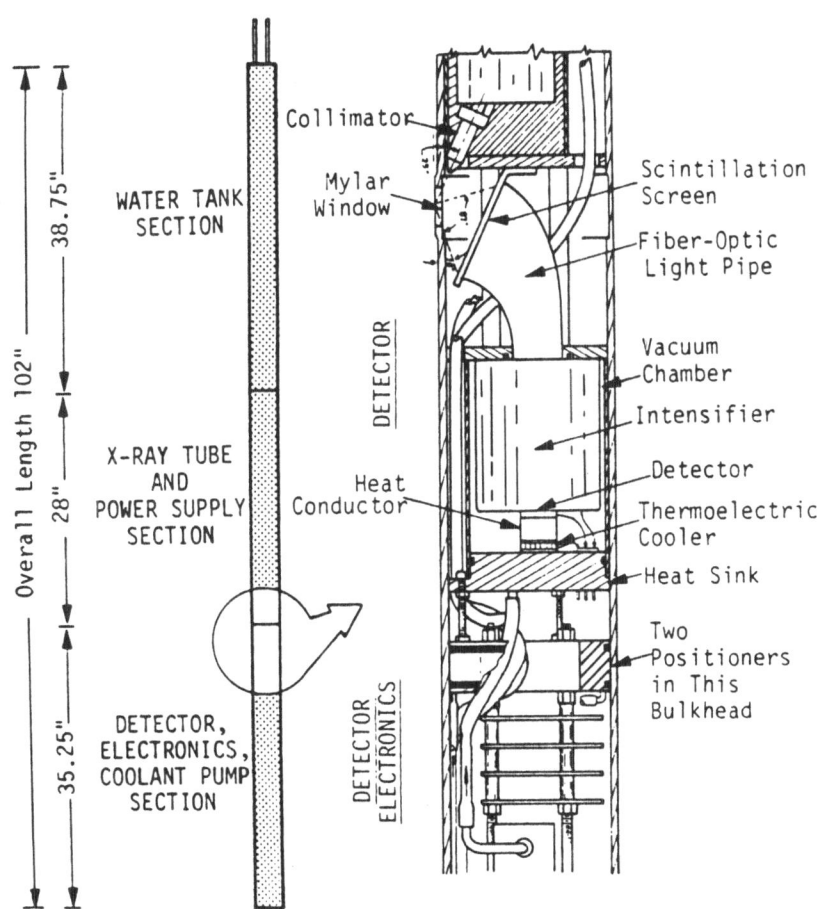

Figure 1. Conceptual Design of the XRD Borehole Probe. The right Side Shows the Arrangement of Some Components.

analysis. A conventional radiation detector would require angular
scanning and would lead to long measurement times. A position sensitive detector based on electro-optical sensors satisfies the requirements of small size and high sensitivity.

Data analysis could be done on-line or off-line. The on-line
mode is obviously more desirable since it would permit one to make
near real time decisions on strategies for the utilization of the
probe in exploration.

DESIGN CHOICES

Following the above analysis, a number of calculations and experiments were performed to establish the technical feasibility of
an XRD probe. The experiments have involved diffraction of X-rays
from single crystals and polycrystalline materials, using configurations closely resembling the geometry to be encountered in the borehole. Experiments were performed both with film and with electro-
optical sensing devices (CCD and Vidicon tubes). The calculations
have involved the time behavior of the heat sink and the development
of numerical methods for the interpretation of diffraction patterns.
The chosen scheme consists of recording the pattern formed on the
surface of a flat detector by the diffraction of a pinhole collimated
X-ray beam. For the correct interpretation of the pattern it is
essential to know the distance from the scattering point on the borehole wall to the surface of the flat detector. It would be cumber-
some to try to measure this distance in the borehole through mechanical means. Instead, it was shown that it is possible to obtain this
distance by a least squares analysis of the shape of the observed
diffraction lines. By shape of the lines it is meant here not the
line profile, but the shape of the conic sections of the Debye-
Scherrer diffraction cones, as they appear on the surface of the
flat detector. The method, which requires that the surface of the
detector not be normal to the incident beam, has been fully described in Reference 2. The method was developed for fitting Debye-
Scherrer lines which are due to a polycrystalline sample, but could
be extended to single crystals by using sequences of Laue spots which
correspond to zone axes.

The experimental and theoretical results obtained, plus consideration of the size of existing components, indicated that it is
technically feasible to build a 3.5 inch diameter diffraction probe
for in-situ borehole analysis. Two preliminary versions of the
mechanical design of the probe have been prepared. They differ
mainly because of the mechanisms used to transfer heat from the X-
ray tube to the heat sink.

Figure 1 shows a partial view of one of the designs. The X-
rays from a 300 watt copper anode, point focus tube are collimated

and pass through a mylar window to reach the borehole surface. The diffracted beams reenter the probe through the mylar window and fall on a scintillation screen. The light generated is directed by a fiber optics bundle on an image intensifier and then on an electro-optical sensor (charge coupled device) which is kept at low temperature for reduction of noise by a small thermoelectric cooler. The X-ray tube runs at a temperature slightly above 100^0 C and is cooled by water contained in a tank at the upper section of the probe. The water, which is fed to the tube by a small pump, vaporizes and is released as steam to the borehole through a pressure relief valve. The high heat of vaporization of water is used to absorb the heat dissipated both by the tube and by the power supply. The volume of water in the tank is sufficient to permit continuous operation of the probe for about three hours. The power supply for the X-ray tube is housed in a section about 28 inches long and is built around a power oscillator. Input to the power supply from the surface is 300 VDC. The possibility of using a fiber optics cable for data transmission has also been considered.

The utilization of a detector system based on electro-optical devices would require considerable development work. For this reason, a testing alternative has been incorporated in the design. This alternative, which is based on the use of standard 35 mm film canisters, would allow one to test the performance of the X-ray tube, power supply, and cooling system, and would permit a preliminary field evaluation of the probe.

ACKNOWLEDGEMENTS

The authors would like to thank the U.S. Bureau of Mines for sponsoring this work and for permission to publish this paper.

REFERENCES

1. D.K. Steinman, J.A. Stokes, N. Vagelatos, D.E. Bryan, and D. Rock, "Future Research in Borehole Assaying Technology", Volume I, Technology Assessment of Borehole Logging Techniques, Final Report on USBM Contract No. J0255018, March 1976.
2. G. Borgonovi, D. Epperson, G. French, G. Houghton, V. Orphan, and D. Sengupta, "A Borehole Probe for In-Situ X-Ray Diffraction Analysis-Technical Feasibility and Application Study", Final Report on USBM Contract No. H0282019, Submitted March 1980.

MEASUREMENT OF CUMULATIVE FATIGUE DAMAGE BY

X-RAY DOUBLE-CRYSTAL AND SCANNING DIFFRACTION METHODS

R. N. Pangborn,* S. Weissmann** and I. R. Kramer***

*The Pennsylvania State University, University Park, PA
**Rutgers University, P. O. Box 909, Piscataway, NJ
***David W. Taylor Naval Ship R & D Center, Annapolis, MD

INTRODUCTION

The ability to predict fatigue failure has long eluded scientists and engineers. Because they can be applied nondestructively, X-ray diffraction methods have been employed to measure fatigue-induced lattice misorientation, lattice strain, residual stress and subgrain formation.[1,2] The changes in the X-ray patterns, however, were found to be restricted principally to the very early and late fractions of the fatigue life. Their invariance during the intermediate portion of the life has hampered the evaluation of cumulative fatigue damage and prediction of ultimate failure.

The primary difference between previous investigations and our recent X-ray diffraction studies is the emphasis we placed on characterizing both the surface and subsurface response to fatigue.[3,4] By evaluating the prefracture damage with depth from the specimen surface a more reliable procedure for failure estimation emerged.

X-RAY DOUBLE-CRYSTAL DIFFRACTOMETRY

A highly-sensitive double-crystal diffraction (DCD) technique was used in the initial analyses of Al 2024 alloy, subjected to push-pull cycling at constant amplitude (R = -1). The method was employed in a modified form designed for application to polycrystalline test specimens.[5] Rocking curves reflected from a large population of grains were sampled since a crystal-monochromated incident beam was used to irradiate nearly the entire specimen gage length. The average halfwidth of the curves was indicative of the mean lattice distortion in the grains of the sample. Rocking curve

analyses were performed at progressively greater depths from the
specimen surface as layers of material were incrementally removed
by electropolishing.[3] The depth profiles of the fatigue damage at
various fractions of life displayed the following features:

(1) The average halfwidth for surface grains increases
     rapidly reaching a near saturation value which is
     maintained until the onset of failure.

(2) A decreasing gradient in halfwidth value from surface
     to bulk exists for all stages of fatigue, and a pla-
     teau level is established at about 200 µm in depth
     which is always lower than the high surface value and
     extends into the bulk material.

(3) This constant-halfwidth plateau level increases in
     elevation by nearly equal amounts for equal increments
     in the fatigue life fraction, revealing a steep en-
     hancement of the subsurface halfwidths during the life.

The steady, linear increase in average halfwidth during the
life measured for grains located in depth became the focus of subsequent research. To analyze the in-depth response nondestructively,
three X-radiations were utilized as shown in Fig. 1. As a result
of their difference in wavelength, chromium, copper and molybdenum
K_α radiations penetrate to successively greater depths, respectively,
in aluminum. From the figure it may be seen that by using Mo K_α
radiation the progressive increase of the halfwidth values for

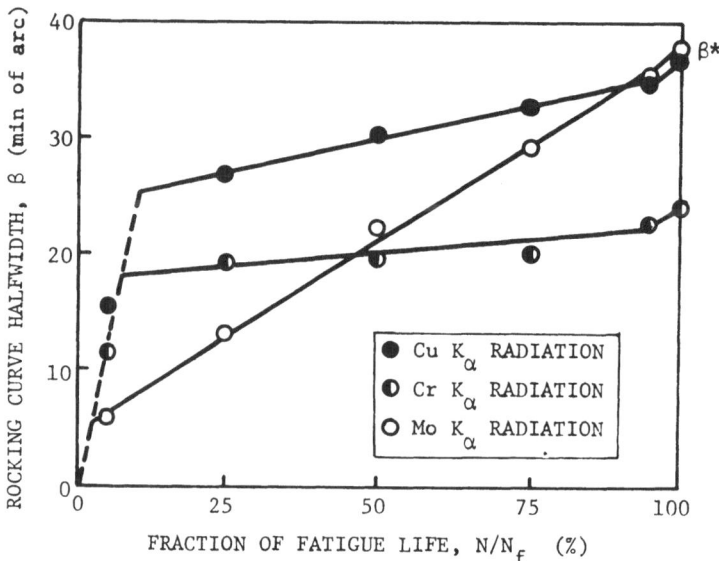

Fig. 1. Dependence of the average halfwidths measured using Cr,
         Cu and Mo K_α radiations as a function of fatigue life.

subsurface metal grains can be readily detected. Due to the steep
slope of the curve, this increase could be easily followed up to a
critical value at failure, β^*, which was independent of the applied
stress amplitude.

SPECTRUM FATIGUE

A spectrum fatigue program was initiated to test the procedure
for estimating fatigue damage. Al 2024 specimens were first subjected to four cycling blocks, each conducted at a different stress
amplitude and representing approximately equal fractions of life.
The amplitude applied during each successive block was higher than
the one before it. After the fourth block the specimens were cycled
to failure at the same amplitude as employed in the last block, and
this remaining fraction of life was determined by reference to the
S-N curve. As shown in Fig. 2, the expended life for the specimens
could be determined graphically by plotting the measured halfwidths
along the calibration curve from prior constant-amplitude testing,
represented by the solid line. Thus, for two identically tested
samples, A and B, the fractions of life expended after the fourth
block were estimated as 0.55 and 0.46, respectively. These values
for the fraction of life used up during spectrum fatigue agreed
favorably with the actual values of 0.58 and 0.42, respectively,
evaluated by cycling both specimens to failure and interpreting
the expended life from the S-N curve data.

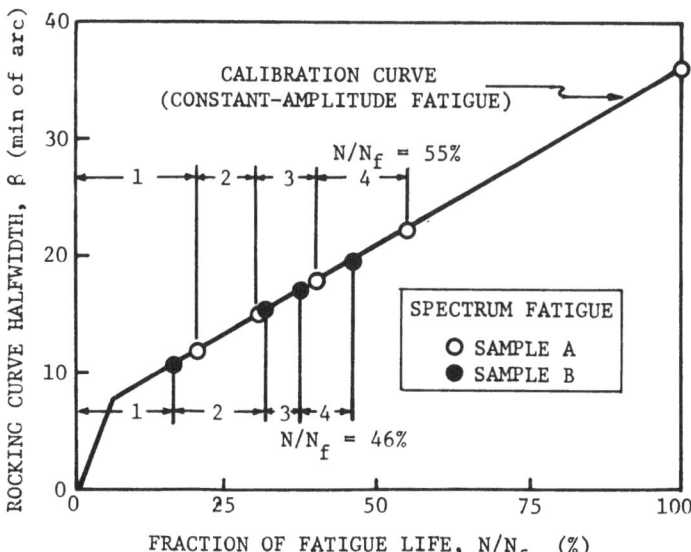

Fig. 2.   Measured halfwidths for spectrum fatigue specimens plotted
          along calibration curve from constant-amplitude testing.

Table 1. Comparison of Cumulative Damage Estimates for
Al 2024 Spectrum Fatigue Samples

Cycling Block	Number of Cycles and Stress Amplitude	Expended Fraction of Life			
		Miner	Kramer	X-Ray DCD Sample A	Sample B
1	36×10^3 cycles @ ±170 MPa	0.20	0.20	0.19	0.15
2	7.5×10^3 cycles @ ±210 MPa	0.40	0.31	0.29	0.30
3	3.0×10^3 cycles @ ±245 MPa	0.60	0.49	0.41	0.38
4	1.5×10^3 cycles @ ±280 MPa	0.80	0.63	0.55	0.46
5	Cycled to failure @ ±280 MPa	Remaining Life			
	Sample A: $N_5 = 3.19 \times 10^3$	X-Ray DCD		0.45	0.54
	Sample B: $N_5 = 4.36 \times 10^3$	Actual (N_5/N_f)	0.42		0.58

The damage estimations based on the X-ray analyses may also
be compared to the predictions of cumulative damage formulations.
Miner's Rule[6] expresses the expended life simply as a summation of
the fractions of fatigue life, f, at each amplitude. Complete
knowledge of the fatigue history is required in the computation
which takes the form:

$$\Sigma f_i = \Sigma (N/N_f)_i = \Sigma [(\sigma_a/\alpha)^{1/P} N]_i \, , \quad i = 1, 2, \ldots, n \qquad (1)$$

where N is the number of cycles run, N_f is the number of cycles to
failure, σ_a is the applied stress amplitude, and α and P are the intercept and slope, respectively, of a log-log plot of the S-N curve.

Kramer[7] recently reported an expression which includes correction terms accounting for the load interaction effects during
spectrum fatigue. This formulation again requires knowledge of
prior cyclic history and takes the form:

$$\Sigma f_i = (\sigma_{a_1}/\alpha)^{1/P} N_1 + (\sigma_{a_2}/\alpha)^{1/P} N_2 [(\sigma_{a_1}/\sigma_{a_2})^{(1/P) f_1}]$$
$$+ (\sigma_{a_3}/\alpha)^{1/P} N_3 [(\sigma_{a_2}/\sigma_{a_3})^{(1/P) f_2} (\sigma_{a_1}/\sigma_{a_2})^{(1/P) f_1 f_2} + \ldots \qquad (2)$$

The X-ray diffraction procedure, by contrast, represents a
direct method for evaluating expended life. The cumulative damage
is determined from the measured halfwidth, β, in combination with
the slope, n, and the critical halfwidth at failure, β^*, of a
previously generated calibration curve. The fraction of expended
life is given by:

$$\Sigma f_i = 1 - (\beta^* - \beta)/n \, . \qquad (3)$$

Table 1 compares the values for the expended life computed
using the two cumulative damage rules and the direct X-ray dif-

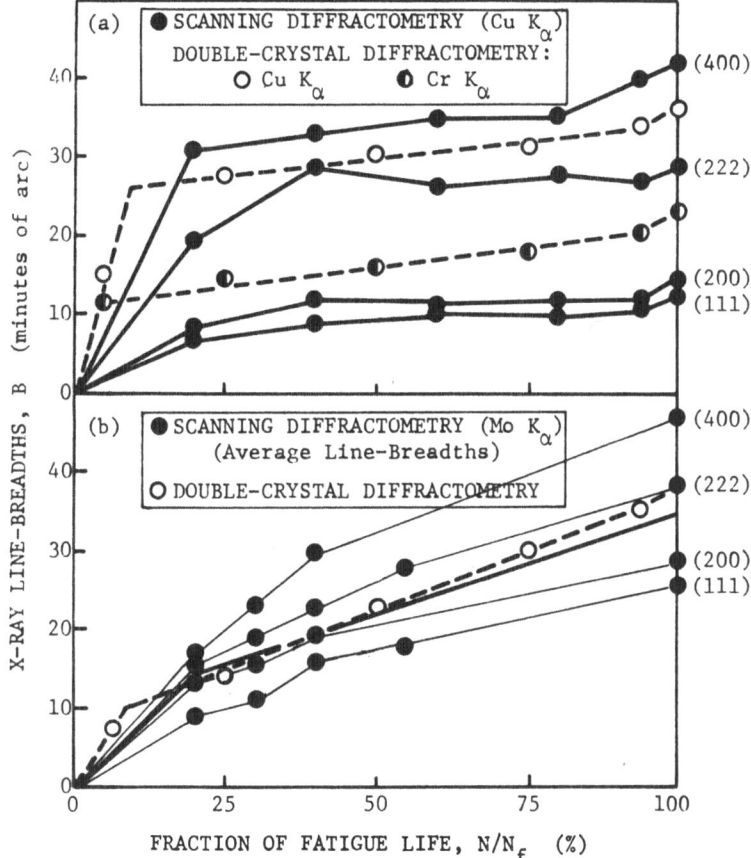

Fig. 3.   Comparison of rocking curve halfwidths from DCD and line-
          breadths from scanning diffractometry. (a) shallow-
          penetrating Cr K$_\alpha$ and Cu K$_\alpha$ radiations, (b) deeply-
          penetrating Mo K$_\alpha$ radiation.

fraction method. A close correlation between the results of X-ray
analysis and the fractions of life predicted by Kramer's equation
is evident. This attests to both the reliability of the X-ray
damage prediction technique and to the advisability of correcting
for prior history in any cumulative damage formulation.

SCANNING DIFFRACTOMETRY

     To permit faster and more convenient analysis, as well as
adaptability to portable instrumentation for on-site application

of the method, the reliability of scanning diffractometry for damage
prediction was investigated. Fig. 3 shows a comparison between the
rocking curve halfwidths measured during the life using double-crystal diffractometry (broken lines) and the line-breadths for selected
(hkl) reflections from scanning diffractometry (solid lines). In
Fig. 3a the halfwidths obtained with DCD using Cr K_α radiation compare favorably with the line-breadths for the low order (111) and
(200) reflections recorded by scanning diffractometry with Cu K_α
radiation. The rocking curve halfwidths are similar in magnitude
to the line-breadths of the higher order (222) and (400) reflections
when Cu K_α radiation was utilized for both diffraction methods.
When Mo K_α radiation was employed for both double-crystal and scanning diffractometry, much steeper dependences of the halfwidths and
line-breadths on the fraction of life were measured. The halfwidths
measured using DCD are nearly coincident with the average line-
breadths of the four (hkl) reflections monitored by scanning diffractometry. This indicates that the extreme precision of double-
crystal diffractometry may be traded for the efficiency of conventional scanning diffractometry without risking the viability of the
damage estimation method.

In summary, reliable evaluation of fatigue damage, even under
spectrum loading conditions may be accomplished by penetrating beyond
the surface layer, for which the diffraction pattern changes saturate
early in the life. This behavior is caused by the surface-sensitivi-
ty of fatigue processes. The steady subsurface and bulk response to
progressive cycling provides a better indicator of accrued fatigue
damage and potential for failure. X-ray diffraction analysis may
be carried out nondestructively by using deeply-penetrating radiation
and automated scanning diffraction equipment.

REFERENCES

1. S. Taira and K. Hayashi, X-ray investigation on fatigue fracture
    of notched steel specimen, Bull. JSME 9:627 (1966).
2. S. Taira, T. Goto and Y. Nakano, X-ray investigation of low-cycle
    fatigue of low carbon steel, in: "Proc. 12th Jap. Congr. on Mat.
    Res.," The Society of Materials Science, Kyoto, 8 (1969).
3. R. N. Pangborn, S. Weissmann and I. R. Kramer, Work hardening in
    the surface layer and bulk during fatigue, Scripta Met. 12:129
    (1978).
4. R. N. Pangborn, S. Weissmann and I. R. Kramer, Fatigue failure
    prediction by double crystal diffractometry and topography, in:
    "Strength of Metals and Alloys," Pergamon Press, NY, 1279 (1979).
5. S. Weissmann and D. L. Evans, An X-ray study of the substructure
    of fine-grained aluminum, Acta Cryst. 7:733 (1954).
6. M. A. Miner, Cumulative damage in fatigue, J. Appl. Mech. 12:
    A-159 (1945).
7. I. R. Kramer, Prediction of fatigue damage, in: "Proc. 2nd Int.
    Conf. on Mech. Behavior of Materials," Boston, 1812 (1976).

AN X-RAY STUDY OF ION-IMPLANTED LIQUID PHASE EPITAXIAL

GARNET FILMS WITH A SINGLE CRYSTAL DIFFRACTOMETER

Po-wen Wang, Kochan Ju and T. C. Huang

IBM Research Laboratory

5600 Cottle Road

San Jose, California 95193

ABSTRACT

A vertical scanning single crystal diffractometer with graphite monochromator and narrow divergent beam controlled by an IBM Series/1 computer was used to study crystal damage in the ion-implanted garnet films including strain effects induced by multi-implantation processes as a function of the types of ions, their energies and doses, growth and annealing temperatures.

INTRODUCTION

Ion-implantation has been used widely in magnetic bubble technology to locally alter the magnetic properties of the liquid phase epitaxial (LPE) magnetic single-crystal $Gd_3Ga_5O_{12}$ garnet (GGG) films (1). For example, uniaxial magnetic anisotropy can be changed to planar anisotropy after ion-implantation.

Implanted ions, when traveling through the lattice, interact with atoms in the LPE film and cause structural damage. In the implanted region, the lattice along the $[111]$ direction undergoes an expansion of about 1 to 2 percent accompanied by a contraction normal to $[111]$. The magnitude and uniformity of the strains caused by these lattice changes are important to the performance of contiguous disk magnetic bubble devices (2).

The observation of lattice expansion by ion-implantation using the X-ray diffraction technique was first reported by North and Wolfe (1). A high resolution double-crystal diffraction method was later used by Speriosu, et al., (3) and Komenou, et al., (4) to investigate the damage in ion-implanted LPE films. This paper presents a rapid and effective single crystal diffraction technique for the study of the strain effects in LPE magnetic garnet films.

EXPERIMENTAL

A vertical scanning Norelco diffractometer with diffracted beam graphite monochromator and vacuum paths (5) was used to record the single-crystal diffraction profiles of the GGG (888) reflection of the ion implanted LPE films on GGG substrates. An incident beam of $0.04°$ divergence was used to avoid beam "walking" during specimen scanning and thereby obtain narrow profiles. Prior to the actual recording, the specimen was rotated around an axis normal to the surface to obtain the maximum intensity of the substrate at a fixed 2θ angle. The specimen was then kept in this position throughout the entire experiment.

Figure 1 shows a typical experimental profile of the GGG (888) reflection of a hydrogen double implanted LPE film and its substrate using an IBM Series/1 computer controlled diffractometer (6) for step-scanning with $\Delta 2\theta = 0.02°$ and $t=1$ second. The total experimental time was less than five minutes. The strongest two peaks are the $CuK\alpha_1$ peaks of the substrate (S) and the non-implanted region of the LPE film (Fn) respectively. The lattice mismatch $\Delta a/a$ between this film and its substrate calculated from the separation of these two peaks was 7.7×10^{-4}. The minimum strain which can be observed with this experimental set up is 4×10^{-4}. Because the Fn reflections had a larger d than the substrate, the film was under a compressive stress. Two small groups of peaks occur at the low 2θ side: the $K\alpha_1$ (Fi) and $K\alpha_2$ peaks from the implanted region of the film. Each of these broadened peaks is a superposition of two peaks (Fi_1, Fi_2) caused by double hydrogen implantation. The corresponding lattice expansion ($\Delta a/a$) were about 7.9×10^{-3} and 7.2×10^{-3} respectively.

Multiple implantations were used to obtain uniform strain in the implanted region. The diffraction profiles resulting from using a second and a third implantations with reduced accelerating energies following the first He^+ implantation, are given in Figure 2. For a single implantation (S) of He^+ at 140 KeV and a dose of 3×10^{15} ions/cm^2, a large amount of non-uniform strain was observed as indicated by the broadened profile between the main peak of the ion-implanted profile (Fi) and the tails of the non-implanted

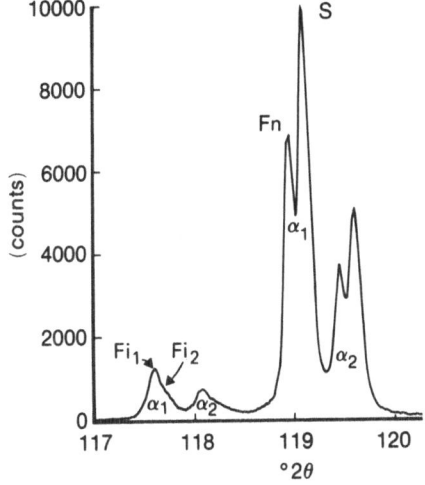 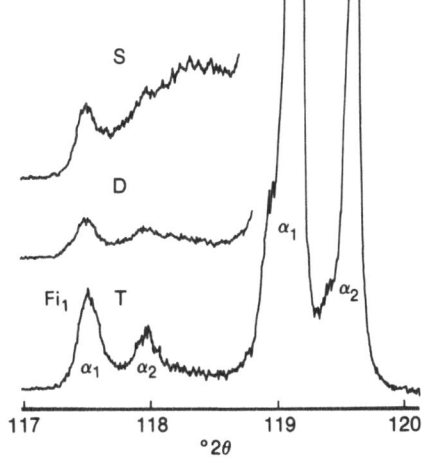

Fig. 1. GGG (888) diffraction profile of a LPE film with CuKα S, Fn and Fi$_{1,2}$ stand for substrate, non-implanted and implanted regions of the LPE film respectively.

Fig. 2. GGG (888) diffraction profiles of multi-implanted LPE films; S, D and T stand for single, double and triple implantations respectively.

and substrate peaks. The non-uniform strain was reduced by a second implantation (D) at an energy of 70 KeV and a dose of 1.4×10^{15} ions/cm^2. A third implantation (T) at an energy of 30 KeV and a dose of 9×10^{14} ions/cm^2 extended the uniformity through a much larger area as indicated by the increase in intensity and sharpeness of the (Fi$_1$) peak.

APPLICATIONS

This single crystal diffraction technique was used to study the strain effects induced by multi-implantation processes as a function of the types of ions, their energies and doses, growth and annealing temperatures.

A typical example of strains for various energy deposition densities (eV/A^3) with hydrogen single implantation in LPE magnetic garnet films is shown in Figure 3. Curves 1 and 2 were measured from two peaks of the hydrogen single-implanted region and curve 3 is from the non-implantation region. They varied almost linearly as a function of the energy deposition density. As the energy deposition density increased above 1.5 eV/A^3, the rate of increase of Δa/a began to reduce and finally no peaks could be observed and an amorphous state was formed. Similar results were also observed from ferromagnetic resonance data. The crystal damages caused by B^+, C^+, O^+ and Ne^+ were more severe; the amorphous state began as early as 1 eV/A^3.

Strain relaxations due to annealing are shown in Table 1. The strain, due to hydrogen double implantation, reduces more rapidly in comparison with other types of ions as the annealing temperature increases. Sharper diffraction profiles and higher intensities for the Fi peaks were observed as the annealing temperature increased. This is the result of non-uniform stress relief and crystalline regrowth due to annealing.

These X-ray diffraction results were correlated with ion-implantation effects on the magnetic properties of the LPE films and used to re-adjust the implantation parameters. For example, in Table 1 the as-implanted strain for Ne^+ is about 30% higher than for other ions. This parameter was used to monitor melt compositions and growth temperature to achieve desired lattice mismatches.

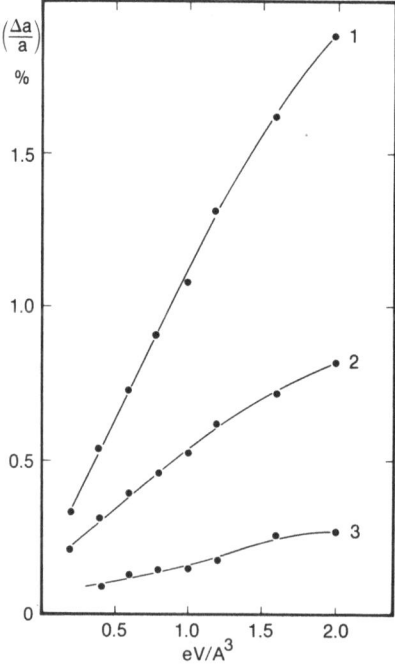

Figure 3.   Strain vs. energy deposition density (eV/A^3) for hydrogen single implantation; curves 1 and 2 for implanted and curve 3 for non-implanted regions.

Table 1. The variation of strains for various annealing temperatures (annealing time 30 minutes).

Ion	Energy (KeV)	Dose $\times 10^{15}$ (ions/cm^2)	Strain = $\Delta a/a$ $(\times 10^{-3})$			
			As-implanted	Annealing temperature (°C)		
				250	350	450
H$^+$	140	23.00	11.1	7.9	6.9	5.6
	60	11.50				
He$^+$	100	4.70	10.8	9.6	9.1	7.4
	40	1.90				
B$^+$	190	0.77	11.7	10.7	9.4	7.3
	75	0.23				
	30	0.09				
Ne$^+$	190	0.22	14.3	12.0	10.7	8.3

ACKNOWLEDGMENTS

The authors would like to thank Drs. B. H. Schechtman, W. Parrish and H. L. Hu for their support of this study.

REFERENCES

1. J. C. North and R. Wolfe, Ion Implantation Effects in Magnetic Bubble Garnet, in: "ion implantation in Semiconductors and Other Materials," B. L. Crowder, ed., Plemun, New York (1973).

2. K. Ju, R. O. Schwenker and H. L. Hu, Damage in Garnet Films Produced by Multiple Ion Implantation, IEEE Trans. Magnetics, 15:1658 (1979).

3. V. S. Speriosu, H. L. Glass and T. Kobayashi, X-Ray Determination of Strain and Damage Distributions in Ion-implated Layer, Appl. Phys. Lett. 34:539 (1979).

Magnetic Properties of An Ion-Implanted Layer in Bubble Garnet Films, J. Appl. Phys. 49:5816 (1978).

5. W. Parrish, X-Ray Analysis Papers, Centrex Publ. Co, Eindhoven, 1965; Chap. 1, "X-Ray and Electron Methods of Analysis," H. van Olphen and W. Parrish, eds., Plenum Press, N.Y., 1968.

6. W. Parrish, G. L. Ayers and T. C. Huang, A Minicomputer and Methodology for X-Ray Analysis, Adv. X-Ray Anal. 23:313 (1980).

RESIDUAL STRESS CHANGE DUE TO ROLLING CONTACT

OF BALL AND ROLLER BEARINGS

Kikuo Maeda, Noriyuki Tsushima, and Hiroshi Muro

NTN Toyo Bearing Co. Ltd.

511 Kuwana, Japan

INTRODUCTION

The life of a rolling bearing is predominantly determined by the contact stress between the rolling elements (ball and roller) and the raceway of inner and outer rings. The contact stress is calculated by assuming Hertzian stress distribution. The maximum Hertzian contact stress Pmax usually ranges from 2000 to 3000 MPa in actual service of rolling bearings. These figures are rather large compared to the compressive yield strength of hardened bearing steel (σ_{ys} = 2000 MPa). Therefore plastic deformation sometimes occurs under the raceway, creating residual compressive stress. Several investigators[1-5] have so far reported about the residual stress due to rolling contact. There were, however, few who referred to the residual stress in used bearings and application of residual stress measurement to failure analysis.

Basic experiments of rolling contact and sliding contact tests with specimens were conducted to correlate the contact stress conditions and residual stress distributions. Application of the result to the presumption of contact stress in used bearings was attempted, and a few supplemental experiments were made.

BASIC EXPERIMENTS

The outlines of the test rigs of rolling contact and sliding contact are shown in Figs. 1(a) and (b). Both specimens were made of SAE 52100 steel. The rolling contact test was carried out with various specimen hardnesses and Hertzian maximum contact stress.

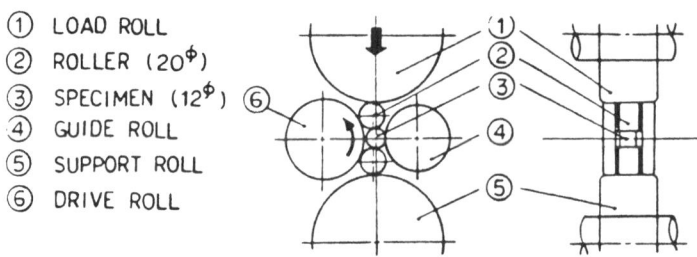

① LOAD ROLL
② ROLLER (20$^\phi$)
③ SPECIMEN (12$^\phi$)
④ GUIDE ROLL
⑤ SUPPORT ROLL
⑥ DRIVE ROLL

Fig. 1(a).   Rolling contact test rig

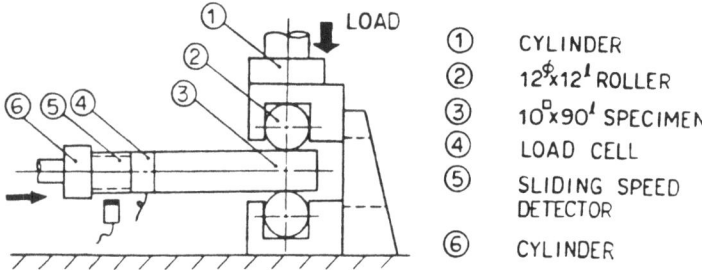

① CYLINDER
② 12$^\phi$x12^l ROLLER
③ 10$^\Box$x90^l SPECIMEN
④ LOAD CELL
⑤ SLIDING SPEED DETECTOR
⑥ CYLINDER

Fig. 1(b).   Sliding contact test rig

The sliding contact test was carried out with various specimen hardnesses, contact stresses and coefficients of friction.

Residual stress on and under the contact surface of the specimen was measured by X-ray.   The X-ray beam was projected in the rolling or sliding direction.

RESULTS

Rolling Contact Test

Fig. 2 shows the typical patterns of residual stress due to rolling contact.   Residual stress due to rolling contact appears when Hertzian maximum contact stress Pmax is larger than 3500 MPa. The increase in contact stress leads to the increase in residual compressive stress peak Zp and in residual stress creation depth Zc.

Comparison with the unidirectional shear stress distribution under the contact surface shown in Fig. 3 shows that compressive residual stress is produced over the region where the shear stress

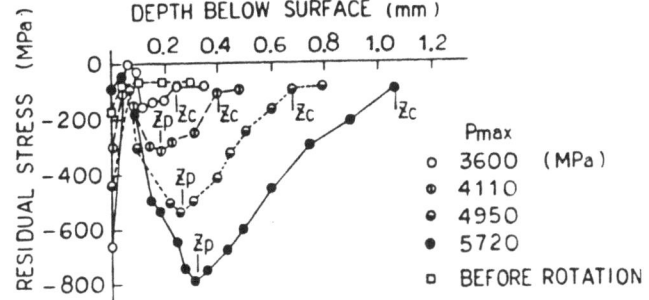

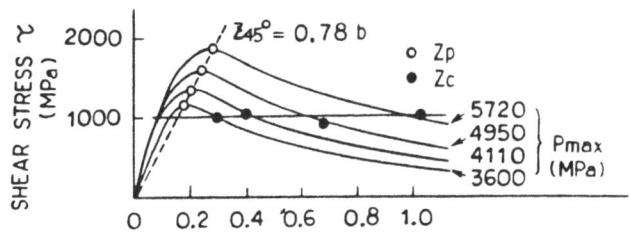

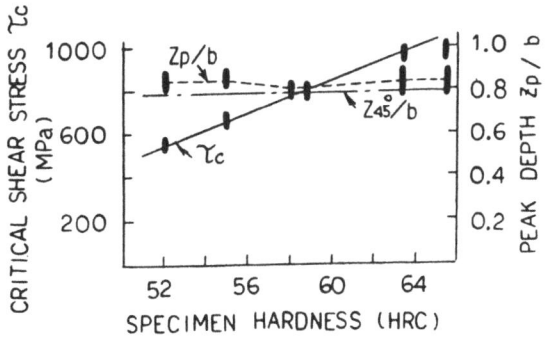

Sliding Contact Test

Fig. 5 shows typical patterns of residual stress due to
sliding contact. Similarly to the rolling contact, increase in
contact stress leads to the increase in Zp and Zc. For the same
contact stress, Zp and Zc are the same, independent of the coefficient of friction.

Fig. 6 shows the variation of Zp and τ_c with specimen hardness. Similarly to the rolling contact, Zp is independent of
specimen hardness and corresponds to $Z_{45°}$. The τ_c value for the
normal hardness region of bearings is 1000 MPa.

Application to the Presumption of Contact Stress

From these experiments, it is proposed that the contact stress
can be obtained from the residual stress measurement under the
raceway surface in the following two ways:

1) Contact stress deduced from residual stress peak depth Zp,
   assuming that Zp coincides with maximum unidirectional shear
   stress depth $Z_{45°}$.

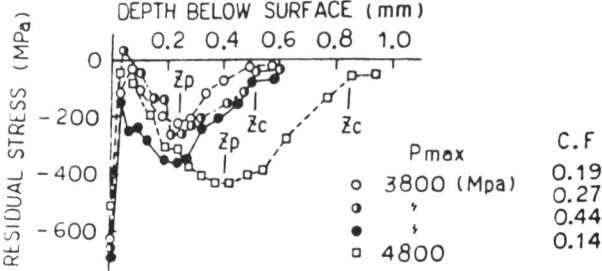

Fig. 5.   Residual stress distributions due to sliding contact

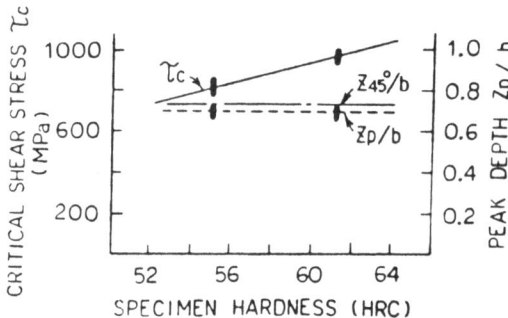

Fig. 6.   Variations of Zp and τ_c with specimen hardness

2) Contact stress deduced from residual stress creation depth Zc, assuming that shear stress at Zc is equal to τ_c, for example, τ_c = 1000 MPa for normal hardness.

Fig. 7(a) shows the results obtained from used bearings which compare the two values of contact stress obtained assuming τ_c = 1000 MPa. The two deduced values coincide for non-failed bearings, but are inconsistent for failed bearings. Namely, for failed bearings, contact stress deduced from Zc is always larger than that deduced from Zp. Decreasing τ_c to 500 MPa in the case of failed bearings improves the correlation as shown in Fig. 7(b).

SUPPLEMENTAL EXPERIMENTS

Rolling Contact With Failed Mating Element

To verify the phenomenon that τ_c is lower in failed bearings than in non-failed bearings, supplemental tests of rolling contact were conducted. The rolling contact test with failed (flaking failed) mating element was done in the same test rig as in Fig. 1(a) under various specimen hardness and contact stress conditions. Fig. 8 shows the results. τ_c is almost constant, in the range of

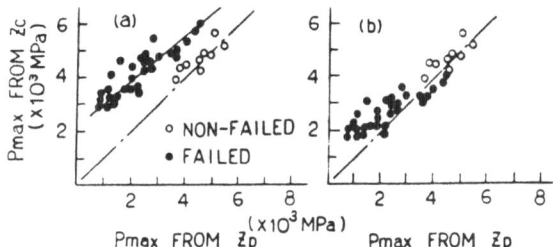

Fig. 7.   Relation between two estimated contact stresses, Pmax from Zp and Pmax from Zc with (a) τ_c = 1000 MPa, (b) τ_c = 500 MPa for failed bearings

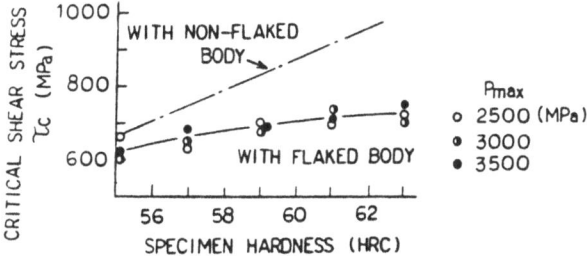

Fig. 8.   Decrease in τ_c in the case of rolling contact with flaked mating element

600 to 700 MPa, independent of specimen hardness. Therefore τ_c for the normal hardness range of HRC 60 to 62 has 1/2 to 2/3 of the value for the rolling contact with non-failed element.

Rolling Contact Under Lubrication With Foreign Particles

The rolling contact test with thrust type bearing 51124 was conducted under lubrication with foreign particles (quenched turning chips of ball bearing steel, 0.1 to 0.2 mm in size) to reproduce the rolling condition after flaking failure. The test rig is shown in Fig. 9. The test was done at very low rotational speed of 120 r.p.m. to eliminate the influences of vibration and temperature rise.

Fig. 10 shows the results. τ_c can decrease remarkably with the increase in cycles of rolling contact. The mechanism for the decrease in τ_c is not clear, but it was confirmed from these experiments that the decrease in τ_c is not due to vibration, temperature rise or edge loading by the flaking edge. It may be related to the fatigue of the material.

REFERENCES

1. J.J.Bush, W.L.Grube, G.H.Robinson, Trans. ASM 54:390 (1961)
2. E.V. Zaretsky, R.J.Parker, W.J.Anderson, J. Lubric. Technol. (Trans. ASME, F), 91:314 (1969)
3. H.Muro, N.Tsushima, Wear, 15:309 (1970)
4. H. Muro, N.Tsushima, K. Nunome, Wear 25:345 (1973)
5. A.P. Voskamp, R.Osterlund, P.C.Becker, O.Vingsbo, Metals Technol. 7:14 (1980)

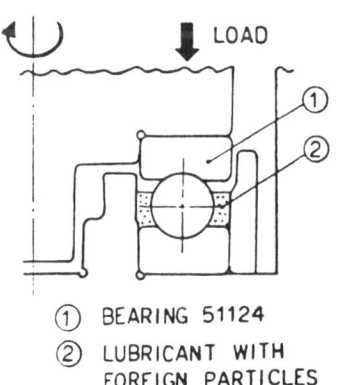

① BEARING 51124
② LUBRICANT WITH
  FOREIGN PARTICLES

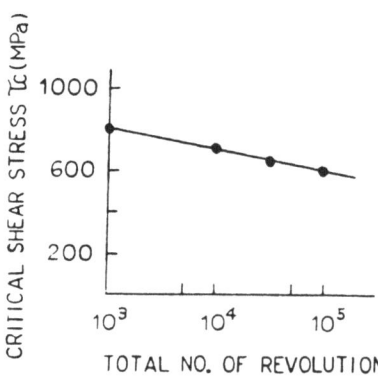

Fig. 9, 10.   Test rig and result of rolling contact test in lubrication with foreign particles

COMPRESSION AND COMPRESSIBILITY STUDIES OF PLUTONIUM AND

A PLUTONIUM-GALLIUM ALLOY*

R. B. Roof

Los Alamos Scientific Laboratory

Los Alamos, New Mexico 87545

INTRODUCTION

Two metal foils, one pure plutonium and the other being a solid solution of 6.5 a/o gallium in plutonium, were examined, in-situ, by X-ray diffraction techniques while under pressure. The purpose was to determine the compression and compressibility of these materials as a function of pressure and to identify the products of any transformation that may occur due to the action of applied pressures.

EXPERIMENTAL PROCEDURES

The pure plutonium foil was prepared by the technique of splat-cooling. The gallium alloy foil was prepared by hot-rolling. The foils were 1-2 mils thick and were wrapped in Al foil to a thickness of 3 mils on each side of the sample. This assembly was sealed between two layers of clear sticky tape. The sample coupon (or sandwich) was placed between the diamond anvils of a high pressure X-ray diffraction camera equipped with a film cassette[1]. Mo Kα X-ray radiation (λ = 0.7107 Å) from a commercial generator (45 kv – 25 ma) and a 100 μm collimator were used to obtain the diffraction film. Exposure times were 400-600 hours.

The Al foil in the coupon acts as the internal pressure standard. It is recognized that the pressure on the sample is not hydrostatic but varies from zero at the edge of the diamond anvil to a maximum at the center of the anvil face. However, with a 100-μm collimator aligned optically to the center of the diamond anvil

---

*Work performed under the auspices of the Department of Energy.

face only a very small region of pressure distribution is examined. The Al samples this small region and a representative value of the pressure is obtained. At high pressure differences in yield strength between a sample and the Al may cause stress or shear gradients that in turn may affect the pressure determination. It is assumed in this work that this effect is negligible within the accuracy to which the diffraction rings can be measured. From the measured experimental lattice constant of Al at each pressure point a volume, V, is determined. From V/V_0 the compression is determined. From a chart of compression vs. pressure previously determined[2] the pressure is read directly.

The X-ray diffraction films obtained were subjected to a program of computer-assisted enhancement[3] in order to improve the intensity-to-background ratio and to enlarge the image of the original film. Measurements of diffraction ring diameters were made on the computer generated photographs. Prior to the start of the experiment a calibration curve was established between sample-to-film distance and the measured diameter of the diffraction rings.

The Bragg equation, $\lambda = 2 \cdot d \cdot \sin\theta$, is the basis for the determination of lattice constants. From the measurement of corrected ring diameters on the film the quantity 2θ is determined. Then θ is found and from the Bragg equation $\underline{d}$ is calculated. For Al, the lattice constant determined from the experimental $\underline{d}$ values is plotted vs. $(\cot^2\theta - 1)$ and the intercept obtained by a least-squares extrapolation to zero is taken as the true value of the lattice constant. This regression function is employed to correct for the truncation of the diffracted Debye cone by the diamond anvils[4].

The slope of this regression line is typically quite small and results in a change in the lattice constant of Al in the range 0 to -0.005 Å. This yields an average volume difference of approximately -0.2% between the uncorrected and corrected volume of Al and indicates that the truncation effects of the diamond anvils border on the negligible. Thus, the experimental error in the determination of the pressure by utilizing the Al lattice constant is ± 0.1-0.2 GPa. Since the 2θ range examined is small (15 to 40°), and the truncation effects practically negligible the lattice constants and standard deviations of plutonium and the plutonium-gallium alloy were determined by a computer code[5] in which extrapolation or regression parameters were set to zero.

EXPERIMENTAL RESULTS

The experimental results for alpha-plutonium are summarized in Table 1. The lattice constants of alpha-Pu at 0 GPa are in good agreement with the usual accepted values[6] of a = 6.183$\pm$1, b = 4.822 $\pm$1, c = 10.963$\pm$1 and β = 101.79°$\pm$1. In a plot of compression vs. pressure a reasonably smooth curve can be drawn through the points

and this is shown in Figure 1. At no time in the pressure range
examined did the alpha-Pu sample exhibit any detectable tendency to
transform to any other phase. Since alpha-Pu is already in the most
dense phase known for this material this observation was expected.

Table 1. Experimental Results of X-ray Diffraction Examination
of alpha-Pu Under Pressure.

Al Lattice Constant	V/V_0	Pressure GPa	alpha-Pu Lat. Con.		V/V_0
4.0499±15	1.0000	0	6.185±	6	
			4.828	5	
			10.980	10	1.0000
			101.81°	9	
4.0480 15	0.9986	0.1	6.185	6	
			4.822	5	
			10.969	9	0.9976
			101.87°	9	
4.0438 5	0.9955	0.3	6.185	6	
			4.808	5	
			10.963	11	0.9933
			102.09°	9	
4.3974 2	0.9922	0.6	6.171	7	
			4.801	7	
			10.938	14	0.9865
			102.30°	12	
3.9741 2	0.9449	4.9	6.044	1	
			4.671	1	
			10.769	2	0.9266
			102.01°	21	
3.8992 14	0.8925	11.25	5.949	14	
			4.592	11	
			10.611	23	0.8848
			101.58°	25	
3.8320 10	0.8471	18.6	5.879	20	
			4.537	16	
			10.487	33	0.8549
			101.24°	36	
3.8142 23	0.8345	20.95	5.863	19	
			4.506	15	
			10.450	32	0.8442
			101.09°	35	

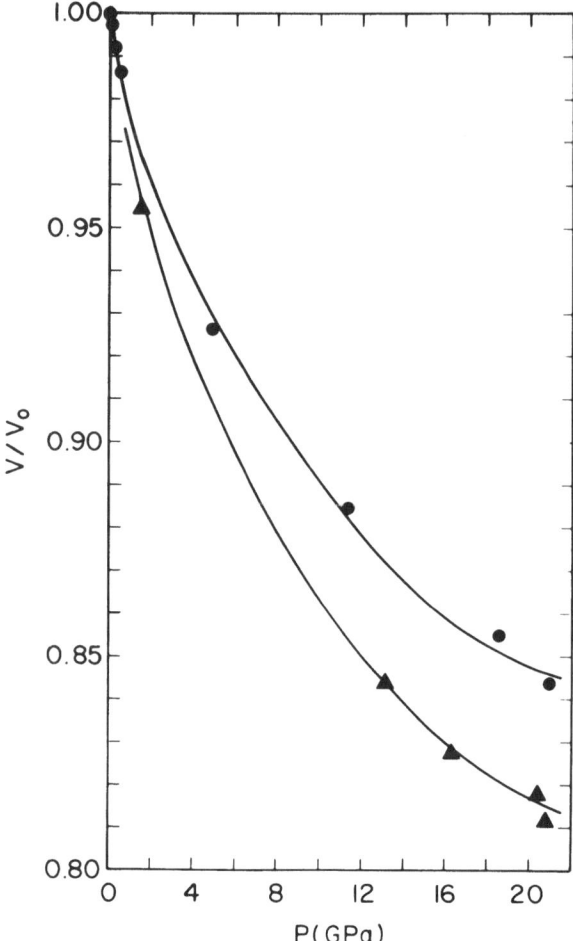

Figure 1.   Compression of alpha-Pu(●) and 6.54
a/o Ga-Pu (▲) as a function of pressure.

The experimental results for the solid solution of 6.54 a/o Ga
in plutonium are summarized in Table 2.   Spectrochemical analysis
indicated that this alloy contained 6.54 a/o Ga.   This Ga content
enables the high temperature delta phase to be retained at room
temperature as indicated by the Pu-Ga phase diagram[7].   When pressure is applied to this material transformation to the alpha-phase
occurs.   At 1.5 GPa a two phase diffraction pattern is observed;
compressed delta-phase and compressed alpha-phase.   The alpha-phase
lattice constants of 6.54 a/o Ga in solid solution substitution in
alpha-Pu at 0 pressure are estimated[8] to be a = 6.298, b = 4.938,
c = 11.141, β = 101.81.   A further increase in pressure results in
complete transformation to the alpha-phase and a reasonably smooth
curve can be drawn through the points as shown in Figure 1.

Table 2. Experimental Results of X-ray Diffraction Examination of 6.54 a/o Ga-Pu Under Pressure.

Al Lattice Constant	V/V_0	Pressure GPa	delta-Pu Lat. Con.	V/V_0	alpha-Pu Lat. Con.	V/V_0
4.0495+7	1.0000	0	4.5909+ 7	1.0000		
4.0243 3	0.9814	1.5	4.5132 48	0.9501	6.246+26	
					4.843 17	
					10.938 42	0.9552
					101.72° 31	
3.8807 6	0.8800	13.1			5.988 22	
					4.658 15	
					10.494 37	0.8445
					101.91° 28	
3.8514 3	0.8602	16.3			5.948 23	
					4.633 15	
					10.418 38	0.8286
					101.80° 29	
3.8181 3	0.8381	20.4			5.918 15	
					4.623 10	
					10.369 24	0.8192
					101.67° 19	
3.8155 2	0.8364	20.75			5.904 14	
					4.613 9	
					10.348 23	0.8129
					101.96° 18	

COMPRESSION AND COMPRESSIBILITY CALCULATIONS

It has been noted[9] that for many solids isothermal compression can be represented by the following linear equation

$$\left[\frac{P v_o^2}{v_o - v}\right]^{\frac{1}{2}} = c_T + S_T \left[P (v_o - v)\right]^{\frac{1}{2}} \tag{1}$$

where the subscript T indicates isothermal conditions and the subscript 0 represents the initial state of the material. P = pressure in units of GPa (10 GPa = 100 kbar). V = volume in units of cm^3/gm. For a least squares fit of the data values of C_T and S_T may be determined. C_T = isothermal bulk sound speed in units of km/sec. S_T = slope constant and is dimensionless. From C_T and S_T two other constants are defined. K_1, the isothermal bulk modulus at zero pressure in units of GPa is equal to $\rho_0 C_T^2$, where ρ_0 = the initial density (19.786 gm/cm^3 for Pu and 17.856 for the Pu-Ga alloy). K_2, the pressure derivative of the bulk modulus at zero pressure is equal to $4 \cdot S_T - 1$, and is dimensionless. Equation (1) may be rearranged[9] to yield an expression for compression

$$\frac{V}{V_o} = 1 - \left[\frac{1}{S_T} + \frac{K_1}{2 S_T^2} \cdot \frac{1}{P} \left[1 - \left[1 + \frac{4 S_T}{K_1} \cdot P \right]^{\frac{1}{2}} \right] \right] \tag{2}$$

The first derivative of this function with respect to P is

$$\frac{dV}{dP} = \frac{V_o K_1}{P^2 \cdot 2 S_T^2} \left[1 - \left[1 + \frac{4 S_T}{K_1} \cdot P \right]^{\frac{1}{2}} \right] + \frac{V_o}{P \cdot S_T} \left[\frac{1}{\left[1 + \frac{4 S_T}{K_1} \cdot P \right]^{\frac{1}{2}}} \right] \tag{3}$$

and compressibility is defined as $(-dV/dP)/V_o$. From the data in Tables 1 and 2 the appropriate quantities involved in Eq. 1 were calculated and are plotted as shown in Figure 2. A least–squares fit of the linear equation through the appropriate points yields for

alpha-Pu $C_T = 1.460 \pm 12$ km/sec $S_T = 2.887 \pm 49$

6.54 a/o Ga-Pu $C_T = 1.209 \pm 48$ km/sec $S_T = 2.778 \pm 128$

It can be shown[10] that

$$c_T = \sqrt{V_L^2 - (\tfrac{4}{3}) \cdot V_S^2} \tag{4}$$

where V_L = longitudinal and V_S = shear velocity in units of km/sec. Values for the longitudinal and shear velocities of alpha-Pu are[11] $V_L = 2.255 \pm 15$ and $V_S = 1.490 \pm 7$ km/sec. Substitution in Eq. (4) yields $C_T = 1.458 \pm 24$ km/sec. Thus, the isothermal bulk sound speed obtained from high pressure X-ray diffraction (1.460 km/sec) is in excellent agreement with the value obtained by calculation (1.458 km/sec) from the longitudinal and shear velocities.

The isothermal bulk modulus, K_1, at zero pressure is calculated for

alpha-Pu $(19.786)(1.460 \pm 12)^2 = 42.175 \pm 0.693$ GPa

6.54 a/o Ga-Pu $(17.856)(1.209 \pm 48)^2 = 26.099 \pm 2.07$ GPa

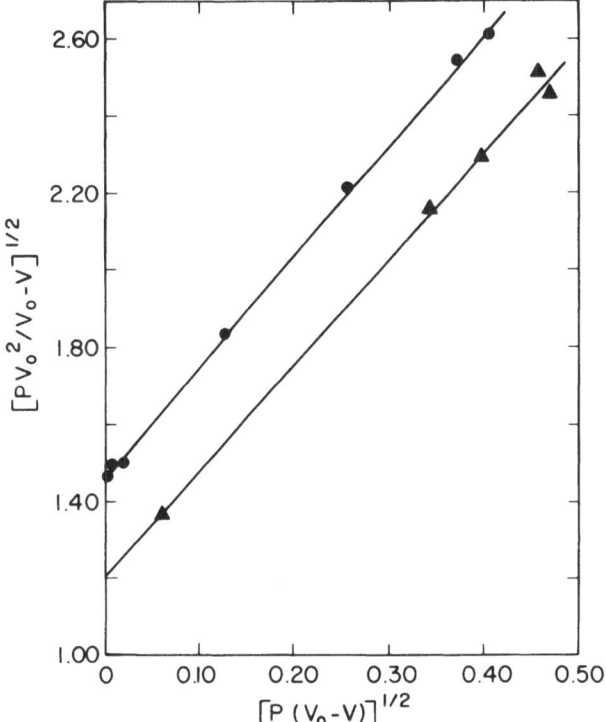

Figure 2. The linear relationship between $[PV_0^2/(V_0 - V)]^{1/2}$ and $[P(V_0 - V)]^{1/2}$. Alpha-Pu ($\bullet$) and 6.54 a/o Ga-Pu ($\blacktriangle$).

The pressure derivative of the bulk modulus at zero pressure, K_2, is calculated for

alpha-Pu (4)(2.887$\pm$ 49) $-$ 1 = 10.548$\pm$196

6.54 a/o Ga-Pu (4)(2.778$\pm$128) $-$ 1 = 10.112$\pm$512

The values of K_1 and S_T given above, coupled with P, were used to calculate the compression according to Eq. 2. The calculated compressions are the solid lines in Figure 1. The values were also used to calculate the compressibility according to Eq. 3. The calculated compressibilities are plotted in Figure 3.

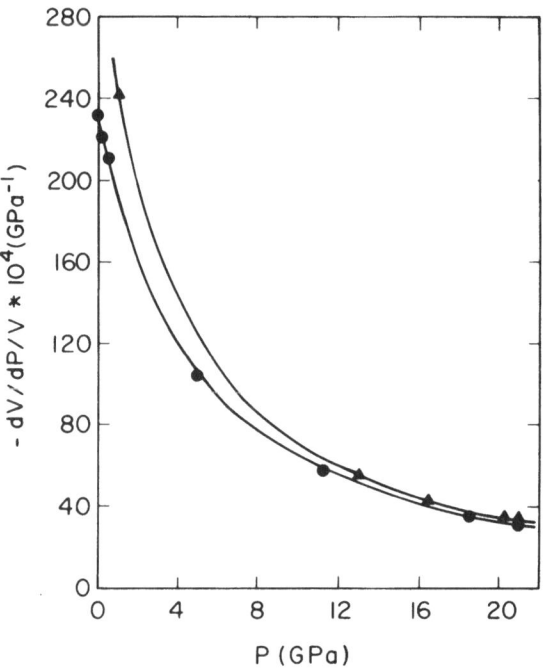

Figure 3. Calculated compressibility of alpha-
           Pu (●) and 6.54 a/o Ga-Pu (▲) as a function
           of pressure.

CONCLUSIONS

Within the pressure range examined (0-20 GPa) alpha-Pu exhibits a smooth compression curve. Any conversion to another phase would be totally unexpected as the alpha-Pu structure is already the most dense phase known for this material.

For 6.54 a/o Ga-Pu low pressures (<1 GPa) are sufficient to convert the material from the initial delta-phase to the transformed alpha-phase. After transformation is complete a smooth compression curve is observed.

The greater the Ga content the greater the degree of compression exhibited by the materials studied. (See Figure 1.)

An emperical linear relationship[9] (see Figure 2) involving pressure and volume was used to calculate compression and compressibility for the materials examined. From this relationship the isothermal bulk sound speed was determined. For alpha-Pu the experimental value of 1.460 ± 12 km/sec obtained from the present work is in excellent agreement with the value of 1.458 ± 24 km/sec obtained by direct sound speed measurement. The value of 1.209 ± 48 km/sec for 6.54 a/o Ga-Pu reflects the effect of Ga addition in "softening" the sample with respect to sound speed values.

The calculated pressure derivatives of the bulk modulus at zero pressure indicate a trend with Ga content. As the Ga content of the alloy is increased the pressure derivative is decreased although the standard deviation on the calculated quantities are such that the decrease is not clear cut.

The compressibility (Figure 3) of the materials below 10 GPa indicates that increased Ga content increased the numerical magnitude of the compressibility. Above 10 GPa however, there appears to be no significant difference in compressibility regardless of the Ga content.

ACKNOWLEDGMENTS

The pure alpha-Pu foil was prepared by R. O. Elliott. The alloy foil was prepared by G. Tate and D. Eash. C. Cox and R. O. Bagley adapted a computer photodensitometer code to the requirements of enhancing the peak-to-background ratio and expanding the scale of the X-ray diffraction high pressure films.

REFERENCES

1.   W. A. Bassett, T. Takahashi, and P. M. Stook, "X-ray Diffraction and Optical Observations on Crystalline Solids up to 300 kbar", Rev. Sci. Instruments, 38:37 (1967).

2.   B. W. Olinger and H. H. Cady, "The Hydrostatic Compression of Explosives and Detonation Products, to 10 GPa (100 kbar) and Their Calculated Shock Compression; Results for PETN, TATB, CO_2 and H_2O," in: Sixth Symposium (International) on Detonation, Office of Naval Research Report No. ACR-221, (1978).

3.   This work was performed by Group M-5, LASL, D. Janney group leader. A two-dimensional digitizing microdensitometer extracts data from the diffraction film for computer enhancement. The enhancement algorithim may make several sequential passes through the data. Diffraction lines which are initially very faint may then display high contrast. While position data are preserved, some intensity data are lost during the enhance-

ment. A positional scale is also synthesized in the computer
and added to the image as a measurement aid.

4. M. Senoo, H. Mii, I. Fujishiro, T. Fujikawa, "Precise Measurement of Lattice Compression of Al, Si, and S1-Si Alloy by
   High Pressure Diffractometry", Jap. J. of App. Physics, 15:871
   (1961).
5. R. E. Vogel and C. P. Kempter, "A Mathematical Technique for
   the Precision Determination of Lattice Parameters", Acta
   Crys. 14:1130 (1961).
6. W. H. Zachariasen and F. H. Ellinger, "The Crystal Structure
   of Alpha Plutonium Metal", Acta Cryst. 16:777 (1963).
7. F. H. Ellinger, C. C. Land, V. O. Struebing, "The Plutonium-
   Gallium System", J. Nuc. Mat. 12:226 (1964).
8. R. O. Elliott, R. E. Tate, R. B. Roof, "Lattice Expansion in
   Dilute alpha-Pu(Ga) and alpha-Pu(Al) Metastable Alloys",
   LA-6922 (1977).
9. B. Olinger and P. M. Halleck, "Compression and Bonding of Ice
   VII and an Emperical Linear Expression for the Isothermal
   Compression in Solids", J. Chem. Phys. 62:94 (1975).
10. A. H. Cottrel, "The Mechanical Properties of Matter", John
    Wiley & Sons, New York (1964), pg. 160, eqs. (6.9) & (6.10).
11. H. L. Laquer, "Sound-Velocity Measurements on Alpha-phase
    Plutonium", in: "The Metal Plutonium", A. S. Coffinberry
    and W. N. Miner, Eds., University of Chicago Press, Chicago
    (1961), pg. 157.

X-RAY DIFFRACTION EVALUATION OF ADHESIVE BONDS AND

STRESS MEASUREMENT WITH DIFFRACTING PAINT

Charles S. Barrett and Paul Predecki

University of Denver Research Institute
Denver, Colorado 80208

ABSTRACT

X-ray diffraction is found effective in disclosing the distribution of stresses over the surface of adhesively bonded lap joints when loads well below the yield point are applied. When a pair of 6061-T6 aluminum strips 1/16" or 1/32" thick and 3/4" wide is adhesively bonded in a single lap joint and loaded in tension, maps giving the distribution of the X-ray-measured stresses show the limits of the bonded area with an accuracy about equal to the width (1 mm) of the irradiated area along the specimen. Attendant bending stresses resulting from the loading are also registered. Stress values can be obtained from the observed diffraction angles by calibration with tensile tests of a single unbonded strip. Similar results are obtained for graphite/epoxy laminates adhesively bonded in a single lap joint to aluminum when diffraction is from the aluminum, but a lower accuracy is obtained when diffraction is from the filled composite.

Another X-ray method was developed for measuring applied (not pre-existing residual) stresses, and for mapping their distribution around a joint. A thin layer of epoxy paint containing a diffracting filler (say aluminum or silver powder) is applied to a specimen and cured. Diffraction from this paint yields shifts in diffraction angle approximately proportional to the magnitude of applied stresses. The limits of the bonded area in single lap joints under load are disclosed. Both methods appear, therefore, to be practicable for mapping the areas that are properly bonded, and presumably also for non-destructive evaluation of bond defects.

INTRODUCTION

X-ray diffraction is widely used for the measurement of residual and applied stresses in metals. Our experiments have shown
that residual and applied stresses can be measured in polymeric
materials, including polymer matrix composites,[1-3] provided suitable diffracting powders are embedded in them before they are
cured. Experiments with graphite fiber reinforced epoxy composites
have led us to conclude that the method could be extended into one
for non-destructively evaluating adhesive-bonded joints. Since
the X-ray measurements can reveal the point to point variation in
stress when a stress is applied to an object, a map of them over
the surface when the joint is loaded should (a) provide evidence
concerning the direct transfer of the applied stress from one adherend to another, even when relatively light stresses are applied;
(b) serve as a nondestructive method for revealing the outer limits
of the bonded area; (c) disclose unbonded patches within the area
that are large enough to significantly alter the stress distribution; and (d) also record any stress components at the surface of
an adherend that arise from the bending of adherends as a result
of the loading. The method should thus provide a direct and quantitative NDE method for comparison with theory[4-6] and should also
reveal deterioration of bond edges with aging when they are exposed to the environment. Also experiments reported here show
that the method can be applied to fillers in a diffracting paint
applied to the surface and cured.

THE METHOD AND THE SPECIMEN DESIGN

The method as applied to single lap joints reported here involved measurement of the diffraction angles by two techniques:
(1) by fitting a parabola to the upper 15% of the $K\alpha_1$ peak in a
standard procedure previously used;[2] and (2) by a more rapid technique that will be here called a "constant angle" technique: measurement of the diffracted beam intensity at a fixed angle on the
low 2θ side of the $CuK\alpha_1$ peak (an angle where the curve of intensity vs 2θ is nearly a straight line). Displacement of the peak
by applied stress then is registered as a change in intensity that
is approximately proportional to the stress change. A modified
version of this technique, (2a), employs measurements at a pair of
fixed angles straddling the peak, and taking either the ratio of
those or their difference.

RESULTS WITH TECHNIQUE (1)

Surface stresses were explored first in a single-lap adhesive
joint by the parabola fitting method, technique (1). Two 6061-T6
aluminum strips were glued together to form a single lap joint with

Hysol EA9309 epoxy adhesive. The sample for this and those for the
other experiments reported below had adherends 1.595 mm thick, 19.7
mm wide, and 95.4 mm long which overlapped each other so as to give
a total sample length of about 152 mm. A 0.0254 mm thick film of
teflon FEP was interposed between the adherends during assembly to
produce a debond. The teflon extended inwards from one edge of the
overlap and across the entire width of the sample. After curing
the epoxy (at 80°C overnight) the sample was inserted in a small,
manually-operated tensile frame that was mounted on a Siemens
horizontal diffractometer. A beam collimated to 1/2 degree divergence irradiated a spot 2 mm wide, 19 mm high on the specimen which
was 107 mms from the slit and on the diffractometer axis. Diffracted rays were passed through a slit 0.610 mm wide at 175 mm from
the specimen and into a graphite crystal monochromator and a scintillation counter. The shift of the 511 + 333 $CuK\alpha_1$ beam ($2\theta \sim$
163°) was measured with 0.2°, 20 sec step-counts, with 13 and 384
lb applied loads (a nominal change of 52.4 MPa, 7.6 ksi).

   On each of the two outer surfaces of the joint the diffraction
angle 2θ was shifted to larger angles by the applied stress at
points within the bonded area corresponding to longitudinal tensile stress (and lateral Poisson's contraction). But prominent
sharp minima in 2θ were seen near the place in each adherend where
it extends beyond the bonded area. Calibration experiments on a
single strip indicated that the surface longitudinal stress at
these minima was -20 MPa (compressive) while the maximum 2θ within
the bonded area corresponded to a stress of +40 MPa.

RESULTS WITH CONSTANT ANGLE TECHNIQUE (2) AND (2a)

   Fig. 1 is a plot for a single lap joint of the same aluminum
alloy (6061-T6) used for the other samples reported here, with
similar sample dimensions. For a series of X-rayed positions along
the strip the plot shows the counts per second (CPS) with a tensile
load of 10 lbs minus CPS for a load of 400 lbs applied to a bonded
pair of 1/16" (1.588 mm) thick strips; 40 second count times were
used, with 2θ held constant at $\sim 0.4°$ less than the $K\alpha_1$ peak. The
joint had a teflon debond built in. The 10 lb load was used to
assure firm seating and centering of the specimen on the diffractometer. Subtraction of the CPS values for the two loads should
eliminate effects from any pre-existing residual stresses. The
plotted intensities represent the effect of an increase in applied
stress in each single strip of 52.4 MPa (7.6 ksi). Increasing
tensile stress at the surface of the aluminum causes an increase
in the 10 lb minus 400 lb CPS, corresponding to an increase in 2θ
and a decrease in interplanar spacing of reflecting planes parallel
to the surface. The incident beam was collimated to 0.25° and was
1 mm wide at the specimen.

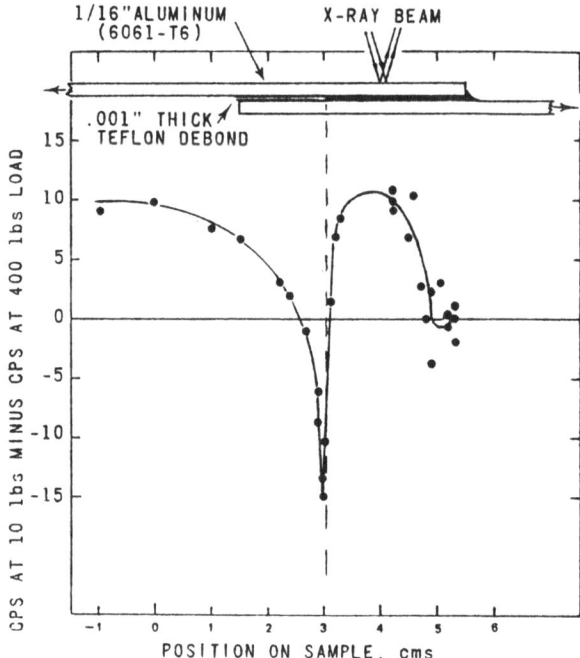

Fig. 1. Diffracted intensities in counts per second (CPS) vs posi-
         tion for a single lap adhesive joint between 1/16" thick
         aluminum strips. Technique (2) with change in load of 390
         lbs (52.4 MPa, 7 ksi). Longitudinal tensile stress in-
         creases upwards on the plot; dips are from sharp (elastic)
         bends located at bond edges.

     The surface stress in the upper adherend of Fig. 1 decreases
steeply as the bond is approached, because of local elastic bending which tends to throw the X-rayed surface into compression.
This bending stress reaches an abrupt maximum at the edge of the
bond as expected,[4-6] and the CPS correspondingly reaches a very
sharp minimum here, showing that the method can locate the edge of
the bond with an accuracy about equal to the width of the X-ray
beam. The bending stresses, which are superimposed on direct
stresses, are largest at the bond edges but also extend with lower
intensity more widely as can be seen with a rubber model under load,
Fig. 2. The prime reason for the bending moments is the eccentricity
in the load path.[4-6] This experiment was repeated with thinner

Fig. 2. Rubber model of an adhesive joint with load applied,
         showing bending of the adherends.

(1/32") strips of 6061-T6, method (2), and similar dips at bond edges were obtained; there were also minor irregularities within the bonded area (possibly from local residual stresses, since only the CPS at a 400 lb load was recorded).

An experiment was performed with a 1/16" thick aluminum adherend joined to a 6-ply unidirectional graphite/epoxy laminate, .032" thick, in which aluminum powder was embedded between the first and second plies nearest the X-rayed surface. Technique (2a) was used and a plot made showing CPS at 10 lbs minus CPS at 400 lbs with a 1 mm wide beam. The plot is not reproduced here because for the aluminum strip the results differ little from those of Figs. 1 and 3. The results from the filled laminate were very irregular, presumably because for this relatively high modulus material a higher stress is needed for precision, but again there was a minimum at the position in the composite opposite the edge of the bond to the underlying strip.

Fig. 3 shows results from constant angle technique (2) on a 1/32" aluminum strip bonded to a 6-ply uniaxial graphite/epoxy laminate. The counts per sec from the aluminum at 0.4° 2θ below the peak 2θ value were recorded at 10 lbs load and those at 400 lbs load were subtracted; thus increasing counts on the plot correspond to increasing 2θ and increasing longitudinal tension at the X-rayed surface. The counts were accumulated in 4 steps of 10 sec each with the specimen rotated about 0.12° between steps, using a 1 mm wide, 1/4" divergent beam.

STRESS MEASUREMENT WITH DIFFRACTING PAINT

We propose another method, which our tests indicate is applicable for measuring applied stresses (though not pre-existing residual stresses), and for mapping stress distributions. A

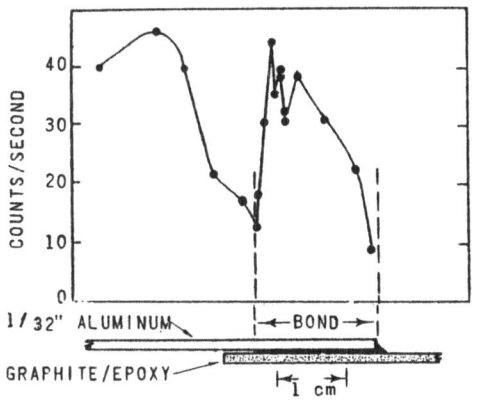

Fig. 3. Diffracted intensities vs position for the aluminum adherend of a single lap adhesive joint between a 1/32" thick aluminum strip and a graphite/epoxy laminate. Counts/sec are for 10 lb load minus those for 400 lb load with technique (2).

tightly adhering paint (e.g. epoxy) containing a diffracting filler
such as aluminum or silver powder can be applied to an object that
diffracts poorly or is amorphous. After curing and suitable calibration experiments, this paint can reveal the applied stresses in
the underlying object by shifts in the diffraction angle.

Fig. 4 shows the results obtained with a 6-ply unidirectional
graphite/epoxy strip thinly coated with an aluminum-filled epoxy
paint, cured,then stressed in tension. Diffraction angles shifted
in proportion to the applied stress up to a stress level that suggests a yield point behavior, as in earlier experiments.[3]

Fig. 5 is a plot of results obtained with a thin epoxy paint
containing silver powder on an aluminum adherend of a single lap
joint. Silver was chosen so that the diffraction peaks from the
underlying aluminum would not interfere (the 333 + 511 Kα_1 peak
has $2\theta_{Ag}$ ~156.7°). Dips close to bond edges were seen with both
technique (2a) and with technique (1), with an applied load of 450
lbs, (128 MPa, 18.5 ksi).

The accuracies in stress analysis attainable with a diffracting
paint are generally somewhat lower than are obtained with the same
filler embedded in a fiber-reinforced composite, and are more subject to stress relaxation. A thin paint with a uniform distribution of filler should be used, one which has an appropriate sensitivity and elastic limit.

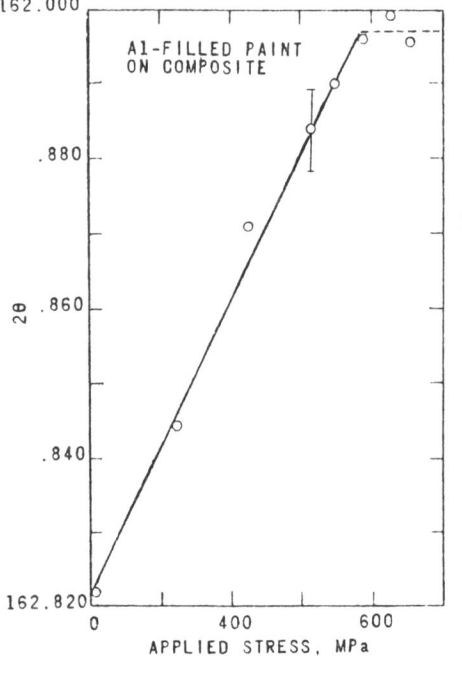

Fig. 4. Applied tensile stress
vs 2θ for aluminum
filled epoxy paint ap-
plied to a graphite/
epoxy composite 0.032"
thick. Technique (1).

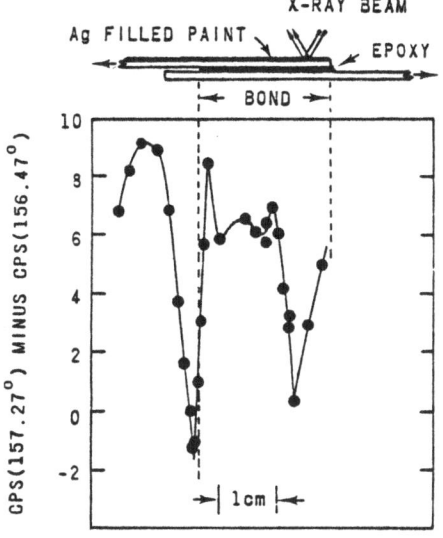

Fig. 5. Diffraction from silver
filler in epoxy paint
applied to 1/32" thick
aluminum strips adhesive-
ly bonded. Technique
(2a), with a load of 450
lbs.

Double-lap joints, tapered-lap, stepped lap and scarf joints
all have far less intense local bending moments than the single-
lap joints studied here,[4-6] so plots of stress distributions in
these cannot be expected to show such marked minima as are seen in
Figs. 1, 3 and 5 (unless unsymmetrically loaded). Experiments
with two double lap joint specimens confirmed this expectation.
These specimens consisted of a pair of 1/32" aluminum (6061-T6)
strips adhesively bonded with epoxy to an inner strip of the aluminum in one case and of 6-ply uniaxial graphite/epoxy in the other.
No well marked minima in diffraction shifts under load were seen
opposite the end of the inner adherend in either specimen, and the
position of the end of the inner adherend was not disclosed with
the accuracy that had been attained in our experiments on single
lap joints. The potentialities of the method for such cases cannot be properly assessed from these two experiments.

ACKNOWLEDGEMENT

This research is supported by the Air Force Office of
Scientific Research under grant #77-3284.

REFERENCES

1. C. S. Barrett and Paul Predecki, Stress Measurement in Polymeric Materials by X-Ray Diffraction, Polymer Eng. & Sci.
16:602 (1976).

2. Paul Predecki and Charles S. Barrett, Stress Measurement in Graphite/Epoxy Composites by X-Ray Diffraction from Fillers, J. Comp. Mat. 13:61-71 (1979).
3. Charles S. Barrett and Paul Predecki, Stress Measurement in Graphite/Epoxy Uniaxial Composites by X-rays, Polymer Composites 1(1) (Sept. 1980), p. 1-6.
4. L. J. Hart-Smith, Adhesive-bonded Joints for Composites-- Phenomenological Considerations, in "Advanced Composites Technology," Technology Conference Associates, P.O. Box 842, El Segundo, Calif. 163-173 (March 14-16, 1978).
5. L. J. Hart-Smith, Analysis and Design of Advanced Composite Bonded Joints, NACA CR-2218 (August, 1974).
6. L. J. Hart-Smith, Advances in the Analysis and Design of Adhesive-Bonded Joints in Composite Aerospace Structures, Presented to Soc. for Advancement of Material and Process, Nat'l Symposium, Anaheim, CA (April 24, 1974).

SEPARATION OF BROAD CRYSTALLINE AND AMORPHOUS X-RAY

DIFFRACTION PEAKS

O. W. Marks*, D. K. Smith#, M. D. Chris*

Hercules Incorporated, Research Center, Wilmington, DE 19899*
Dept. of Geosciences, Pennsylvania State University,
University Park, PA 16802#

INTRODUCTION: Separating overlapped peaks is a part of many x-ray diffraction analyses, for example, polymer crystallinity. Natta [1] defined a method for polypropylene in 1957. His method was computerized at the Hercules Research Center in 1960 with an automatic "curve follower" which punched paper tape for the computer. A later method deviated from Natta's method by approximating the amorphous curve with a fixed shape and a height chosen to best fit the diffraction data from $2\theta = 7.5$ through 10. degrees. Neither of these methods worked on "smectic" polymer samples, i.e., composed of very small crystallites. Also, a different computer program was used for each different polymer, so a general purpose computer program was developed using a peak profile method. This method has been used on polymer mixtures and copolymers of ethylene, propylene, and butene; and on cellulose, modified cellulose, and catalysts. The selection of a profile function is discussed in the next section. In later sections, the background, the fitting procedure, and computer input and output are discussed.

PROFILE FUNCTION: Like in Salazar's and Johnson's methods [2,3], the peak shape is defined as a variable sum of a normal and a Cauchy distribution. The diffraction pattern for amorphous polypropylene becomes remarkably normal in shape when corrected intensity is plotted versus log d-spacing. (see figure 1)

If we could assume that the instrumental abberations and spectral distribution are fit by a Cauchy function of width W, and the sample dispersion is fit by a normal distribution of width w' then, a normal distribution of Cauchy peaks would describe the composite diffracted profile:

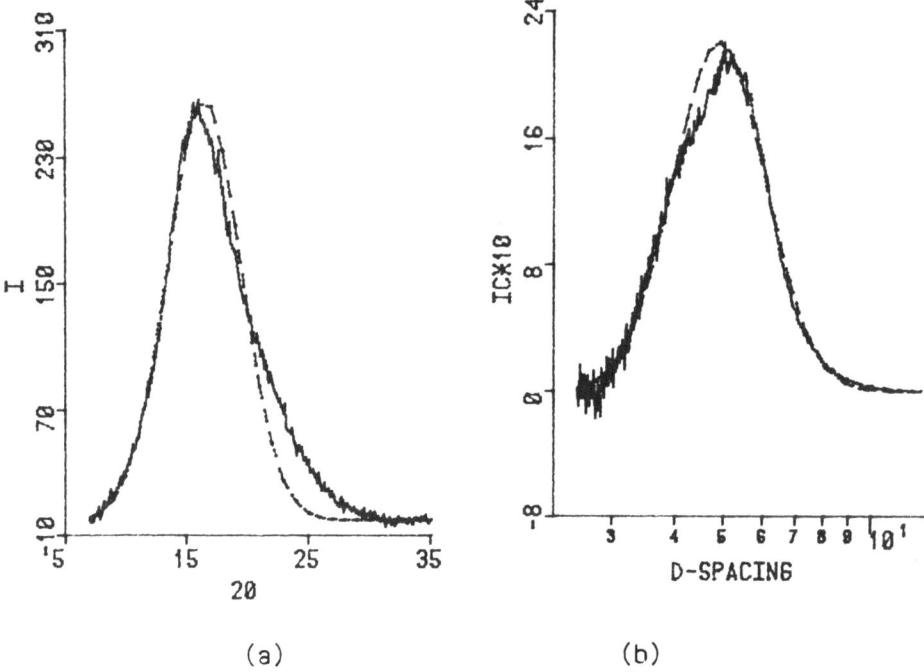

(a)   (b)

Figure 1. Amorphous polypropylene. (a) Intensity versus
diffraction angle. (b) Corrected intensity versus log d-
spacing. The dashed line is a normal curve with same height
and width.

$$I_x = N \int exp[-\ln 2(y/w')^2]/\{1 + [(y-x)/W]^2 \} \, dy \qquad (1)$$

where N is a normalizing factor for the integration, and I_x is
the intensity at $x = \log d/d_0$.

The integration in (1) would lead to an excessive amount of
computation even for the current high speed computers when many
peaks times many points multiplied by many iterations of a fitting
procedure to determine the nonlinear parameters are considered.
However, some numerical experiments have shown that the maximum
deviation between (1) and:

$$I_x = (1 - W/w)exp[-\ln 2(x/w)^2] + (W/w)/[1+(x/w)^2] \qquad (2)$$

is about one percent. The total width w is chosen so that the
width of (1) equals the width of (2). If w = W the function is all
Cauchy and if w >> W (broad peak) the function is mostly normal.

BACKGROUND: The background of the diffraction pattern has some
curvature which is neatly described by a constant plus a reciprocal
diffraction angle term:

$$I = a + b/\theta. \tag{3}$$

A recently purchased instrument equipped with a theta compensating
slit has a background with a positive slope. The procedure now
allows for a choice of base line types between (3) and
$$I = a + b\,\theta$$

Fitting Procedure: The equation used by the fitting procedure is:

$$I_\theta = a + b/\theta \text{ or } b\theta$$

$$+ \frac{1}{C_\theta} \sum_{j=1}^{m} \sum_{i=1}^{n_j} h_j h_{ij} \{(1 - \frac{W}{w_j w_{ij}}) \exp[\ln\frac{1}{2}(\frac{x - x_{ij}}{w_j w_{ij}})^2]$$

$$+ \frac{W}{w_j w_{ij}} \quad \frac{1}{1 + (\frac{x - x_{ij}}{w_j w_{ij}})^2} \} \tag{4}$$

where I_θ = computed intensity, a and b are baseline parameters, θ
= diffraction angle, x = log d-spacing at θ, x_{ij} = log d-spacing
position of i-th peak in j-th fraction, C_θ = intensity correction
at θ, h_j = peak height factor for j-th fraction, h_{ij} = relative
height of i-th peak in j-th fraction, W = width of the Cauchy
dispersion, w_j = peak width factor for j-th fraction, w_{ij} =
relative width of i-th peak in j-th fraction.

The variables x_{ij}, h_{ij} and w_{ij} may be either parameters
to be adjusted by the nonlinear least squares fitting procedure or
they may be constant. At least one h_{ij} and one w_{ij} for each
fraction must be constant, or else the procedure has too many
unknowns. The h_j and w_j are always parameters, but they can be
fixed for the first fraction by setting the upper and lower limits
equal to the desired value. This has been used to set the shape of
the amorphous profile. Another parameter, which is not shown in
equation (4), is an adjustment of the diffraction angle for zero
location. If there is no standard peak in the sample pattern, the
upper and lower limits on this parameter are set to some
insignificant value to stop the adjustment.

An algorithm based on the work of Marquardt [4] is used to determine all the parameters in equation (4) simultaneously. This algorithm provides for upper and lower limits on the parameters. If, in any iteration, a parameter estimate moves outside its limit, it is set at the limit for the next iteration. This is vital for the peak positions which tend to vary too much in the early steps of the procedure. The fitting is based on minimizing the sum of the squares of the differences between calculated and measured uncorrected intensity of the diffraction pattern.

Marquardt's algorithm is very robust. It starts a calculation using the steepest descent method guaranteeing convergence. As the calculation proceeds, the method changes progressively to the Newton-Raphson method which has quadratic convergence. If the error exceeds that of the previous iteration then it backs off toward the steepest descent method for the next iteration, thus maintaining the convergence guarantee. The algorithm uses partial derivatives to compute new estimates of the parameters and converges rapidly.

COMPUTER INPUT AND OUTPUT: The FORTRAN program uses "NAMELIST" input to set the fitting controls, parameter limits, raw-data input medium, and output options. This means "default" variables to be changed are selected by specifying their names.

One input record of d-spacing or diffraction angle position, relative height, relative width, fraction assignment, and description is given for each peak. A minus sign on the position, height, or width indicates that value is an initial estimate for the fitting procedure, except for the first peak of each fraction. The height and width values for the first peak in a fraction are the initial estimate of the fraction parameters h_j and w_j. The values for h_{1j} and w_{1j} are set to one. If the height h_j or h_{ij} is zero, the program generates the initial estimate. It also generates the initial estimate of the baseline parameters.

Printed output consists of peak record input, fraction area percents, parameters resulting from the fitting procedure, and a table of final values for each of the peaks, i.e., the position, height = $h_j h_{ij}$, width = $w_j w_{ij}$, and individual peak area percents. An example of plot output is shown in figure 2.

REFERENCES

[1] G. Natta, P. Corradini, M. Cesari, "Quantitative Analysis of Crystallinity of Polypropylene", Rend. Classe sci. fix. nat. e nat., 22, 11-17 (1957).

[2] J. M. Salazar, J. C. G. Ortega, and F. J. B. Calleja, "On the Separation of Crystalline and Diffuse X-ray Scattering in Semicrystalline Polymers", Anales de Fisca, 73, 244-247 (1977)

[3] D. J. Johnson, Advances in X-ray Analyses, 24 (1981).

[4] D. W. Marquardt, "An Algorithm for Least-Squares Estimation of Nonlinear Parameters", J SIAM 11, 431 (1963).

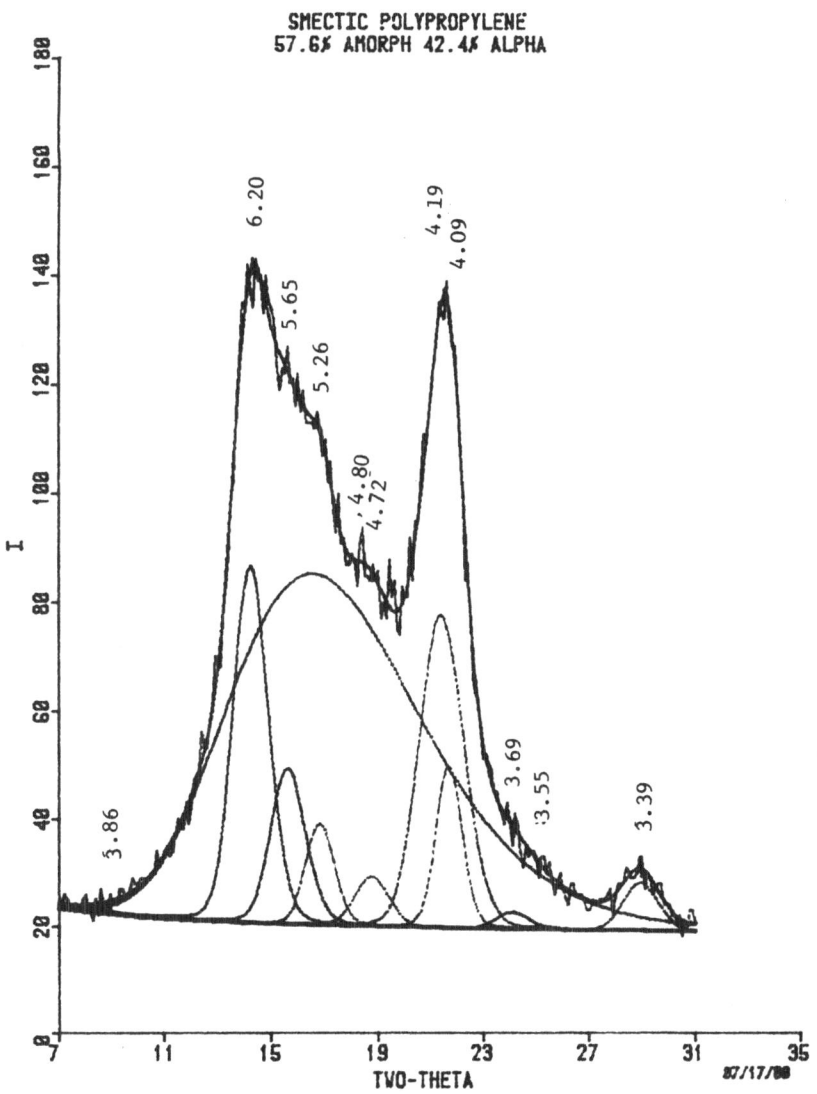

Figure 2.   Computer plot showing individual peaks.

ENERGY-DISPERSIVE DIFFRACTION ANALYSIS OF THE STRUCTURE OF

METALLIC GLASSES

C.N.J. Wagner, D. Lee, S. Tai and L. Keller

Materials Science and Engineering Department
University of California
Los Angeles, California 90024

ABSTRACT

The intensities of x-rays scattered by amorphous $Fe_{80}P_{13}C_7$ and $Fe_{40}Ni_{40}P_{14}B_6$ samples have been measured as a function of photon energies E at fixed scattering angles $2\theta_i$ using a Li-drifted Si detector and polychromatic x-rays generated by a 50KV full-wave rectified generator. The coherently scattered intensity per atom was calculated for free-standing samples as well as samples contained in a Be or pyrolytic graphite cell, after the evaluation of the energy dependence of the primary beam spectrum by an iterative process. The interference functions were then calculated from the data obtained in transmission and reflection, and compared with those measured with the conventional variable 2θ technique. Good agreement between energy-dispersive diffraction (also called variable wavelength technique) and variable 2θ diffraction was observed in all cases.

INTRODUCTION

The goal of any scattering experiment on amorphous or liquid samples is to determine the scattered intensity as a function of the length of the diffraction vector $\vec{K} \equiv \vec{k} \equiv \vec{Q} \equiv \vec{q} \equiv \vec{s}$, i.e.,

$$K = (4\pi/\lambda)\sin\theta = (4\pi/hc)(\sin\theta)E \tag{1}$$

where λ is the wavelength, 2θ is the diffraction angle, E is the energy of the x-rays, h is Planck's constant, and c is the velocity of light. In the past, it has been customary to vary K by scanning through 2θ using a monochromatic radiation. This conventional technique may now be called the variable 2θ method.

245

With the development of modern solid state detectors (Li-drifted Si or Ge, and hyperpure Ge) with an energy resolution of ~200eV for 8KeV photons, it became possible to vary K by measuring the energy E of polychromatic x-rays scattered at a fixed angle $2\theta_i$. This technique, called energy-dispersive x-ray diffraction technique or variable λ method, has been introduced by Giessen and Gordon (1968) and later applied by Prober and Schultz (1975) to determine the structure of liquid Hg. In the case of the diffuse diffraction pattern, characteristic of amorphous solids or liquids, the limited energy resolution of the solid state detector is still sufficient to determine accurately the scattered intensity as a function of K. This has been clearly demonstrated by Egami (1978, 1980) who developed a rigorous data reduction procedure and applied it successfully to the evaluation of the structure and the structural relaxation of metal-metalloid glasses.

In this paper, we will apply the energy-dispersive x-ray diffraction method (also called variable λ method) to the analysis of the structure of metal-metalloid glasses. The results will be compared with the data obtained with the conventional variable 2θ method using monochromatic Ag or Mo radiation. Both the variable λ and 2θ methods have been employed with the reflection and transmission geometries. In addition, a simulation experiment was carried out where the transmission sample was held in a flat container in order to evaluate the data reduction procedures when a container scattering has to be measured in a separate experiment.

EXPERIMENTAL PROCEDURES

A Li-drifted Si detector has been mounted on a conventional, horizontal diffractometer (Siemens), and the scattered x-rays, produced by a high intensity W or Cr target x-ray tube with full-wave rectified generator at 50KV, were measured at nine different scattering angles $2\theta° = 5.66, 11.32, 17.02, 22.76, 34.44, 46.49, 59.12, 72.60$, and $87.37°$. These angles were chosen so that the quantity $(4\pi/hc)\sin\theta = 1.0135 \sin\theta$, when E is expressed in KeV, represents a simple number varying from 0.05 to 0.7 for the different 2θ angles chosen, which also ensured a sufficient range and overlap in K because of the limited range of available photon energies (from 16 to 35KeV). The lower limit was chosen to avoid the W-L lines, and the double pulses from the Cr-K-lines and the fluorescence radiation of the sample (Ni and Fe). The raw data of the intensity, scattered at fixed 2θ angles, are shown in Fig. 1. The same set-up was also used in the conventional variable 2θ technique using an Ag or Mo tube in conjunction with a narrow energy window of the single channel analyzer, set about the energy of the $K\alpha$ line (Wagner, 1978). Measurements were made on free-standing samples of the metallic glasses $Fe_{80}P_{13}C_7$ and $Fe_{40}Ni_{40}P_{14}B_6$, as well as on samples enclosed in a Be or pyrolytic graphite cell. Both the reflection and transmis-

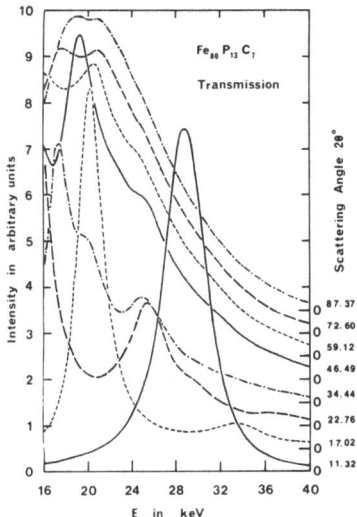

Fig. 1.   Scattered intensity
(in arbitrary units)
of an amorphous Fe_{80}
$P_{13}C_7$ alloy as a function of the photon en-
ergy E.

Fig. 2.   Ratio of the measured
and calculated intensity
[eqs. (5), (8) and (9)].
Also shown is the spec-
tral intensity of the
primary beam.

sion geometries (Wagner, 1978) were employed to measure the x-ray
intensities, scattered by the sample and cell, and the cell, respectively.

The intensity $I_M^{SC}(E,\theta)$, scattered by the sample S and cell C,
as a function of E at different scattering angles $2\theta_i$ is given by:

$$I_M^{SC}(E,\theta) = I_M^S(E,\theta) + I_M^C(E,\theta) \tag{2}$$

where $I_M^S(E,\theta)$ and $I_M^C(E,\theta)$ can be expressed in the following way
(Egami, 1978)

$$I_M^x(E,\theta) = N_x \{C(E)I_p(E)P(E,\theta)A_{x,xy}(E,\theta)[I_a(K) + I_{MS}(K)]^x$$
$$+ C(E')I_p(E)P(E,E',\theta)A_{x,xy}(E,E',\theta)[I_{inc}(K')]^x\} \tag{3}$$

N_x is the number of atoms in the irradiated volume of medium x, $C(E)$
is the detection efficiency, $I_p(E)$ is the spectral distribution of
the primary beam, $P(E,E',\theta)$ is the polarization factor (Egami, 1980),
$A_{x,xy}(E,E',\theta)$ is the absorption correction of the radiation scattered in medium x and absorbed in medium x and y (Wagner, 1978),
$I_a(K)$ is the coherent scattering per atom, $I_{MS}(K)$ is the multiple
scattering, and $I_{inc}(K')$ is the incoherent scattering per atom. The
value of K is given by eq. (1) whereas $K' = (1.0135 \sin\theta)E'$, E'

being the initial energy (in KeV) of the incoming x-ray photon which is reduced to E after the Compton scattering, i.e., $E' = E + \Delta E = E/(1 - 0.00392 \sin^2\theta \; E)$.

What we are interested in is the intensity $I_M^S(E,\theta)$ scattered by the sample alone, which can be readily evaluated with the following relation:

$$I_M^S(E,\theta) = I_M^{SC}(E,\theta) - I_M^C(E,\theta)A_{C,SC}(E,\theta)/A_{C,C}(E,\theta) \qquad (4)$$

where $I_M^S(E,\theta)$ is given by eq. (3). All factors in eq. (3) can be calculated or taken from tables except $I_p(E)$.

The task remains to establish the values of $I_p(E)$ as a function of the photon energy E. This can be most readily done by using directly the x-ray scattering of the amorphous sample itself. Since $I_a(K)$ will tend towards $<f^2>$ for large values of K, say $K > 15\text{Å}^{-1}$ we can set $I_a(K) = <f^2>$ in eq. (3) and solve for $I_p(E)$. However, it is more convenient to combine the three factors N_S, $C(E)$, and $I_p(E)$, i.e., $I_o(E) = N_S C(E) I_p(E)$. Thus eq. (3) reduces to the following expression for large values of K:

$$I_M^S(E,\theta) = I_o(E)\{P(E,\theta)A_{S,SC}(E,\theta)<[f(K)]^2>$$

$$+ [I_o(E')/I_o(E)]P(E,E',\theta)A_{S,SC}(E,E',\theta)I_{inc}(K')\} \qquad (5)$$

neglecting the multiple scattering. For large values of E and 2θ, the Compton shift becomes appreciable. Nevertheless, in the first iteration step we will assume that $I_o(E) = I_o(E')$, which is a reasonable approximation at lower energies E and scattering angles 2θ. We can calculate the function $I_o^1(E)$ as follows:

$$I_o^1(E) = I_M^S(E,\theta)/[PA_{S,SC}<f^2> + P'A'_{S,SC}I'_{inc}] \qquad (6)$$

where $P' = P(E,E',\theta)$ and $A'_{S,SC} = A'_{S,SC}(E,E',\theta)$. Then we evaluate the quantity $[I_M^S(E,\theta)]_c^1$ by introducing the energy shift of the Compton scattering into $I_o^1(E)$ of eq. (5). The ratio of $I_M^S(E,\theta)/[I_M^S(E,\theta)]_c^1$ is shown in Fig. 2. It is obvious that we underestimated $I_o(E)$ when using the assumption that $I_o(E) = I_o(E')$.

In the second iteration we make the approximation $I_o(E')/I_o(E) = I_o^1(E')/I_o^1(E)$. It can be readily shown that with this assumption

$$I_M^S(E,\theta)/[I_M^S(E,\theta)]_c^1 = I_o(E)/I_o^1(E). \qquad (7)$$

We then calculate the function $I_o^2(E)$ with the relation

$$I_o^2(E) = I_o^1(E)I_M^S(E,\theta)/[I_M^S(E,\theta)]_c^1. \qquad (8)$$

We continue the process with the assumption that $I_o(E')/I_o(E) = I_o^{n-1}(E')/I_o^{n-1}(E)$ where n is the nth iteration. Consequently, $I_o(E)$ can be found by applying the recursive formula.

$$I_o(E) \equiv I_o^{n}(E) = I_M^S(E,\theta) \prod_{j=1}^{n-1} I_o^j (E)/[I_M^S(E,\theta)]_c^j . \tag{9}$$

Usually four to five iterations are sufficient to ensure that the ratio $I_M^S(E,\theta)/[I_M^S(E,\theta)]_c$ modulates uniformly about one.

With the primary beam spectrum thus established, it is now possible to evaluate the intensity per atom $I_a(K)$ by matching the individual sections of $I_a(K)$, determined from the different $2\theta_i$ angular settings, to each consecutive one, starting with the data of the 2θ run used to evaluate the primary beam spectrum $I_o(E)$. The interference function $I(K)$, defined as

$$I(K) = \{I_a(K) - [<f^2> - <f>^2]\}/<f>^2 \tag{10}$$

can be calculated from $I_a(K)$, and is shown in Fig. 3 for the range $3.3 < K < 20\overset{\circ}{A}^{-1}$. It is clearly seen that the match between segments of the different 2θ settings is satisfactory, and that modulations about one are still visible at $20\overset{\circ}{A}^{-1}$.

RESULTS AND DISCUSSION

The variable λ or energy-dispersive x-ray diffraction method has been employed to determine the interference functions $I(K)$

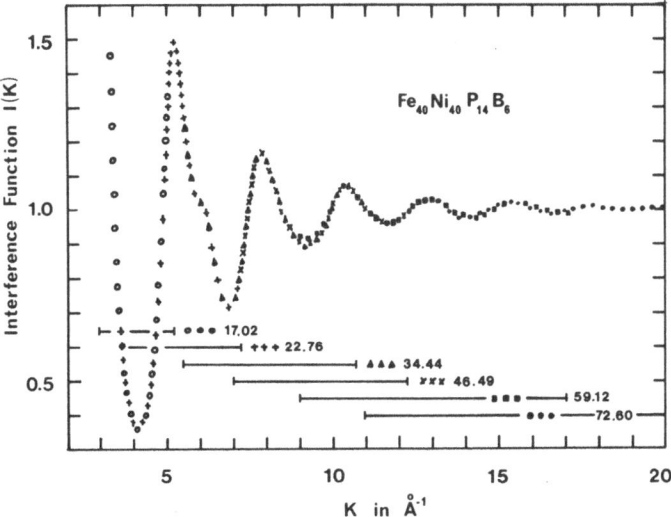

Fig. 3. Matching of the different segments of the interference function [eq. (10)], determined from the scattered intensities at different 2θ angles.

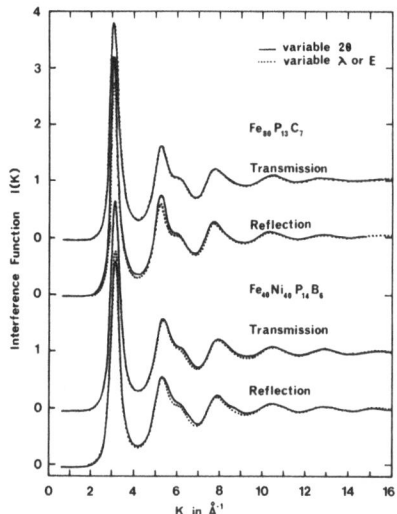

Fig. 4. Interference functions
I(K) of amorphous Fe_{80}
$P_{13}C_7$ and $Fe_{40}Ni_{40}P_{14}B_6$
alloys.

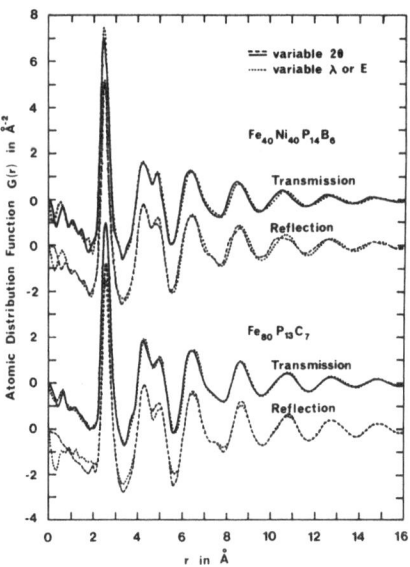

Fig. 5. Fourier transforms G(r)
[eq. (11)] of the inter-
ference function given
in Fig. 4.

[eq. (10)] of a $Fe_{80}P_{13}C_7$ metallic glass, prepared by the piston and
anvil technique, and an amorphous $Fe_{40}Ni_{40}P_{14}B_6$ alloy, prepared by
melt-spinning (Metglas 2628), which are given in Fig. 4. The same
functions have also been determined with the conventional variable
2θ technique using $AgK\alpha$ radiation in conjunction with the Si-solid
state detector except for the reflection run of the $Fe_{80}P_{13}C_7$ sample where $MoK\alpha$ radiation was employed. It is evident that there is
an excellent agreement between the two methods, the variable λ and
the variable 2θ techniques, and the two diffraction geometries, the
transmission and the reflection techniques. Only the run taken with
$MoK\alpha$ radiation produced an I(K) which is somewhat higher in the first
and second peaks. However, when the same sample was also measured
with $AgK\alpha$ radiation, as all the others were, excellent agreement was
again observed between the variable λ and 2θ methods. Our data on
$Fe_{40}Ni_{40}P_{14}B_6$ are also in excellent agreement with those of Egami
(1978).

The quality of the interference functions I(K) can be judged
from the magnitude of the ripples in G(r), shown in Fig. 5, the
Fourier sine transform of K[I(K)-1] at small values of r. Since
G(r) is defined as:

$$G(r) = 4\pi r[\rho(r) - \rho_o] = (2/\pi) \int K[I(K) - 1](\sin Kr)dK \qquad (11)$$

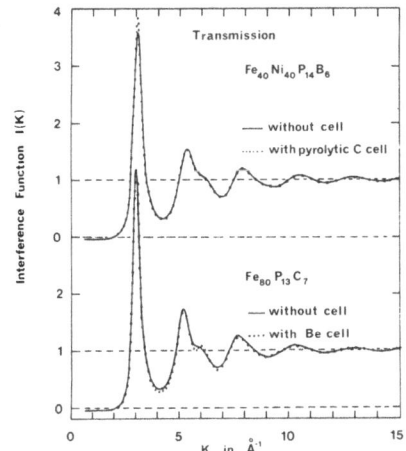

Fig. 6.   Comparison between the
          interference functions
          I(K), obtained from
          free-standing samples
          and samples enclosed
          in a cell.

where $\rho(r)$ is the weighted atomic
distribution function of the alloy
(Wagner, 1978) and ρ_0 is the average atomic density, it is readily
seen that $G(r)$ tends towards $-4\pi r \rho_0$
when r goes to zero. Thus, $G(r)$
should be a straight line at $r < r_0$,
the value of the hard sphere diameter of the smallest atom in the
alloy. The size of the modulations
at low r in Fig. 5 is reasonably
small, and this has been accomplished by applying only the data
reduction as outlined above. No
variable background function has
been introduced (D'Antonio et al,
1977) which can be chosen by minimizing the ripples in the process
of repeated Fourier transformations
and inversion of slightly altered
$K[I(K) - 1]$ and the corresponding
$G(r)$, respectively.

        When the amorphous samples were enclosed in a cell, i.e., two
50 μm thick Be sheets or two 125 μm pyrolytic graphite sheets, the
data reduction became somewhat more involved (Lee, 1980) but good
agreement could be found between the data taken with and without
cell as shown in Fig. 6.

        It should be emphasized that all the data reported here have
been obtained with a conventional 50KV full-wave rectified generator and a Li-drifted Si detector, which impede with an efficient
generation and detection, respectively, of photons with energies
above 30KeV. Therefore, a new experimental set-up has been developed
for the variable λ method consisting of a 100KV constant potential x-
ray generator with a W-target x-ray tube inclined at 45° to the diffraction plane, a primary beam collimator 1 m in length, and a hyperpure Ge detector. The specimen (up to 25 x 25 x 5 mm) can be mounted
on a newly designed high temperature, high vacuum enclosure capable
of reaching 1500°C (Wagner and Mardesich, 1981). This experimental
set-up permits us to use photons with energies between 20 and 60 KeV
to determine the interference functions of liquid and amorphous samples. It is anticipated that the new variable λ experimental set-up
will greatly reduce the counting time for a given counting statistics
and also somewhat simplify the data reduction procedure because of
a more uniform primary beam spectrum $I_p(E)$.

ACKNOWLEDGEMENT

    The research leading to this paper has been sponsored by the grant DMR 78-09929 from the National Science Foundation.

REFERENCES

D'Antonio, P., Moore, P., Konnert, J.H., and Karle, J., 1977, Electron Diffraction of Amorphous Materials, Trans. Amer. Cryst. Assoc. 13:43-66.

Egami, T., 1978, Structural Relaxation in Amorphous $Fe_{40}Ni_{40}P_{14}B_6$ Studied by Energy-Dispersive X-ray Diffraction, J. Materials Science, 13:2587-2599.

Egami, T., 1980, Structural Study by Energy Dispersive X-ray Diffraction, in: "Metallic Glasses," eds. H.J. Guntherodt and H. Beck, Springer Verlag, Heidelberg, New York (in press).

Giessen, B.C., and Gorden, G.E., 1968, X-ray Diffraction: New High-Speed Technique Based on X-ray Spectrography, Science 159, 973-975.

Lee, Dokyol, 1980, The Structural Study of Metallic Glasses Using Variable Two Theta and Variable Energy X-ray Diffraction Techniques, Ph.D. Thesis, University of California, Los Angeles.

Prober, J.M., and Schultz, J.M., 1975, Liquid Structure Analysis by Energy-Scanning X-ray Diffraction: Mercury, J. Appl. Cryst. 8:405-414.

Wagner, C.N.J., 1978, Direct Methods for the Determination of Atomic Scale Structure of Amorphous Solids (X-ray, Electron, and Neutron Scattering), J. Non-Cryst. Solids, 31:1-40.

Wagner, C.N.J., 1980, Diffraction Analysis of Metallic, Semiconducting and Inorganic Glasses, J. Non-Cryst. Solids (in press).

Wagner, C.N.J. and Mardesich, N., 1981, A Novel Design of a High-Temperature Furnace for X-ray or Neutron Diffraction Studies (to be published).

INTERNAL STANDARDS FOR QUANTITATIVE X-RAY PHASE ANALYSIS:

CRYSTALLINITY AND SOLID SOLUTION

G.J. McCarthy and R.C. Gehringer, Department of Chemistry, North Dakota State University, Fargo, ND 58105, D.K. Smith,[a] V.M. Injaian,[b] D.E. Pfoertsch,[c] and R.L. Kabel,[b] Departments of Geosciences,[2] Chemical Engineering,[b] and Materials Research Laboratory,[c] The Pennsylvania State University, University Park, PA 16802.

INTRODUCTION

Quantitative phase analysis by X-ray diffraction (QTXRD) has been an established tool of analytical chemistry for more than four decades. Despite its age, this tool remains ascendant as the only universally applicable method for determining the manner in which elements are combined into crystalline phases in multiphase solids. QTXRD is entering its second renaissance. The first came with the introduction of the counter diffractometer in the late 1940's. The specimen preparation and data collection processes were exacting and tedious, but reasonably accurate analyses could be obtained. The second came with the introduction of computer controlled diffractometers, whose software packages include QTXRD routines, in the late 1970's. With the tedium of data collection and analysis greatly reduced, we can expect even more widespread adoption of this tool in the general analytical laboratory. The accuracy of the analyses will benefit from the improved precision made possible with the automated instruments, but will still be heavily dependent on specimen preparation and a factor of which the analytical chemist may not be aware, the choice of standards for the phases of interest. This factor, the appropriateness of analyte standards, is the subject of this paper.

There are five classes of QTXRD methods:

1. Direct (or external standard) method in which the intensity of selected reflections of the analyte in a multiphase specimen is compared to the intensity of the same reflec-

tions in the pure analyte phase. After factoring in the
absorption of the analyte (known) and the matrix (calculated
from elemental analyses or measured) the amount of analyte
is obtained (1,2).

2. Internal standard methods in which a fixed amount of a
    reference phase is added to the multiphase specimen and
    either with calibration curves (1,2) or a reference inten-
    sity ratio (2-4) the amount of analyte is obtained.

3. Spiking (or doping) method in which known amounts of the
    analyte are added to the specimen (1,2,5,6).

4. Dilution method in which the amount of analyte is obtained
    from plots of analyte intensity after known additions of a
    (preferably amorphous)diluent (7).

5. The "no standards" method of Zevin (8) in which the number
    of phases is known and an equal number of samples having
    very different ratios of these and only these phases is
    available.

A common feature of the first three methods is the need for standard
versions of the analyte. While it may seem to be a simple matter
of acquiring a laboratory chemical, commercially available mineral,
ceramic, metal or alloy etc., it will be shown here that the selection
of the analyte standard can strongly influence the accuracy of the
result.

This is not a new consideration. Chung (4) cautions that the
reference intensity internal standard technique requires that "the
perfection or imperfection in the crystal structure of the component
sought is of the same order" as that of the analyte standard. In
his review of QTXRD methods for clays and soils, Brindley (9)
emphasizes the importance of matching composition of the analyte
standard to that of the analyte. Yet one does not find this consideration featured in the standard reference on this subject by
Klug and Alexander (1). And, nowhere is there a discussion of just
how much error can be introduced by the selection of an inappropriate
analyte standard.

This paper describes four sets of experiments, performed with
established X-ray laboratory procedures, that provide insight into
the magnitude of these errors. In each case, an intensity ratio
will be measured. This is the ratio of the integrated or peak
intensity of the analyte standard to that of an admixed reference
phase such as α-Al_2O_3, NiO or ZnO. The procedure allows one to make
intensity comparisons among solids without concern for instrumental
drift. Such an intensity ratio is utilized in the internal standard

QTXRD methods. The question to be addressed is to what extent do
different versions of a particular phase (having, for example,
different origins, thermal history, impurities) give variable X-ray
intensity. Because in each of the three methods cited earlier the
amount of a phase is proportional to its intensity, such variations
could result in proportional errors depending on which version of
the phase is chosen for the analyte reference.

I. REFERENCE INTENSITY RATIO OF Co_3O_4 PREPARED BY DECOMPOSITION OF COBALT SALTS*

A supported Co_3O_4 catalyst is prepared by soaking the support
with a solution of a Co salt, drying and then calcining the impreg-
nated support in air. During a study of the CO oxidation activity
of unsupported Co_3O_4 prepared from different salts over a range of
calcining temperatures and times, it was noted that, in spite of
efforts to use reproducible specimen preparation and experimental
conditions, the intensity of diffraction appeared to vary from sample to sample. This was tested quantitatively by comparing the intensity of the most intense reflection from the Co_3O_4 samples to
that from an internal intensity standard.

Experimental

The reference intensity standard used by the U.S. National
Bureau of Standards (NBS), α-Al_2O_3 (Linde A corundum), was chosen as
the internal standard. The (311) reflection of Co_3O_4 was compared
to the (113) of corundum in a 50:50 weight mixture.

The Co_3O_4 was prepared by decomposing Co nitrate and carbonate
in air from 200 to 900°C (Co_3O_4 reduces to CoO near 910°C in air)
for 6 and 24 hours (h). The products were ground in an agate mortar
and pestle and passed through a 400 mesh sieve. The 50:50 mixture
of Co_3O_4 and corundum was homogenized by grinding and dusted onto
a sticky glass slide. This type of specimen preparation was selected
to minimize preferred orientation.

Intensity data were acquired on a Seimens diffractometer
equipped with a graphite monochromator using CuKα radiation. Two
specimens of each sample were prepared and each was scanned three
times after small adjustments of the specimen position in the X-ray
beam. Integrated intensities were measured with a planimeter from
strip charts. The mean of three measurements was used. Altogether,
six values of reference intensity ratio (I/Ic) were obtained for

---

*Abstracted from the M.S. thesis in Chemical Engineering of
V.M. Injaian, The Pennsylvania State University (1979).

each Co_3O_4 preparation and were plotted against calcining temperature. The typical spread in I/Ic values was 10% of a given value.

Particle size data were collected with a scanning electron microscope procedure developed by Johnson, et al (10). The theoretical I/Ic was calculated by the method of Hubbard, et al (11).

Results and Discussion

The salts decomposed to give phase pure Co_3O_4 when calcined above 200°C. After grinding and sieving, the mean particle sizes of the Co_3O_4 ranged from 2-6 μm with a few percent at most of particles above 10 μm in diameter. The corundum had a narrow distribution of sizes around a mean of 0.6 μm.

Figs. 1 and 2 show the variation of I/Ic with calcining temperature and Table 1 gives the values read from the curves in the figures for 300, 600 and 800°C. In both decompositions, the I/Ic values generally increased with increasing calcination temperature until a maximum value was reached. As expected, increasing calcination temperatures increased crystallinity until the sample was fully crystalline. However, as shown in Table 1, there are significant differences in I/Ic values between the nitrate and carbonate decompositions at higher temperatures.

The calculated I/Ic value is 4.39. In the nitrate preparation, I/Ic's fall below this value while in the carbonate preparation I/Ic's match the calculated value (within experimental error) over the range 500-750°C, but exceed it slightly near 800°C.

As a check on the effect of calcining time on I/Ic, the 758°C firing in the nitrate preparation was continued to 24 h. The I/Ic value was within the range of experimental error of the 6 h calcination. Thus, after 6 h the Co_3O_4 sample had approached the same level of crystallinity achievable for that temperature after much longer calcinations.

We suggest that the substantial differences in I/Ic between the two preparations are due to variable oxygen stoichiometry, Co^{2+}/Co^{3+} ratio and accompanying defects in this spinel structure oxide. During the decomposition stage of the calcination, the nitrate produces an oxidizing atmosphere and the carbonate a reducing atmosphere. The important point is that choice of one precursor salt over another could give a 20% difference in reference intensity ratio that would be reflected in the amount of Co_3O_4 found in a QTXRD phase analysis. It is also evident that a Co_3O_4 standard prepared from the carbonate at 300°C would not be representative of a Co_3O_4 specimen that has been treated at 800°C. The latter has an I/Ic 50% greater than the former.

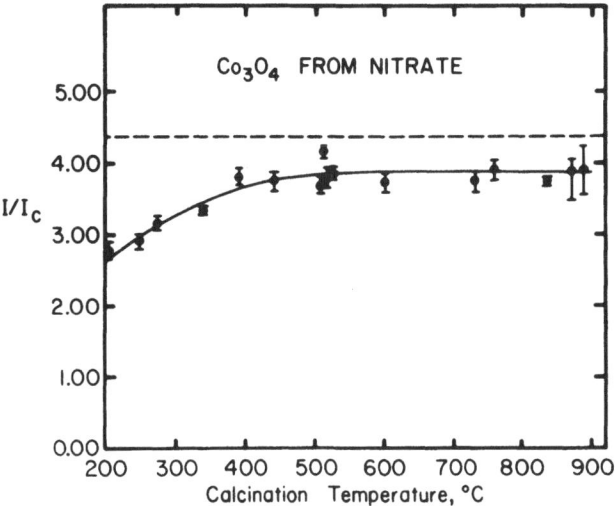

Fig. 1.   Reference intensity ratio for Co_3O_4 synthesized by cobalt
          nitrate decomposition in air. Dashed line is the
          calculated ratio.

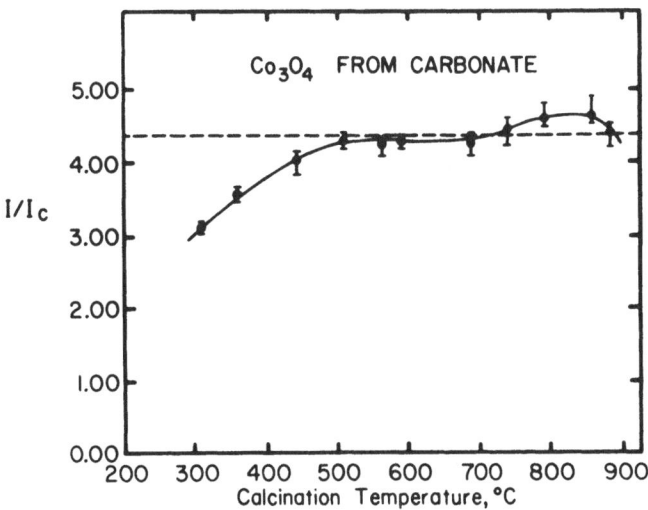

Fig. 2.   Reference intensity ratio for Co_3O_4 synthesized by cobalt
          carbonate decomposition in air.

Table 1

I/Ic Values for Co_3O_4 Preparations

Firing	$I/I_{corundum}$		% Change
Temperature	Co_3O_4 From Nitrate	Co_3O_4 From Carbonate	From Nitrate
300°C	3.25	3.05	− 6%
600°C	3.85	4.30	+ 12%
800°C	3.85	4.60	+ 20%
% Change from 600 to 800°C	None	+ 6%	

II. REFERENCE INTENSITY RATIO OF Y_2O_3 PREPARED BY DECOMPOSITION OF YTTRIUM NITRATE

In order to further explore the relationship between calcination temperature and I/Ic, we repeated the above experiment with
a salt, $Y(NO_3)_3$, that would decompose to give an oxide, Y_2O_3, with
a unique cation valence, and thus a fixed stoichiometry, over the
chosen temperature range. The decomposition was performed from
500-1400°C for 6 h. After grinding and sieving, particle sizes
for the 1215°C preparation had a mean of 0.8 μm and a standard
deviation of 0.3 μm. The (222) reflection of Y_2O_3 was referenced
to the (113) of corundum.

I/Ic data are plotted in Fig. 3. The calculated I/Ic for
Y_2O_3, 9.38, is also shown. The 1123°C firing was continued for
an additional 18 h and the I/Ic value was virtually identical to
that of the 6 h firing. As with the Co_3O_4 preparations, I/Ic
values rise with increasing temperature and level off at a value
that should be indicative of full crystallinity. The curve drawn
through the mean I/Ic values falls somewhat above the calculated
value, but is generally within the scatter in the measurements.

All products gave pure Y_2O_3 diffractograms with no evidence
of a diffuse scattering maximum from a noncrystalline Y_2O_3 phase.
It is the range of I/Ic values that is of interest. If the
crystallinity of Y_2O_3 in an unknown were typical of Y_2O_3 prepared
at 500°C and the specimen prepared at 1200°C were used as a standard,
the analysis would be in error by about 60% of the actual amount
present.

III. EFFECT OF HEAT TREATMENT ON CRYSTALLINITY OF NaCl, CaF_2, AND ZnO

In order to further explore the effects of heat treatment on

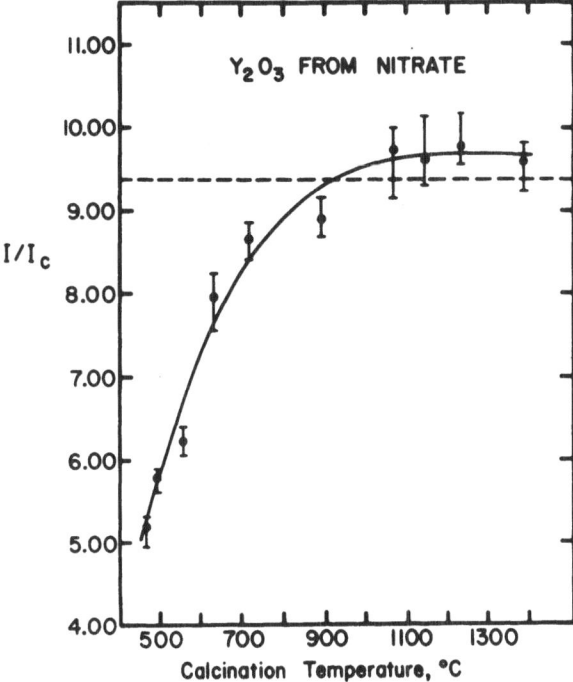

Fig. 3. Reference intensity ratio for Y_2O_3 synthesized by $Y(NO_3)_3$ decomposition in air. Dashed line is the calculated ratio.

crystallinity, as reflected in the I/Ic ratios, we studied these three air stable, stoichiometric salts.

Experimental

The salts were heat treated as described in Table 2. The reagent from the bottle and the heat treated sample were handled identically. After grinding and sieving, the mixtures of salt plus corundum were loaded into aluminum mounts using the side-drifted procedure recommended by the NBS (12) for minimizing preferred orientation. Both peak and integrated intensities were obtained with the scaler-timer of a late model Philips diffractometer using Cu radiation. For the (113) of corundum typically 9,000 counts peak and 3,000 counts integrated were collected. Full width at half maximum (FWHM) and resolution of the α_1/α_2 doublet between 60 and 80° 2θ were also examined as additional measures of changes in crystallinity with heat treatment.

Results and Discussion

A diffractogram taken at high count sensitivity indicated that each of the salts was phase pure before and after the heat treat-

Table 2

I/Ic Before and After Heat Treatment

Phase	Reflection Used	Specimen Treatment	FWHM (degrees)	I/Ic[a] Integrated	Peak
NaCl	(220)	Bottle	0.173	3.10	2.79
		melted at 840°C 5 Hrs.	0.195	3.04 (-2)[b]	2.43 (-13%)
CaF$_2$	(220)	Bottle	0.187	4.74	3.15
		960°C 18 Hrs.	0.171	4.20 (-11%)	4.38 (+39%)
ZnO	(102)	Bottle	0.179	1.46	1.22
		960°C 18 Hrs.	0.187	1.38 (-5%)	1.12 (-8%)

[a](113) reflection of corundum
[b]Percent change from original intensity ratio

ment.   Data are presented in Table 2.   Each value of I/Ic represents
3-6 repetitions of the measurement.   In all cases, the standard
deviation of the measurements was within 1% of the mean.

Melting NaCl caused both a drop in peak and integrated intensity and a broadening of the peak.   The fired CaF$_2$ seemed to be more
crystalline than the fresh reagent based on its FWHM and peak I/Ic,
but it did show a small but significant drop in its integrated I/Ic.
The latter may be due to microabsorption effects from increases in
particle sizes (not measured in this study) rather than from loss
in crystallinity.   The CaF$_2$ case was the only one where a marked
improvement in α_1/α_2 resolution was observed after firing.   After
firing the ZnO, small reductions in both its peak and integrated
I/Ic and an increase in FWHM were recorded.

The use of peak intensities in QTXRD often looks attractive
to the analyst from the standpoint of counting time and in cases
of partial overlap with the best analytical reflections.   However,
the data in Table 2 confirm the well known fact that peak intensities
are more sensitive to crystallinity than are integrated intensities.
Only in the case of CaF$_2$ could a QTXRD error outside of the usually
accepted range arise from inappropriate choice of reference if
integrated intensities are used.

IV.   SOLID SOLUTION EFFECTS ON REFERENCE INTENSITY RATIO

Two systems with extensive solid solution ranges were chosen
for study, one simple and one with a complex defect mechanism.   The
corundum structure oxides Cr$_2$O$_3$ and Fe$_2$O$_3$ are completely miscible

and have a simple substitutional mechanism. In the second, CeO_2-UO_{2+x}, approximately 50 mole % of U oxide can be dissolved in CeO_2 while still retaining the cubic fluorite structure (13), but the U for Ce substitution is accompanied by partial oxidation of the U from 4+ to 5+ or 6+ and the addition of charge balancing oxygen interstitials to the structure. Both solid solutions are of interest in nuclear waste ceramics (14).

Experimental

Cr_2O_3, Fe_2O_3 and four intermediate mixtures were given identical grinding, pelletizing, and 36 h 1200°C firings with one intermediate regrinding. In the preparation of (Ce, U)O_{2+x} phases, appropriate volumes of Ce and $(UO_2)^{2+}$ nitrate solutions were mixed, dried at 110°C, denitrated at 600°C and fired at 1200°C for two and 24 h. All solid solutions were single phase after these treatments. For the reference intensity measurements, ZnO was chosen as the internal standard for the Cr_2O_3-Fe_2O_3 system and NiO for the CeO_2-UO_{2+x} system. The NBS' specimen preparation and scalar-timer recording of intensity data, as described above, were also used in this study. From 2 to 4 reloadings of the specimen holder were made for each sample and the reference reflections from each specimen were counted at least three times. Typical total counts for the standards were: ZnO (100) reflection: 13,000 peak and 3500 integrated; NiO (III) reflection: 75,000 integrated.

Results and Discussion

Cr_2O_3-Fe_2O_3. To a first approximation, one might expect the reference intensity ratio of any member of a solid solution to be a linear combination of the end member values. Fig. 4 shows that this is not the case for I/I_{ZnO} in the Cr_2O_3-Fe_2O_3 system. Both peak and integrated values have large positive deviations from linearity. At their maximum deviations, the values are about 15% greater for the integrated and 45% for the peak I/I_{ZnO}. Thus, in this particular system, substantial error would be introduced by using the pure end member as a QTXRD standard for a solid solution phase in a multiphase unknown. Even when the elemental composition of the phase in the unknown is determined (e.g. by electron microprobe analysis), trying to approximate the reference intensity ratio from end member values could also introduce considerable error. Clearly, the better procedure is to synthesize the appropriate solid solution composition. But, the conditions chosen for the synthesis are also important, as shown in the next example.

CeO_2-UO_{2+x}. The data from this system demonstrate conclusively that not only the composition but also the heat treatment can affect the reference intensity ratio of solid solutions. Table 3 gives the results for two different heat treatments of the same

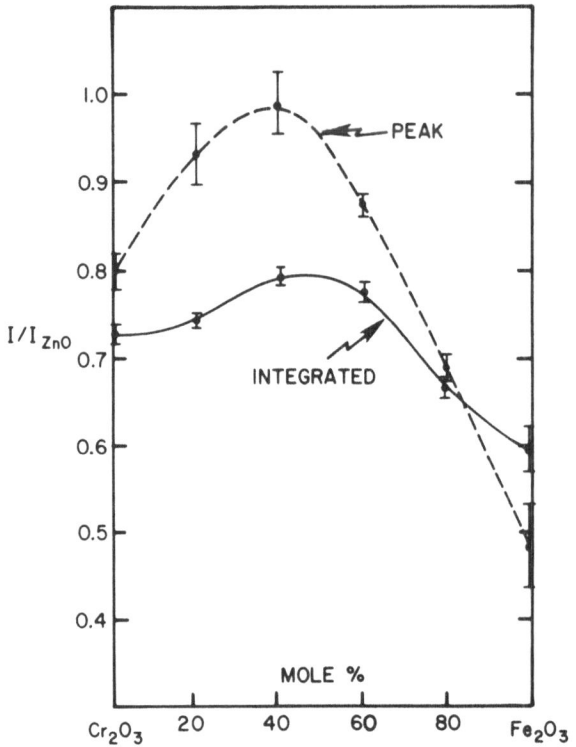

Fig. 4. Reference intensity ratio for Cr_2O_3-Fe_2O_3 solid solutions.

six compositions. The samples fired for 24 hours had about half the I/I_{NiO} of those fired for only 2 hours. The change in I/I_{NiO} with composition is also remarkable. The value for pure CeO_2 decreases by about 30% with only 10 mole % substitution of UO_{2+x}. Further substitution leads to only a small decrease followed by a small increase.

In the research project for which this data was obtained, the nuclear waste ceramics had also been prepared from nitrate solutions and had final firings of 1200°C for about 2 hours. The importance of using the same firing duration for the fluorite structure solid solution phase standard is obvious. If a standard fired for 24 hours had been used, the QTXRD analysis to be made with the reference intensity method of Chung (3,4) would have been in error by about 50% of the amount present.

CONCLUSIONS AND RECOMMENDATIONS

The results presented here demonstrate that differences in

Table 3

I/I_{NiO} Values for the CeO_2-UO_{2+x} System

Composition	1200°C/2 h	1200°C/24 h
CeO_2	1.96 (7)[b]	0.87 (8)
$(Ce_{0.9}U_{0.1})$[a]	1.33 (3)	0.63 (5)
$(Ce_{0.8}U_{0.2})$	1.22 (2)	0.55 (7)
$(Ce_{0.7}U_{0.3})$	1.32 (3)	0.59 (7)
$(Ce_{0.6}U_{0.4})$	1.33 (3)	0.60 (9)
$(Ce_{0.5}U_{0.5})$	1.24 (5)	0.60 (3)

[a] metal ratios in $(Ce_{1-y}U_y)O_{2+x}$ [b] standard deviation of 3-4 reloadings of the specimen holder

mode of preparation of analyte standards, especially heat treatment, give a wide range of variations in reference intensity ratio. The examples cited had variations ranging from 2% for NaCl before and after melting to 100% for $(Ce, U)O_{2+x}$ with the 2 hour compared to the 24 hour firing. These variations are primarily a function of crystallinity differences, but differential microabsorption effects, due to increases in particle sizes with higher firing temperatures, should also be a contributing factor. Reference intensity ratio can also be a sensitive function of composition in solid solutions. In all cases, the differences were smaller when integrated intensities rather than peak intensities were used.

We have the following recommendations for the selction of analyte and internal intensity standards:

•Match the composition of the analyte standard to the analyte. This often requires sophisticated microphase chemical analysis of the analyte.
•In this chemically matched analyte standard, approach the crystallinity observed in the specimen by:
    •comparable thermal treatment of analyte standard to that of the specimen,
    •matching of the breadth and α_1/α_2 splitting for selected reflections in analyte and analyte standard.
•To better assess the accuracy of an analysis, measure reference intensity ratios for several versions of an

analyte standard covering the range of compositions and
heat treatments appropriate to the specimen.
•In choosing an internal intensity standard, approximate
the linear absorption coefficient (μ) of the standard, the
analyte and the multiphase unknown as closely as possible.
The NBS will issue intensity Standard Reference Materials
covering a wide range of μ in the near future (15).

ACKNOWLEDGMENTS

The research in III and IV was supported by the Department
of Energy through Battelle, Pacific Northwest Laboratories.
Research in I and II was supported by the National Science Founda-
tion, Continental Oil Company and Exxon Research and Engineering
Company. We thank J.G. Pepin for synthesizing the CeO_2-UO_{2+x}
solid solutions.

REFERENCES CITED

1. H.P. Klug and L.E. Alexander, X-Ray Diffraction Procedures,
    2nd Ed. Wiley Interscience, New York (1974).
2. L.E. Alexander, in Adv. in X-Ray Analysis, Vol. 20, Plenum
    Press, New York, pp. 1-13 (1977).
3. F.H. Chung, Adv. in X-Ray Analysis, Vol. 17, Plenum Press,
    New York, pp. 106-115 (1974).
4. F.H. Chung, J. Appl. Cryst. 7, 519-525 (1974).
5. L.E. Copeland and R.H. Bragg, Anal. Chem., 30, 196-208 (1958).
6. S. Popivic and B. Grzeta-Plenkovic, J. Appl. Cryst., 12,
    205-208 (1979).
7. N.H. Clark and R.J. Preston, X-Ray Spectrometry, 3, 21-25 (1974).
8. L.S. Zevin, J. Appl. Cryst., 10, 147-150 (1977).
9. G.W. Brindley, in The Identification and Crystal Structures
    of Clay Minerals, G. Brown (Ed.) Mineralogical Soc. London
    pp. 489-516 (1972).
10. G.G. Johnson, Jr., E.W. White, D. Strictler and R. Hoover,
    Proc. Symp. Electron Microscopy of Microfibres, HEW Publ. No.
    (FDA) 77-1033, pp. 76-82 (1977).
11. C.R. Hubbard, E.H. Evans and D.K. Smith, J. Appl. Cryst., 9,
    169-174 (1976).
12. M.C. Morris, H.F. McMurdie, E.H. Evans, B. Paretzkin and J.H.
    deGroot, Mono. 25, National Bureau of Standards, Washington,
    D.C. pp. 2-3 (1979).
13. J.G. Pepin and G.J. McCarthy, submitted for publication.
14. G.J. McCarthy, Nucl. Techn. 32, 92-105 (1977).
15. C.R. Hubbard, Accuracy in Powder Diffraction, NBS Spec. Publ.
    567 pp. 489-502 (1980).

APPLICATION OF GANDOLFI X-RAY DIFFRACTION TO THE CHARACTERIZATION

OF REACTION PRODUCTS FROM THE ALTERATION OF SIMULATED NUCLEAR WASTES

C. A. F. Anderson, M. E. Zolensky, D. K. Smith[1],
W. P. Freeborn and B. E. Scheetz[2]

[1]Department of Geosciences and [2]Materials Research
Laboratory, The Pennsylvania State University,
University Park, PA 16802

Abstract: Accurate phase characterization of the alteration pro-
          ducts of rad-waste requires the separation and identifi-
          cation of scattered individual grains from among the
          bulk product. These grains are typically 5 to 100 µm
          in size. Bulk x-ray powder diffraction will normally
          not detect these minor phases, and even if the phase can
          be detected, it often may not be identifiable. The use
          of the Gandolfi technique with the individual particle
          not only facilitates the identification, but also allows
          the assignment of the identification to the specific
          grain.

Research into the hydrothermal stability of the leading candidate rad-waste forms results in complex assemblages of both crystalline and amorphous products. The rad-waste forms contain up to
forty elements, and the starting materials are rarely homogeneous.
Individual reaction products are isolated under a binocular microscope sometimes after characterization in a scanning electron
microscope. The position of the particle in the product material
often indicates whether it formed by solid state or solid-fluid
reaction or from quenching of the liquid phase. Micro reactions
lead to a large diversity of phases which may form, and this diversity represents local equilibrium not bulk equilibrium. It is these
interactions which must be defined to completely understand the
potential decomposition of the rad-waste form. The effectiveness
of this microdiffraction approach is shown by the long list of compounds which have been identified.

The byproducts of commercial power reactors and the defense effort of the United States Government represent some 10^5 cubic meters of high-level, long-lived radioactive waste (Bartlett, 1976). Research efforts at Sandia Laboratories (Johnstone, 1979), Battelle Pacific Northwest Laboratories (Mendel, 1977), Catholic University (Simmons et al.), Canberra University (Ringwood, 1979) and The Pennsylvania State University (McCarthy, 1977) are focused upon attempts to develop waste forms for reprocessed wastes that can suitably immobilize this hazardous waste for upwards of 10^4 years; until the radiation levels contained within the waste drops to background levels. In addition to the reprocessing alternatives, spent fuel elements from a once-through cycle of commercial reactors has become the leading waste form because of the moratorium placed upon reprocessing of wastes in the United States.

The most widely accepted mode of terminal storage today is via deep geological isolation (KBS, 1978). Dispersal of the radionuclides by intruding groundwaters is the most likely mechanism of repository failure that is currently postulated. Therefore, the presence of intruding groundwaters into a repository during the thermal period (the first 500 to 600 years [McCarthy et al., 1978]) coupled with a repository failure in which the pressure of the overburden rock was entirely transferred to the groundwater represent the most extreme circumstances that are likely to be encountered. Closely linked to the success of any candidate waste form is its stability (resistance to alteration) under potential "worst-case" storage conditions.

Research into the hydrothermal stability of the leading candidate waste forms, supercalcine-ceramic, glass and spent fuel elements both alone and in the presence of evaporite assemblages and crystalline silicate geomedia were shown to result in a complicated assemblage of both crystalline and noncrystalline phases (Scheetz et al., 1979, 1980; Freeborn et al., 1979; Komarneni et al., 1979; White et al., 1979; McCarthy et al., 1979). Gandolfi x-ray diffraction techniques coupled with qualitative chemistry obtained with the aid of a scanning electron microscope equipped with energy dispersive x-ray analyser have allowed a significant advance to be taken in the interpretation and understanding of these complicated (typically 35 to 40 oxide components) phase assemblages.

Most of the reaction products analyzed in these studies were aggregates of crystals in the range of 5 to 100 μm in size or were layered alteration rinds formed either on fragments of the reactants or adhering to the gold reaction capsules. Gandolfi x-ray diffraction techniques have allowed the characterization of minute single phase grains that were hand-selected from these runs. Careful separation of the intergrown phases facilitate the identification of specific phases which are present and provides information on the spatial aspects of the chemical reactions. The small particles are

often single phase. The separation of grains was accomplished under
a binocular microscope at 40X to 100X magnification with the aid of
a fine needle. Selected crystals were then mounted onto the ends of
low-scattering Lindeman glass capillaries which had been drawn to a
diameter comparable to that of the specimen by first touching them
to a hot Pt wire. Cenco Tackiwax, clear nail polish, or collodion
in amyl acetate was used to cement the larger samples to the capil-
lary. Tackiwax has the disadvantage of producing diffraction lines
at 4.12 Å and 3.75 Å; however, it remains flexible allowing the
sample to be accurately positioned at the end of the glass fiber.
The nail polish produced no pattern but dries quickly, although the
drying time can be extended by dilution with acetone. Collodion in
amyl acetate is used as a cement for the smallest particles.

Wherever possible samples consisting of several individual
crystals were used, but occasionally certain run products would be
scarce, and few or only one crystal would be all that could be
mounted.

The preferred orientation problem inherent in the x-ray identification of single crystals is essentially eliminated by the sample
motion of the Gandolfi camera, which simultaneously rotates the
samples about two axes approximately 45° to each other bringing
almost all Bragg planes into a position to diffract (Gandolfi,
1967). Because of the small size of the samples, precise centering
of the sample in the camera is essential.

To minimize air scattering of the x-ray beam and allow longer
exposure times, the camera was evacuated to a pressure of about 10^{-4}
to 10^{-3} torr during the film exposures. Very fine collimators
(0.2mm) were also routinely used to minimize unnecessary scattering
of the x-ray beam. Contrary to the findings of McCrone and Delly
(1973), a rule-of-thumb for exposure time cannot be simply stated
because the exposures are so dependent upon the size, density,
symmetry and orientation of the sample, film type, and operating
conditions. For example, the 57.3 mm cassette utilized with $CuK\alpha$
radiation at 40kV and 20 mA commonly required exposure times of 96
hours, although the average duration of exposures was 48 hours. The
114.6 mm cassette affords greater accuracy but requires four times
the exposure time of the 57.3 mm cassette.

According to the camera's manufacturer, Officina Elettrotecnia
di Tenno, the lower limit of sample size is 30 μm; however, 5 μm
particles are routinely x-rayed while acceptable patterns have been
obtained from 2 μm particles as suggested by McCrone and Delly (1973).
For the smallest and least dense samples, a brass insert was designed
and constructed at The Pennsylvania State University X-Ray Crystallography Laboratory which decreases the radius of the 57.3 mm cassette
to 28.65 mm and correspondingly reduces the exposure time by four

times. The decreased camera radius, however, has the drawback of reduced resolution and accuracy.

To prevent darkening of the film from L fluorescence from heavy elements such as uranium, the surface of the film was covered by a thin sheet (25 μm) of aluminum foil during the exposures. All films must be corrected for shrinkage after processing using the Straumanis (1949) technique to achieve precise cell parameters.

The following list of characterized phases and prototype structures from the reaction products of simulated nuclear wastes can be used to illustrate the effectiveness of this approach in unraveling complicated phase relationships in complex oxide systems:

Elements: gold, Au; graphite, C; palladium, Pd.

Borates: $Na_2B_4O_7$; $Na_2B_2O_5 \cdot 4H_2O$; tincalconite, $Na_2B_4O_7 \cdot 5H_2O$, borax, $Na_2B_4O_7 \cdot 10H_2O$.

Carbonates: calcite, $CaCO_3$.

Chlorides: $NdCl_3 \cdot 6H_2O$; $KCaCl_3$; $NaMgCl_3$; sylvite KCl.

Molybdates: powellite, $(Ca,Sr,Ba)MoO_3$; $Na_2MoO_4 \cdot 2H_2O$; $NaLa(MoO_4)_2$; $NaNd(MoO_4)_2$; $NaGd(MoO_4)_2$; $Nd_5Mo_3O_{16}$; $[HoVWO_6]$; $[GdVWO_6]$.

Oxides: corundum, Al_2O_3; hematite, Fe_2O_3; RuO_2; cerianite, CeO_2; uraninite, UO_2; $UO_{2.25}$; U_3O_7; αU_3O_8; $UO_3 \cdot 0.5H_2O$; $UO_3 \cdot 0.8H_2O$; $UO_3 \cdot H_2O$; schoepite, $UO_2 \cdot 2H_2O$; ianthinite, $UO_2 \cdot 5UO_3 \cdot 10.5H_2O$; spinel, $(Fe^{2+},Zn,Ni)(Fe^{3+},Cr)_2O_4$.

Phosphates: hydroxyapatite, $Ca_5(PO_4)_3OH$; wilkeite, $Ca_5(P,Si,S)_3O_{12}(Cl,OH,F)$.

Silicates: nesosilicates -- willemite, Zn_2SiO_4.
            inosilicates -- acmite-augite, $(Na,Ca)(Fe^{2+},Fe^{3+})Si_2O_6$; diopside, $CaMgSi_2O_6$; hypersthene, $CaFeSi_2O_6$.
            sorosilicates -- hardystonite, $Ca_2ZnSi_2O_7$.
            phyllosilicates -- talc, $Mg_3Si_4O_{10}(OH)_2$; nontronite, $Na_{0.33}Fe_2^3(Al,Si)_4O_{10}(OH)_2 \cdot nH_2O$; sauconite, $Na_{0.33}Zn_3(Al,Si)_4O_{10}(OH)_2 \cdot 4H_2O$.
            tektosilicates -- quartz, SiO_2; albite, $NaAlSi_3O_8$; K-spar, $KAlSi_3O_8$; Sr-paracelsian, $SrAl_2Si_2O_8$; analcime-pollucite, $(Na,Cs)AlSi_2O_6 \cdot H_2O$; phillipsite, $(K_2,Na,Ca)(Al_2Si_4)O_{12} \cdot 4-5H_2O$; mordenite, $(Ca,Na_2,K_2)(Al_2Si_{10})O_{24} \cdot 7H_2O$; wellsite, $(Ba,Ca,K_2)(Al_2Si_3)O_{10} \cdot 3H_2O$.
            other silicates -- truscottite, $Ca_2Si_4O_9(OH)_2$; zeophyllite, $Ca_4Si_3O_7(OH)_6$; xonotlite, $Ca_6Si_6O_{17}(OH)_2$; weeksite, $(Na,K,Rb,Cs)_2(UO_2)_2Si_6O_{15} \cdot 4H_2O$; haiweeite, $Ca(UO_2)_2Si_6O_{15} \cdot 5H_2O$; boltwoodite, $(Na,K)_2(UO_2)_2Si_2O_6(OH)_2 \cdot 5H_2O$; uranophane, $Ca(UO_2)_2Si_2O_7 \cdot 6H_2O$.

Sulfates: barite, $BaSO_4$; zippeite, $K_4(UO_2)_6(SO_4)_3(OH)_{10} \cdot 4H_2O$.

Sulfides: bornite, Cu_5FeS_4; pyrrhotite, $Fe_{1-x}S$; wurtzite, ZnS.

This research was sponsored through subcontracts from the Department of Energy administered through the Office of Nuclear Waste Isolation, Battelle Memorial Institute, Columbus, Ohio, and the Basalt Waste Isolation Program, Rockwell Hanford Operations, Richland, Washington.

REFERENCES

Bartlett, J. W. (Task Leader), 1976, "Alternatives for Managing Wastes from Reactors and Post-Fission Operations in the LWR Fuel Cycle," Vol. 4, ERDA-76-43.

Freeborn, W. P., Zolensky, M., Scheetz, B. E., Komarneni, S., McCarthy, G. J., and White, W. B., (submitted 15 July 1979), "Hydrothermal Interaction Between Calcine, Glass Spent Fuel and Ceramic Waste Forms with Representative Shale Repository Rocks", Office of Nuclear Waste Isolation, Battelle Memorial Institute, Columbus, Ohio.

Gandolfi, G., 1976, Miner. Petrog. Acta, 13:67-74.

Johnstone, J. K., Headly, T. J., Hlava, P. F., and Stohl, F. V., 1979, in: "Scientific Basis for Nuclear Waste Management", Vol. 1, G. J. McCarthy, ed., Plenum Press, New York, p. 211-217.

Kä RW-Bränsle-Säkerhet, 1978, "Handling and Final Storage of Unprocessed Spent Nuclear Fuel", Vol. 1: General, AB Teleplan SOLNA, Karnbraslesakerhet, FACK, S-102 40 Stockholm, Sweden, pp. 53-60.

McCarthy, G. J., 1977, "Advanced Waste Forms Research and Development Comprehensive Progress Report", The Pennsylvania State University, FRDA-COO-2510.

McCarthy, G. J., White, W. B., Roy, R., Scheetz, B. E., Komarneni, S., Smith, D. K., and Roy, D. M., 1978, Nature, 273:216-217.

McCarthy, G. J., Scheetz, B. E., Komarneni, S., Smith, D. K., and White, W. B., "Hydrothermal Stability of Simulated Radioactive Waste Glass", Advances in Chemistry Series (in press).

McCrone, W. C. and Delly, J. G., 1973, "The Particle Atlas", 2nd ed., Ann Arbor Science Publishers Inc., pp. 119-129.

Mendel, J. E. et al., 1977, "Annual Report on the Characteristics of High Level Waste Glasses", BNWL-2252, Battelle Pacific Northwest Laboratories, Richland, Washington.

Ringwood, A. E. and Kesson, S. E., 1979, in: "Ceramics in Nuclear Waste Management", T. D. Chikalla and J. E. Mendel, eds., Proceedings of an International Symposium Held in Cincinnati, Ohio on April 30-May 2, 1979, Sponsored by The Nuclear Division of the American Ceramic Society and DOE, pp. 174-178.

Simmons, J. H., Macedo, P. B., Bankatt, A., and Litorivtz, T. A., 1979. Nature, 278:729-731.

Straumanis, M. E., 1949, J. Appl. Phys., 20:726.

THE CHARACTERIZATION OF ALPHA AND INTERMEDIATE

ALUMINUM OXIDE MIXTURES BY SEMI-AUTOMATED XRD

John R. Burleson

AC Spark Plug Division
General Motors Corporation
Flint, Michigan 48556

ABSTRACT

Intermediate aluminum oxides (aluminas) have broad X-ray diffraction (XRD) peaks. Many of the phases have similar XRD patterns
with overlapping of major peaks. To characterize a multi-component
alumina mixture several regions of multiple overlapping peaks are
examined. Each region is integrated to a single sum without regard
for individual peaks. The integrated regions are applied to various
ratio functions obtained from single component standards. By using
an iterative technique, the functions converge to yield weight fractions for up to six alumina phases.

INTRODUCTION

The intermediate aluminas are poorly crystallized. Many of
the phases have similar XRD patterns with most major peaks overlapping. There exist many minor variations among the aluminas.

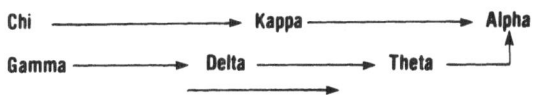

Direction of increasing stability with increasing temperature

Fig. 1. Condensed System of Aluminas[3]

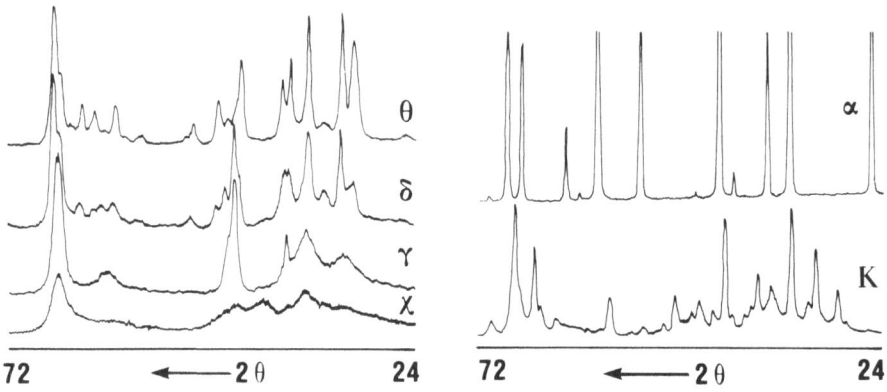

Fig. 2. XRD Patterns of Six Aluminas. All patterns were run with
        identical machine parameters. (Copper Kα)

This results in four different gamma and three delta phases being
entered among the eighteen alumina patterns published.[1,2] For
simplification, I limited the eighteen to a consistent and progressive set of six phases. See Figure 1. One nominal standard for
each phase was acquired from Alcoa[4]. Their diffraction patterns
are shown in Figure 2.

DESCRIPTION OF METHOD

        This method was developed using a manual General Electric
XRD-700 X-ray diffractometer. It is equipped with a graphite
monochromator and copper target X-ray tube. The machine is set
up for medium resolution optics. Its scaler/timer is interfaced
to a Wang 600 desk calculator. The Wang 600 collects and processes the intensity data.

TABLE I REGIONS OF A MAJOR OVERLAP

Region I.D.	Copper Kα 2θ Range	Do All Six Phases Contribute?	Amt. of Contribution of Each Phase
A_{28}	28.0-32.2	No	Variable
A_{32}	32.2-34.0	No	Variable
A_{67}	63.5-71.0	Yes	Uniform
A_{55}	54.7-55.0	No	Background

Only alpha and kappa aluminas have major peaks relatively free of overlap problems. Both phases have a major peak near $43°2\theta$. These peaks rest in a broad, poorly defined valley formed by peaks of the other four phases. The region is scanned at $0.2°2\theta$ per minute with intensity data collected continuously and recorded every six seconds. After the data set is collected, it is analyzed to determine the positions of secondary valleys on either side of the alpha and/or kappa peak(s). The area above and between the secondary valleys is integrated. The weight fractions are obtained directly from calibration curves.

With the multiple overlapping of major peaks, a different approach is necessary for measuring chi, gamma, delta, and theta aluminas. Since the broad peaks could not be measured individually, I selected three regions with multiple overlappings of major peaks from several phases. The several peaks in a region are integrated to a single sum, without regard for any individual peak. One region was chosen to have a somewhat uniform intensity for 100% samples of all six phases. The two remaining regions were chosen with delta and theta in mind. See Table I and Figure 3. The regions are scanned at $0.4°$ per minute.

The weight fraction of the gamma phase is obtained by performing an area balance about region A_{67}.

$$A_{67} = A_{67\gamma} + A_{67\delta} + A_{67\theta} + A_{67\chi} + A_{67\alpha} + A_{67\kappa}$$

or noting that $\omega_i = A_{67i}/A_{67i}(100)$

$$\omega_\gamma = \frac{A_{67}}{A_{67\gamma}(100)} - \omega_\kappa R_\kappa - \omega_\alpha R_\alpha - (\omega_\delta R_\delta + \omega_\theta R_\theta + \omega_\chi R_\chi)$$

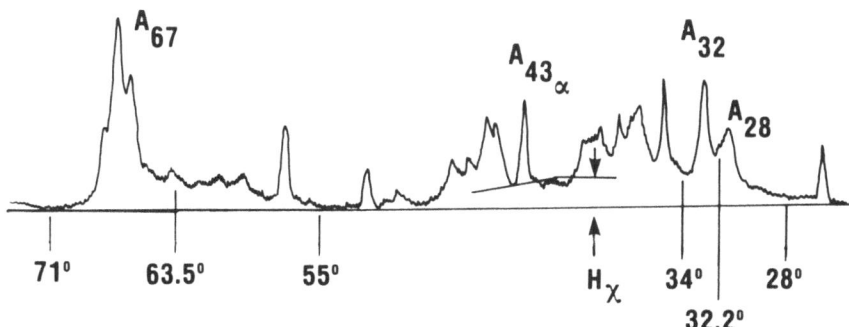

Fig. 3. An XRD Pattern of a Mixture of Aluminas. (Copper Kα)

TABLE II   VALUES OF Ri

$R_\gamma = 1.000$ $R_\theta = 1.119$ $R_\alpha = 0.648$

$R_\delta = 1.096$ $R_\kappa = 1.174$ $R_\chi = 0.648$

where ω_i is the weight fraction of the ith phase and R_i is the ratio $A_{67i(100)}/A_{67\gamma(100)}$ for 100% samples of γ and the ith phase. Values of R_i are given in Table II. While all the phases contribute some-what uniformly to the A_{67} region, this is not true of the A_{28} and A_{32} regions. The net contributions to these regions by kappa and alpha are subtracted away. Of the remaining four phases, chi and gamma have a relatively small contribution to A_{28} and A_{32}. Based on initial estimates of the chi and gamma weight fractions, their con-tributions are subtracted away. The ratios of A_{28}/A_{32} for 100% samples of delta and theta are quite different. Using these ratios, the remaining amount of the regions A_{28} and A_{32} are assigned to delta and theta. Algebraically

$$A_{28\delta+\theta} = A_{28} - A_{28\gamma} - A_{28\chi} - A_{28\alpha} - A_{28\kappa}$$

$$A_{32\delta+\theta} = A_{32} - A_{32\gamma} - A_{32\chi} - A_{32\alpha} - A_{32\kappa}$$

Noting that for $i = \gamma, \chi, \alpha,$ or κ $A_{ji} = \omega_i \cdot A_{ji(100)}$, and defining D_i from 100% samples

$$D_\delta = A_{28\delta(100)}/A_{32\delta(100)} = .748$$

$$D_\theta = A_{28\theta(100)}/A_{32\theta(100)} = 1.123$$

ω_δ and ω_θ can be expressed as

$$\omega_\delta = \frac{A_{28\delta+\theta} - D_\theta A_{32\delta+\theta}}{A_{28\delta(100)} - D_\theta A_{32\delta(100)}}$$

$$\omega_\theta = \frac{A_{28\delta+\theta} - D_\delta A_{32\delta+\theta}}{A_{28\theta(100)} - D_\delta A_{32\theta(100)}}$$

It was necessary to analyze the nearly amorphous chi phase with its very broad, weak peaks differently. As an estimate of the chi concentration, the height of one of its major peaks is measured.

The peak chosen is at approximately $43°2\theta$. Its height is also the height of the valley between the alpha and kappa peaks integrated earlier. From the count data at $28°$ and $55°$ the background under the chi peak is calculated. Corrections from contributions by each other phase are necessary, weighted by each weight fraction.

$$\omega_\chi = H_{43}/H_{43\chi}(100) - \Sigma K_i \omega_i$$

Where H_{43} is the peak height and K_i is the correction coefficient. Now, each ω_i can be calculated as well as each phase's contribution to A_{67}. Using the area balance about A_{67}, any remainder is assigned to $A_{67\gamma}$. With the initial assumption that ω_γ is $1.00 - (\omega_\alpha + \omega_\kappa)$, an iteration about region A_{67} can be run until all six weight fractions converge. Note that the sum of the weight fractions is not forced to unity. Experience has shown that samples with low crystallinity sum to values less than 1.00. Highly crystalline samples will sum slightly higher than 1.00. This reflects the amorphous content of both the samples and standards.

Coefficients used in the program must be corrected for long term intensity variations. The updating is done by using the slope of the alpha alumina calibration curve (m_α) to adjust the various 100% areas from the initial values ($T = 0$) to the current values ($T = t$).

$$A_{ji(100)_t} = A_{ji(100)_0} \frac{m_{\alpha_0}}{m_{\alpha_t}}$$

STATISTICS

For gamma, delta and theta, repeatability on the same sample is in the range of ± 5 wt. % at one standard deviation. For samples having 0.5 to 4 wt. % alpha alumina one standard deviation for alpha is 0.13 wt. %. This includes multiple testing of multiple samples from lots of blended material. The lower detectable limit for alpha is 0.2 wt. %.

REFERENCES

1. JCPDS, Powder Diffraction File, Alphabetical Index Inorganic Materials, 29:15 (1979)
2. K. Wefers and G. Bell "Oxides and Hydroxides of Aluminum", p. 38, Alcoa Research Laboratories, Technical Paper No. 19, 1972
3. ibid. P. 43
4. Thomas L. Francis, Alcoa Research Laboratories, Private Communication, Jan., 1977

EXPERIMENTAL EVALUATION OF PEAK HEIGHT

APPROXIMATION FOR X-RAY DIFFRACTED

INTEGRATED INTENSITY METHOD

Charles P. Gazzara

Army Materials and Mechanics Research Center
Watertown, Massachusetts   02172
DRXMR-EM

INTRODUCTION

The mathematical description of an X-ray peak, diffracted from a powder or polycrystalline material, using physically meaningful parameters has been of interest for many years. With the popularity of computers, this need to characterize a diffraction peak has intensified.

A key problem which persists is how to describe the instrumental diffracted profile and therefore the observed diffracted characteristic peak with subsequent combinations of K_α doublets and mixed overlapping peaks. Many attempts have been made at finding a "true" function to fit the observed diffracted peak; however, a practical solution has yet to be found. The reason is twofold. First, many investigators since Smith[1,2] have tried to find one function that duplicates the X-ray powder peak, finding similar failures with either Gaussian, Cauchy (pure or modified), or Lorentz functions. These results are summarized by Khattak and Cox,[3] who claim to have found a reasonable fit using a cross between a modified and a pure Lorentz function. Secondly, the employment of a composite of three or four[4] or seven[5] Lorentzians adequately describe an X-ray diffracted peak but require the use of untenable computer programs for pattern-fitting crystal structure refinements, or for quantitative phase analyses where many compounds or mixtures are involved. Similar problems arise using convolution techniques involving multiple functions or angle-dependent Fourier coefficients.

The intent of this paper is to experimentally test, in part, a simple but effective alternative to the previously mentioned attempts to characterize a K_α X-ray diffracted peak.[6] Simply stated, the essence of this method is to fit a Gaussian function to the top half and a Cauchy function to the bottom half of the diffraction peak. An application of this procedure to an X-ray diffraction scan of a mixture of silicon and α and β silicon nitride is reviewed[7] in which the integrated intensity is approximated by the peak height for one set of experimental conditions (i.e., a single beam and receiving slit). The degree of correlation of the X-ray peak heights to compositional levels is taken as confirmation of the validity of the proposed method. Another check on the accuracy of the curve fit is the improvement in the accuracy of lattice constant measurements of silicon using the peak positions of the K_{α_1} and K_{α_2} unresolved doublets convoluted from the Gaussian-Cauchy curves.[8]

The final endorsement of this method is the experimental evidence presented in this study, using silicon powder. In this case the measurement of integrated intensities are performed using variations in slit systems for different commercially available X-ray units for testing the effectiveness of the peak height approximation as well as the peak shape-fitting technique.

PROCEDURE

The relationship between the X-ray diffracted peak height y_T and the integrated intensity of the peak was theoretically developed[6] for one set of experimental conditions whereby the upper half of a reflection was matched to a Gaussian curve while the bottom half of the low angle side of a CuK_{α_1} (531) reflection of silicon powder was fit to a Cauchy distribution.

These mathematical functions were used to generate constant intensity curves of y_T versus 2θ angles for different levels of $\omega_{h\frac{1}{2}}$ (see Figure 7[6]). In this case, $\omega_{h\frac{1}{2}}$ (peak width at half height of K_α peak) is assumed to be constant over the 2θ range considered. One major advantage of this analysis was the ability to determine this change in peak height for an unresolved K_α doublet which is usually what is detected in practice. A relationship whereby $\omega_{h\frac{1}{2}}$ can be abstracted from the half width of the K_α doublet $\omega_{T\frac{1}{2}}$ was worked out and given in Figure 6. Therefore, the usual procedure is to measure $\omega_{T\frac{1}{2}}$ for two (in the case of Si using CuK_α radiation) or more peaks, read the respective values of $\omega_{h\frac{1}{2}}$ from Figure 6, and obtain the decrease in peak height for the corresponding angle 2θ from Figure 7.

In this study this procedure is tested for different combinations of beam and receiving slits on three commercial X-ray systems using the change in height of the silicon (111) peak at 28.5° 2θ, and the (200) peak found at 47.7° 2θ.

Diffraction scans of the 400 mesh silicon powder were performed for combinations of beam and receiving slits, and their integrated intensities measured by planimetering the diffractograms. The criterion for the accuracy involves first computing the change in peak height from the (111) to (200) reflection of silicon $\Delta y_{T\ CALC}$ assuming constant intensity, using Figures 6 and 7[6]. Then, the ratio of the measured integrated intensities, $I(111)/I(200)$, was multiplied by the inverse ratio of the measured peak heights, $y_T(200)/y_T(111)$, to yield the observed reduction in peak height, $\Delta y_{T\ OBS}$. The difference, $\Delta y_{T\ CALC} - \Delta y_{T\ OBS}$, in percent, is called % deviation$_c$, and is taken as the measure of accuracy for this method. (The subscript c denotes that a constant value of $\omega_{h\frac{1}{2}}$ is assumed.)

As will be seen for the Norelco X-ray system, the condition that $\omega_{h\frac{1}{2}}$ remain constant is not met for case 2. In this case, therefore, $\Delta y_{T\ OBS}$ is multiplied by $\omega_{T\frac{1}{2}}(200)/\omega_{T\frac{1}{2}}(200)'$ where $\omega_{T\frac{1}{2}}(200)'$ is the apparent doublet peak width, at half height, read from Figure 6[6]. starting with the observed value of $\omega_{T\frac{1}{2}}(111)$. Or,

$$\Delta y_{T\ OBS}' = \Delta y_{T\ OBS} \times \omega_{T\frac{1}{2}}(200)/\omega_{T\frac{1}{2}}(200)'$$

and % deviation$_v$ = $\Delta y_{T\ CALC} - \Delta y_{T\ OBS}'$. (The subscript v indicates that $\omega_{h\frac{1}{2}}$ is variable.) The justification for calculating $\Delta y_{T\ OBS}'$ assumes that the convolution of the instrumental functions do not significantly affect the shape of the diffracted X-ray peak with a variation in the slit systems, e.g., the diffracted X-ray peak will remain essentially Gaussian with the convolution of a Gaussian instrumental function.[6]

RESULTS

Values of $\Delta y_{T\ OBS}$ and $\Delta y_{T\ OBS}'$ were measured and the % deviation$_c$ and % deviation$_v$ calculated and plotted in Figure 1 for various configurations of beam and receiving slits.

The results for the General Electric system, seen in Figure 1a, are not surprising. Since the functions chosen to match the (531) reflection of Si were obtained with a beam slit of 1°, the present experimental results would be expected to possess a good fit with the same beam slit. These results do not change in going from % deviation$_c$ to % deviation$_v$, confirming the existence of a constancy in $\omega_{h\frac{1}{2}}$ between the two reflections investigated. Some departure from the ideal case (i.e., % deviation$_c$ = 0) exists with the 3° beam slit, but improves in going to % deviation$_v$, except for a very narrow receiving slit. As expected, the increase in % deviation$_c$ is high for a 0.4^o beam slit, with some improvement in correcting for the change in peak width, but also increasing the scatter in % deviation$_v$ with a variation of the receiving slit.

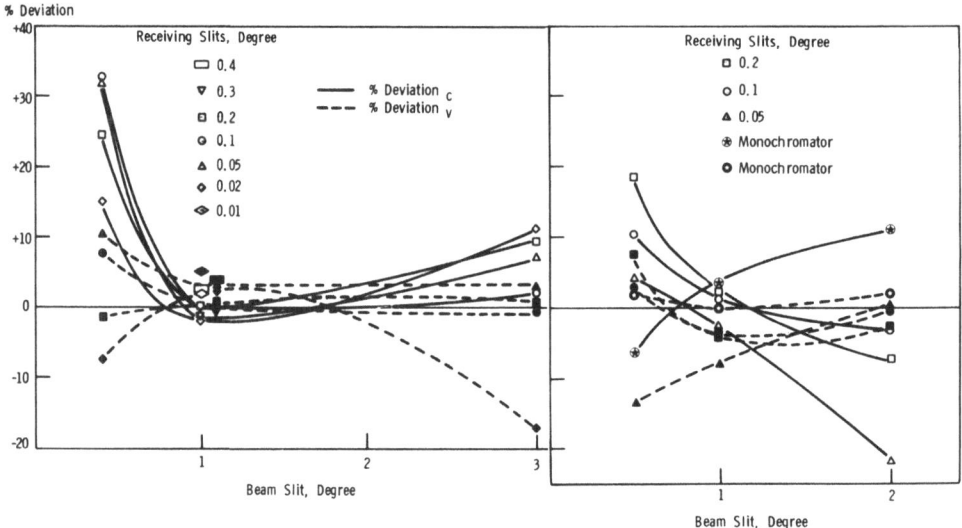

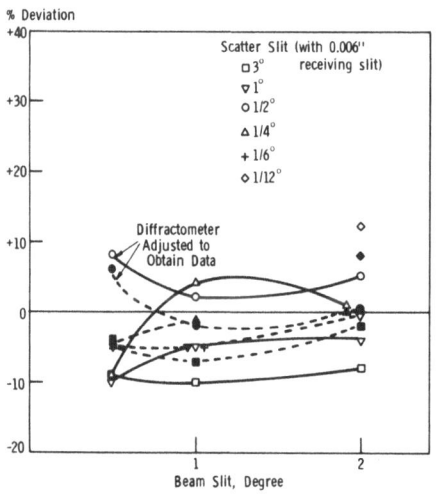

a. General Electric XRD-5 X-ray system.

b. Siemens Crystalloflex IV Type F X-ray system.

c. Norelco Vertical Diffractometer X-ray system, with a conventional slit system.

d. Norelco Vertical Diffractometer X-ray system, with one receiving slit.

Figure 1. Percent deviation of experimental X-ray diffracted (111), (200) silicon powder intensities from theoretical values versus slit configurations.

The plots for the Siemens system (Figure 1b) show a similar behavior for a small beam slit ($\frac{1}{2}^{O}$). For a beam slit of 1^{O} with 0.1^{O} and 0.2^{O} receiving slits, the behavior of % deviation$_c$ is excellent, with a decrease of approximately 5% in the value of % deviation$_v$. A reversal in this behavior occurs for a beam slit of 2^{O}. In this case, the % deviation$_c$ values are too low, and increase to a level in fairly good agreement with the calculated values, or % deviation$_v \approx 0$. The monochromator yields an interesting situation. The value of % deviation$_c$ is within the experimental limits for a beam slit of 1^{O}, but is unreasonably high at 2^{O}, while the values of % deviation$_v$ are excellent for all the beam slits.

The Norelco system warrants some consideration due to the special arrangement of the slit geometry. In the conventional case (case 1, Figure 1c), a scatter slit, which probably should be termed a defining slit, provides a high degree of resolution of the diffracted beam. In fact, the CuK$_{\alpha 1}$ and CuK$_{\alpha 2}$ silicon peaks are resolved to a higher degree than with either of the other two X-ray systems. This condition, however, is achieved at the expense of a loss in integrated intensity of the diffracted peak. With the slit arrangement involving a 2^{O} beam slit and removing the 0.006" receiving slit (proceeding from case 1, with a 0.006" receiving slit and a 1^{O} scatter slit to case 2 with a 0.154^{O} receiving slit), the integrated intensity is increased by a factor of eight times.

From Figure 1c it may be seen that good agreement of the experimental results with the calculations exists, particularly with the 2^{O} beam slit and the larger scatter slits.

With the Norelco system, in the case 2 mode, only one receiving slit was installed in order to increase the diffracted X-ray intensity. As can be seen in Figure 1d, this results in a marked departure from the calculations. While the agreement at a beam slit setting of $\frac{1}{4}^{O}$ is good, the increase in % deviation$_c$ is extremely rapid with increasing beam slit size. Although, the % deviation$_v$ fit is far from ideal, it demonstrates that even with an unusual diffracting geometry, reasonable results are possible.

CONCLUSIONS

This study has given further justification to the use of a simplified method of approximating the shape of an X-ray K$_\alpha$ diffraction peak and approximating its integrated intensity using the peak height and breadth. The advantages of this technique are obvious with search systems applications.

It should be cautioned, however, that before this technique is freely used, some areas should be considered. Namely, a test of the Gaussian-Cauchy curve fitting should be performed and its effect on the accuracy of the resulting peak heights determined. In fact, before the procedure is applied, the shape of the characteristic line should be matched to the Gaussian-Cauchy functions.

Particular care should be exercised when trying to apply the suggested method to an X-ray system involving a variation from the standard diffracted beam geometry, as in case 2.

Thanks to Mr. T. Sheridan for collecting most of the X-ray data.

REFERENCES

1. D. K. Smith, A Fortran Program for Calculating X-ray Powder Diffraction Patterns, Lawrence Radiation Laboratory, Livermore, California, UCRL-7196 (April 1963).

2. D. K. Smith, Computer Simulation of X-ray Diffractometer Traces, Norelco Reporter, XV; no. 2: 57 (1968).

3. C. P. Khattak, and D. E. Cox, Profile Analysis of X-ray Powder Diffractometer Data: Structural Refinement of $La_{0.75} Sr_{0.25} CrO_3$, J. Appl. Cryst., 10: 405 (1977).

4. D. Taupin, Automatic Peak Determination in X-ray Powder Patterns, J. Appl. Cryst., 6: 266 (1973).

5. T. C. Huang, and W. Parrish, Accurate and Rapid Reduction of Experimental X-ray Data, Appl. Phys. Lett., 27: 123 (1975).

6. C. P. Gazzara, Peak Height Approximation for X-ray Diffracted Integrated Intensity, in "Advances in X-ray Analysis." R. W. Gould, C. S. Barrett, J. B. Newkirk, and C. O. Ruud, ed., 19: Kendall Hunt, Iowa, 735 (1976).

7. C. P. Gazzara, and D. Messier, Determination of Phase Content of Si_3N_4 by X-ray Diffraction Analysis, Amer. Cer. Soc. Bull., 56: 777 (1977).

8. C. P. Gazzara, The effect of the K_α Doublet Diffracted Peak Position on the Precision of the Lattice Constant, in "Advances in X-ray Analysis." H. F. McMurdie, C. S. Barrett, J. B. Newkirk and C. O. Ruud, ed., 20: Plenum Press, New York, 161 (1977).

9. F. W. Jones, Measurement of Particle Size by the X-ray Method, Proceedings of the Roy. Soc. (London), 166A: 16 (1938).

THE APPLICATION OF AN AUTOMATED SINGLE CRYSTAL ORIENTER

FOR LARGE SPECIMENS

A. V. Karg, J. M. Walsh and J. M. LaGrotta

Materials Engineering and Research Laboratory
Pratt & Whitney Aircraft
Middletown, Connecticut 06457

ABSTRACT

A special purpose, automated, four circle goniometer arrangement has been developed for the rapid orientation of large specimens
of single crystal turbine hardware. This system rapidly establishes
the complete orientation of these large components and generates a
fully documented stereographic projection. The system is also capable of generating detailed information on crystal quality including
rocking curves in several dimensions. X, Y translational capability
of the specimen makes it possible to characterize crystallographic
defects such as sub-grain boundaries.

INTRODUCTION

One of the major advances in materials technology for advanced
gas turbine engines during the past decade has been the development
of superalloy single crystal high pressure turbine airfoils.[1-3] By
completely eliminating grain boundaries, component intergranular
cracking is prohibited. Elimination of grain boundaries also permits greater flexibility of alloy design, since the designer does
not have to be concerned about the impact of modifications upon
grain boundary properties. These crystals typically grow with [001]
parallel to the major stress axis of the part; however, crystal orientation is required to assure uniformity of properties from part to
part. The Laue technique has been used extensively for measuring deviations of the stress axis from the [001] orientation. While a diffractometer method would provide some advantages over the film method
for this orientation task, there are no commercially available instruments that can accomplish this on the large turbine airfoil crystals that are of interest. A diffractometer-based system was

283

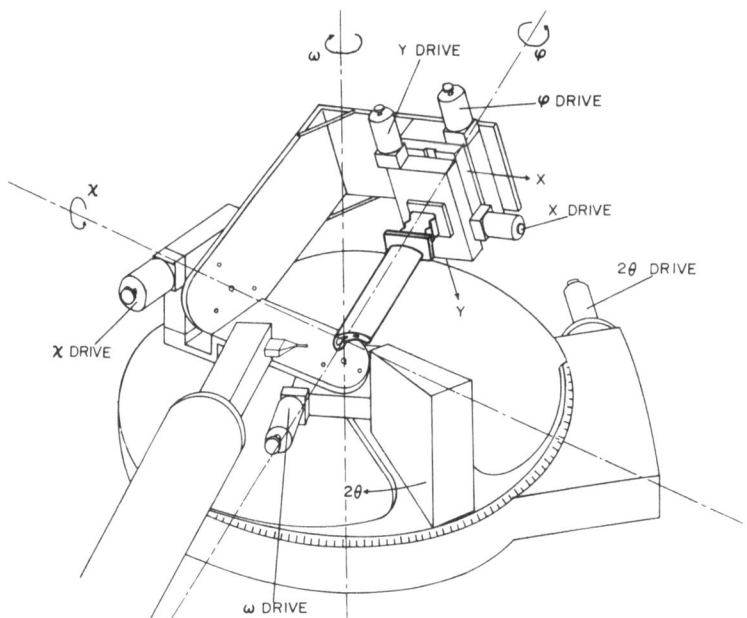

Fig. 1. Apparatus for automatic orientation of large single crys-
         tals with turbine blade in analysis position.

therefore developed that would perform the required orientation
task and accommodate the large crystals of interest. In addition
to the orientation task, elements were incorporated into the design that allow detailed characterization of the quality of these
same crystals.

EQUIPMENT

     The Rapid Automated Single Crystal Aligner (RASCAL) system
mounts on a horizontal goniometer and provides for (1) rotation
about the diffractometer axis (ω), (2) tilt about an axis perpendicular to the diffractometer axis (χ), (3) rotation about the crystal
axis (φ) and (4) X, Y translation of the crystal surface relative
to the X-ray beam. The mechanical arrangement is shown in the
sketch in Figure 1. The axial motions are accomplished with 100
oz-inch stepping motor driven rotary positioning tables with one-
half arc minute resolution. X, Y translation of the specimen is
accomplished with stepping motor driven linear positioning tables.
A 100 oz-inch stepping motor is also coupled directly to the two
theta shaft of the spectrogoniometer. The system was designed to
mount on a General Electric XRD-7 X-ray unit with a #6 spectrogoniometer. The standard geometry did not provide sufficient space
between the detector and the X-ray tube to accommodate the large
specimens of interest and the necessary crystal manipulation equipment. The diffraction geometry was reversed by rotating the

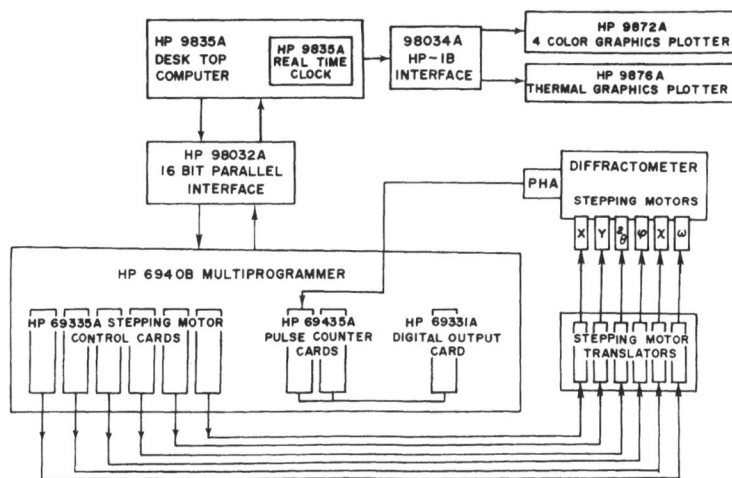

Fig. 2. System configuration for automated crystal orienter.

spectrogoniometer 90° relative to the X-ray source, i.e. with 2θ
increasing in a clockwise direction. The primary beam collimator
support normally mounted to the base of the G.E. goniometer was removed to provide free motion of the ω bracket and the collimator
was mounted directly on the X-ray tube. This arrangement provided
the necessary space.

While many variations of the system are possible, the current
embodiment was optimized for a specific application. This application involves the orientation of cubic crystals of a nickel superalloy up to 4 inches in diameter and 7 inches in length where the
[001] is typically within 20° of the crystal growth axis.

Automation is accomplished through the use of a Hewlett Packard
9835A desk-top computer. An HP 6940B multiprogrammer links the control and detection electronics to the computer. Data are displayed
on an HP 9876A thermal graphics printer and an HP 9872A four color
graphics plotter. A block diagram of the system is shown in Figure
2. The PHA output is routed to a counter circuit in the multiprogrammer and X-ray intensity is integrated against a real-time clock
in the computer.

Versatile software has been developed in HP enhanced BASIC to
perform a variety of single crystal characterization tasks. All
tasks are handled in an interactive question and answer format on
the microcomputer terminal which provides extensive prompting of
the operator for experimental parameters. The special function
keys of the 9835A were programmed in such a way as to give the
operator manual control of the entire system at any point.

A 0.040-inch round primary beam collimator was used in all the work reported here, since analysis was desired in very specific areas of these crystals.

ORIENTATION

The orientation task is accomplished in the following manner. Since the crystals of interest tend toward (001) orientations, the (002) reflection is checked first. With two theta tuned to the (002) reflection of the material of interest, an intensity vs. angle scan is made of each of the axes ω, χ and 2θ. The ω and χ axes are typically scanned $\pm2°$ in 0.4° increments and 2θ is typically scanned $\pm0.5°$ in 0.1° increments. Depending on the degree of precision required in the orientation, 2θ can be held fixed. If preselected intensity threshold criteria are met, the scan ranges and increments are reduced by a preselected convergence factor (typically 0.3X) and a second iteration is made of each axis. Successive iterations are conducted until the nth iteration produces an intensity within a preselected intensity increment of the $n-1^{th}$ iteration. If the intensity threshold criteria are not met after the first iteration is completed, then the system reverts to a search mode. The search mode generates an ω/χ spiral with a pitch of some multiple of the peaking mode scan increment (typically 3X). When preselected intensity criteria are met in the search mode, the peaking mode is resumed until the intensity convergence criteria are satisfied. If the given limits for the ω/χ spiral are reached and search mode intensity criteria are not satisfied, the system returns to the zero position, an "out of range" message is given, and a new sample is called for. Experience has shown that the initial ω/χ peaking scans will often find the tail of a peak even when the misorientation is a few degrees and therefore avoid search mode. Upon finding (002), the (111) locus is computed and a 90° segment of the (111) locus is systematically searched by a combination of φ/χ motions. This search is conducted in the region providing the maximum degree of freedom which is in the range of $\chi = +90°$ to $\chi = +180°$. Once a (111) has been located, the poles are all transformed to coincide with the starting position of the specimen and the (100), (010) and (011) poles are generated. The spherical coordinates of these poles are then transformed to rectilinear coordinates and a fully documented stereographic projection is plotted.

The crystals of interest provide very high intensities; therefore, surface condition is not critical for doing orientation determinations. An as-cut surface from an abrasive cutoff wheel provides sufficient intensity for orientation. This is an important advantage over the Laue method. Another advantage is that an operator with no knowledge of crystallography can routinely orient these crystals.

CRYSTAL QUALITY CHARACTERIZATION

In addition to handling the orientation task, the system provides a very flexible arrangement for characterizing the quality of
these large crystals. Utilizing the automated X-Y translation capability, the oriented specimen can be rapidly surveyed for subgrain
boundaries by traversing the crystal across the X-ray beam and
noting discontinuities in the monitored X-ray signal as boundaries
are encountered. Detailed information on crystal quality and sub-
grain development can also be generated readily through ω and χ
rocking curves. An even more comprehensive picture of crystal quality can be generated in the form of three-dimensional crystal quality maps. These maps are established by generating a spiral scan
of two angles and representing various intensity levels by different
symbols and colors on the HP graphics plotter. As with the orientation work, the crystal quality data are generated automatically
based on a simple dialogue with the microcomputer terminal.

REFERENCES

1. B. H. Kear and B. J. Piearcey, Tensile and Creep Proper-
    ties of Single Crystals of the Nickel-Base Superalloy
    Mar-M200, Trans. TMS-AIME, 239, 1209-1215 (1967).
2. F. L. VerSnyder and M. E. Shank, The Development of Colum-
    nar Grain and Single Crystal High Temperature Materials
    Through Directional Solidification, Mater. Sci. Eng.,
    6, 213-247 (1970).
3. J. Mayfield, Single Crystal Technology Use Starting,
    Aviation Week & Space Tech., 69-73, October 1, 1979.

ON STREAM ANALYSIS OF LEAD AND ZINC ORE FRACTIONS USING

ENERGY DISPERSIVE AND WAVELENGTH DISPERSIVE TECHNIQUES

S. K. Kawatra

Metallurgy Department
Michigan Technological University
Houghton, MI 49931

and

J. L. Dalton
Canada Centre for Mineral and Energy Technology
Ottawa, Ontario

ABSTRACT

On-line X-ray fluorescence analysis is used in the mineral processing industry to monitor the composition of the solids contained in various slurry streams. This study compares wavelength-dispersive and energy dispersive techniques by using a slurry recirculation system employing both an X-ray tube excitation-wavelength dispersive system, and an isotope-excitation energy dispersive system. The results showed the less costly energy dispersive system yields accurate information that can be used to control milling operations.

INTRODUCTION

Compositional analysis of process streams in the minerals industry has been achieved on-line through X-ray fluorescence analysis (XRFA) for a number of years (1-8). Slurry samples, taken from various points in the mineral concentrator, are pumped to an air-conditioned control room, in which a wavelength-dispersive spectrometer is located. Depending upon the manufacturer, each slurry stream may be analyzed for as many as eight elements, and up to 14 streams may be analyzed by one spectrometer.

The greatest drawback of XRFA in mineral concentrators is the high cost of such a system. Many efforts have been made to try to

reduce this cost. In Australia, immersible probes for on-line
analysis (3,4) have been designed for immersion in flotation cells.
Each probe contains a radioisotope source and an X-ray detector.
Such a sensor is advantageous in that its simplicity eliminates the
need for a great deal of extraneous equipment. The major disadvantage is that it is only capable of measuring one element. Another
alternative is the use of an energy-dispersive detector (EDX) such
as a Si(Li) X-ray detector used with radioisotope excitation. This
system is an attractive means of on-line XRFA, but has not received
any attention in scientific literature. The technique has been in
use for analyzing discrete samples for some time (8).

The type of instrumentation used in on-line XRFA is adequately
described in references 1 to 7. For the relation of chemical composition of the mineral product to X-ray fluorescence intensities,
Kay uses linear regression techniques, with interelement and density effect considerations at Inco's Clarabelle mill (9), so also,
Gowans uses linear programming at the Sullivan concentrator of
Cominco (1). Smith emphasizes that calibration requires carefully
controlled experiments and the application of relatively sophisticated mathematical techniques (6).

Thus, the development of a less expensive technique of on-line
XRFA involves concentrating on two areas:

   i.   The use of an energy-dispersive X-ray detector with ra-
        dioisotope excitation.

   ii. The mathematical relationships between X-ray fluorescence
        intensities and chemical composition.

EXPERIMENTAL

A test rig having both wavelength-dispersive and energy-dispersive instrumentation was constructed so that a comparison between the two techniques could be made on the same slurry. A lead-
zinc ore was used in this study. The composition of the ore fractions is given in Table 1.

The test circuit is shown in Fig. 1. The closed loop consisted of a three-litre receiving tank, a variable-speed drive
peristaltic recirculating pump, a density gauge assembly, a con-
stant-flow head tank, a single-window flow cell in an assembly
with isotope-excitation and a Si(Li) X-ray detector, and a single-
window flow cell mounted to a modified Philips model 1540 X-ray
spectrometer. Details of the density gauge and the two single-
window flow cell assemblies are listed in Table 2.

Density gauge assembly utilized a 10mCi Co-60 point source
isotope collimated through a 0.95cm diameter aperture. As shown

Table 1. Analysis of Lead-Zinc Ore Products

Sample	% Moisture	% Lead	% Zinc	% Iron	% Sulphur
75-1423 Brunswick Head	0.12	3.69	9.31	29.04	35.10
75-1424 Brunswick Pb Conc.	0.15	36.32	5.97	21.48	31.62
75-1425 Brunswick Comb. Tail	0.11	1.01	1.88	34.18	37.45
75-1426 Brunswick Zn Conc.	0.24	1.50	51.66	9.00	30.53
75-1427 Cominco Zn Conc.	0.26	6.04	48.27	10.84	29.96
75-1428 Cominco Pb Conc.	<0.02	65.83	4.02	8.99	17.56
75-1429 Cominco Zn Ro. Conc.	1.05	1.84	10.20	33.26	24.69
75-1430 Cominco Tail	1.01	0.46	0.42	34.59	20.52
75-1431 Cominco Feed	0.02	6.81	6.77	23.30	17.72

in Fig. 2, the single-window flow cell in the isotope-excitation
assembly utilized two radioisotopes, Gd-153 and Am-241. Such a
configuration allowed a comparison between the two to be made. The
characteristics of these sources are detailed in Table 3. A schematic of the Philips 1540 X-ray spectrometer, modified to accept a
slurry flow cell in a vertical alignment, is shown in Fig. 3.

Experimental procedure entailed preparing mineral slurries to
simulate concentrator conditions. Changes in solids was accomplished by simply mixing various quantities of concentrate and
tailings to simulated feed and tailings lines, reagent grade chem-
icals and feed to a concentrate line, and pyrite and feed to a
tailing line. All recorded changes in measurements were compared
to those made on water. Painstaking care was taken to ensure consistency. A typical slurry run consisted of approximately 20 ore

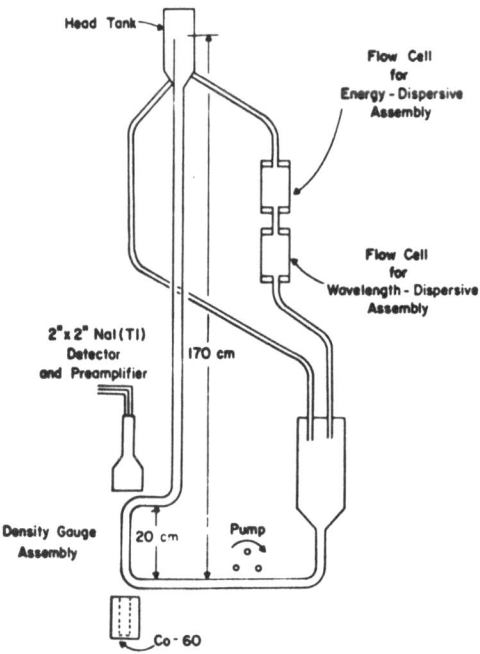

Fig. 1.   Experimental Test Rig

additions to the circuit and required eight to ten hours. The fol-
lowing measurements were recorded: PbL_α, ZnK_α, WL_α, and FeK_α radiation with the Philips spectrometer and PbL_α, PbL_β, ZnK_α, FeK_α
Compton and Rayleigh scatter radiation with the EDX system.

RESULTS AND DISCUSSION

     A multiple linear regression technique was used in analyzing
the data. The mathematical model that was derived from the data
was as follows:

Percentage of Element in Solids =

$K_0 + K_1 I_{Pb} + K_2 I_{Zn} + K_3 I_{Fe} + K_4 I_{Sc} + K_5 I_{Pd}$
where,

K_0 to K_5 were calibration constants

I_{Pb}, I_{Zn}, I_{Fe}, and I_{Sc} were the X-ray intensities from lead, zinc,
iron and scatter radiations, respectively

I_{Pd} was the transmitted gamma-ray intensity from the density gauge

The measured intensities of I_{Pb} were either I_{PbL_α} or I_{PbL_β} and
I_{Sc} were either Compton or Rayleigh counts.

Table 2.   Details of Analytical Instrumentation

1. Density gauge

    Radiation path length
      through slurry   -20 cm
    Isotope   -10 mCi Co-60 point source (AECL)
    Detector   -2 in x 2 in NaI (Tl) (Harshaw)
    High voltage supply   -Ortec Model 456
    Analyzer   -Baird-Atomic Model 530
    Discriminator settings   -Lower Level, 1.0 MeV
                                -Upper Level, Threshold

2. Single window flow cell with isotope excitation and energy
      analysis system

    X-ray window   -3 mil polypropylene, 3.2 cm in
                                 diameter
    Excitation   -1) 100 mCi Am-241 annular source
                                 (Inax Instruments)
                                -2) 100 mCi Gd-153 annular source
                                 (Inax Instruments)
    Collimation   -Inax Instruments dual assembly,
                                 sample aperture 1.6 cm detector
                                 aperture, .33 cm material-pure tin
    Detection   -Si(Li) X-ray detector, active
                                 diameter 6 mm, depth 3 mm, Ortec
                                 Model 7116-06215
    High voltage supply   -Ortec Model 456
    Amplifier   -Ortec Model 730
    Analyzer   -Tracer-Northern TN-1700
                                 Pulse Height Analyzer

3. Single window flow cell with X-ray tube excitation and wave-
      length analysis system

    X-ray window   -3 mil polypropylene, 3.2 cm in
                                 diameter
    Spectrometer   -Modified Philips 1540 X-ray spec-
                                 trometer fitted with a Philips
                                 NR-100 tungsten target X-ray tube
    Analyzing crystal   -LiF (220)
    Detection   -Philips PW 1964 Scintillation
                                 Detector
    High voltage supply   -Philips PW 4025
    Electronics   -Philips single channel analyzer
                                -PW 1365 Pulse shaper
                                -PW 4280 Amplifier/Analyzer
                                -PW 4231 Scaler
                                -PW 4261 Timer

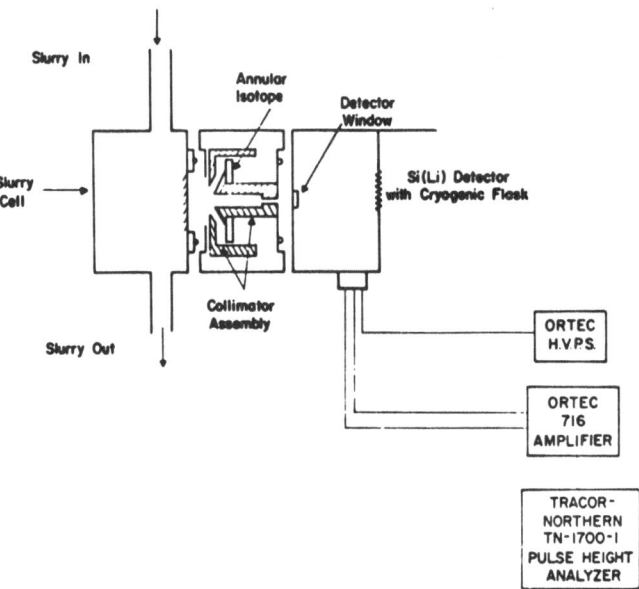

Fig. 2.   Schematic Diagram of the Energy Dispersive X-ray System

Table 3.   Characteristics of Radioisotopes Used in Study (ref. 10)

Isotope	Half-Life	Gamma Radiation (keV)
EDX system		
Am-241	485 years	13.945
		17.740
		20.774
		60.0
		small amounts of radiation
		in the range 100-700
Gd-153	240 days	41.5
		47.0
		70.0
		99.0 (complex)
Density gauge		
Co-60	5.2 years	1,173
		1.332

Data from the Philips spectrometer was highly consistent and straight-forward as far as analysis was concerned. Lead, zinc, and iron were always determined as a function of the intensities of PbL_α, ZnK_α, FeK_α, WL_α - Rayleigh Scatter, and the density gauge.

However, the EDX system was more complicated as two isotopes were used, both Rayleigh and Compton scatter measurements were taken, and both lead L_α and L_β radiations were measured.

The equations that resulted from the data analysis clearly showed that both energy dispersive and wavelength dispersive techniques are acceptable for on-line analysis since most of the percentage error was less than five percent.

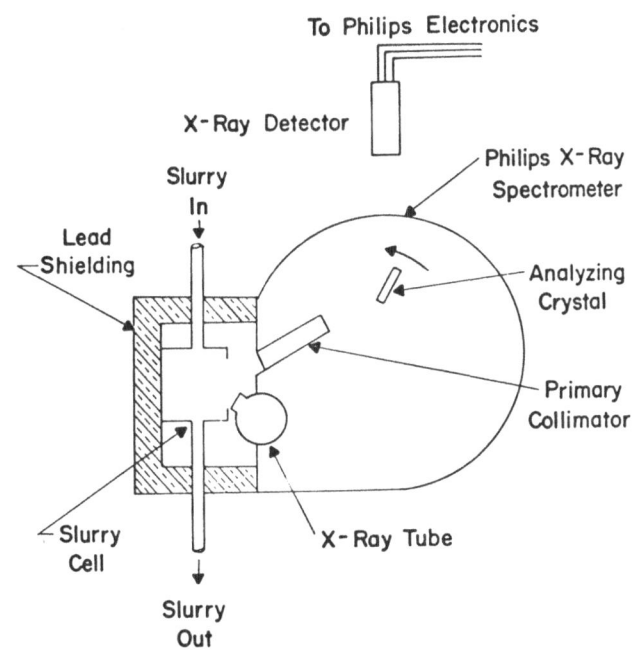

To Philips Electronics

X-Ray Detector

Slurry
In

Lead Shielding

Philips X-Ray Spectrometer

Analyzing Crystal

Primary Collimator

Slurry Cell

X-Ray Tube

Slurry Out

Fig. 3.   Schematic Diagram of the Wavelength
          Dispersive X-ray System

Standard deviations using these equations ranged from 0.02 to 1.02 and corresponding percentage errors ranged from 1 to 5 percent. The only exceptions were the lead and zinc determinations in the two tailings runs with the Gd-153 isotope were rather high. It was not readily apparent why these values are high.

In comparing the Gd-153 isotope to the Am-241, it was seen that there is very little difference between the two. Mathematically determined errors for the two isotopes were almost the same.

It should also be noted that the Compton scatter provides a better correction factor than the Rayleigh. Further analysis of the data showed that the wavelength-dispersive technique employing an X-ray tube with a crystal gave yields more precise results than the energy-dispersive technique employing an isotope and a solid state detector. However, it must be reiterated that the energy-dispersive technique gave the desired precision for control on mineral processing operations. Moreover in a plant, the precision was restricted not by the technique, but by the variation of plant parameters such as particle size (4) and entrained air in mineral slurries (5).

The energy-dispersive system described has many advantages over the wavelength-dispersive system. First of all, it is portable. This allows the system to be used on almost any stream of interest in a plant without much trouble in relocating it. The EDX system is also less expensive than the wavelength-dispersive system. It does not require a central control room necessitating the use of long sampling lines which add to the cost of the system. Thus, an energy-dispersive system may be feasible for assay or process control work where the wavelength-dispersive system cannot always be justified.

CONCLUSIONS

An energy-dispersive system employing an isotope and a solid state detector is capable of giving the desired precision for control of a mineral processing plant. It is more economical than a wavelength-dispersive system due mainly to the fact that it is highly portable and does not require the costly equipment a stationary system does. Such a system offers a highly accurate means of control at a reduced cost to various types of mineral processing operations.

REFERENCES

1. Gowans, W.K., Proceedings of the Seventh Annual Meeting of the Canadian Mineral Processors, Ottawa, Canada, pp. 143-166, Jan. 1975.
2. Kawatra, S.K., Int. J. Miner. Process. V. 3, pp. 41-50, Feb. 1976.
3. Fookes, R.A., Gravitis, V.L., Watt, J.S., Wenk, G.J., Wilkinson, L.R., McColl, I.G. and Baughen, H.J., Proc. Australas. Inst. Min. Metall., No. 239, pp. 93-100, Aug. 1971.
4. Kawatra, S.K., "On-Line Determination of Copper and Lead in Mineral Slurries," Ph.D. thesis, University of Queensland, 1975 (unpublished).
5. Kawatra, S.K., Can. J. Spectrosc., V. 21, No. 1, pp. 5-10, Jan./Feb. 1976.
6. Smith, H.W., Min. Eng. (NY), V. 26, No. 11, pp. 11-13, Nov. 1974.
7. Lederer, C.M., Hollander, J.M. and Perlman, I. Table of Isotopes, John Wiley & Sons, Inc., New York; 1968, p. 273.
8. Burkhalter, P.G., Anal. Chem., V. 43, No. 1, pp. 10-17, Jan. 1971.
9. Kay, K.R. and Paterson, J.G., CIM. Bull., V. 68, No. 754, pp. 65-69, Feb. 1975.

ON-STREAM ANALYSIS OF FLOAT PROCESS

SLURRIES BY XRF

J. Ereiser, G. Frank, J.W. Guy, D. Litchinsky

Bondar-Clegg & Co. Ltd.

6 Bexley Place, Ottawa, Ontario, Canada

ABSTRACT

An on-stream XRF analysis system for monitoring the concentrations of the elements in float process slurries is described. The system consists of multiple XRF stations monitored
by a central computer. The stations utilize high resolution
Si(Li) detectors and radioisotope x-ray sources for stability and
for good separation of adjacent elements such as Cu and Zn.
Results obtained from an on-stream slurry analysis of Copper
and Lead Tails and Copper and Lead Concentrates are presented and
these are compared to assay lab results.

INTRODUCTION

Ores are a dispersion of valuable mineral in gangue or waste
matrix. Concentration of the mineral is usually low and must be
increased by grinding and floatation methods. The process of
concentration depends heavily on ore grade and heterogeniety and
tight control on critical processes must be maintained for
efficient separation and maintenance of the product quality. One
of the more important areas for process control is the monitoring
of the mineral analysis. Traditionally chemical analysis has been
performed in the assay lab using 'grab samples' obtained on a regular basis by mill employees. These samples are then analyzed in
the traditional way either by gravimetric methods, X-ray
Fluorescence or Atomic Absorbtion. The difficulty with this
type of process monitoring is the large error that is introduced
by intermittent sampling and the time lag between the sample
being taken and the analysis results being obtained by the mill
control room. Recently (1 , 2) on-stream x-ray fluorescence

analysis systems have been introduced to redress these problems
and in addition allow more process control capabilities for improved recovery and higher product quality. These on-stream
methods of monitoring chemical concentrations provide a fast,
accurate and effective method of monitoring because several orders
of magnitude more of the stream is seen than in discrete sampling
methods. These systems reduce operating costs by improving recovery and in addition, reduce the number of assay lab staff
especially during night shift and over weekends.

INSTRUMENTATION

     Figure 1 shows an on-stream XRF system, model CE800, which
is produced by Bondar-Clegg & Company Ltd. Slurry sample flows
are cut from the main stream and move past an x-ray fluorescence
analysis station composed of a special slurry flow cell, an I-125
radioisotope source and a cooled Si(Li) detector. The key to
accurate analysis is the sample presentation at the flow cell.
X-rays have limited penetration in mineral slurries and consequently only a small fraction of the process stream is viewed by
the XRF system. Entrapped air, sample segregation and particle
size result in sample presentation problems and thus errors in
the mineral analysis. Figure 2 shows a thief sampler system used
by B.C.C. for cutting samples from a slurry stream. Very good
results with regard to reduced air entrapment and sample
segregation are obtained with this method and down time due to
blockages is low. Data is accumulated by an instrument package
in the mill control room and interpreted by a PDP 11 computer
which in addition may be coupled to other instruments for limited
process control.

     The advantages of the system are; continuous monitoring of
up to 16 remotely located streams each with an independant 1024
channel MCA, high resolution cooled Si(Li) detectors for good
separation of adjacent fluorescence lines such as Cu and Zn,
use of stable monoenergetic I-125 radioisotope X-ray sources
having an incident energy of approximately 27 KeV, automatic
detection of electronic faults and slurry leakages at the
window between the detector and the slurry flow.

DATA INTERPRETATION

     Both percent solids and percent element are calculated
using the equations shown in TABLE 1. A,B, & C are constants
derived from linear regression. CRo is a count rated measured
for a standard at a DAY 0, DAYS is the elapsed number of days
since DAY 0 and the 86.6 value in the percent solids equation is
related to the decay constant of the I-125 radioisotope source.
Scatter is the compton backscatter integrated intensity and the
time is the measurement time in seconds.

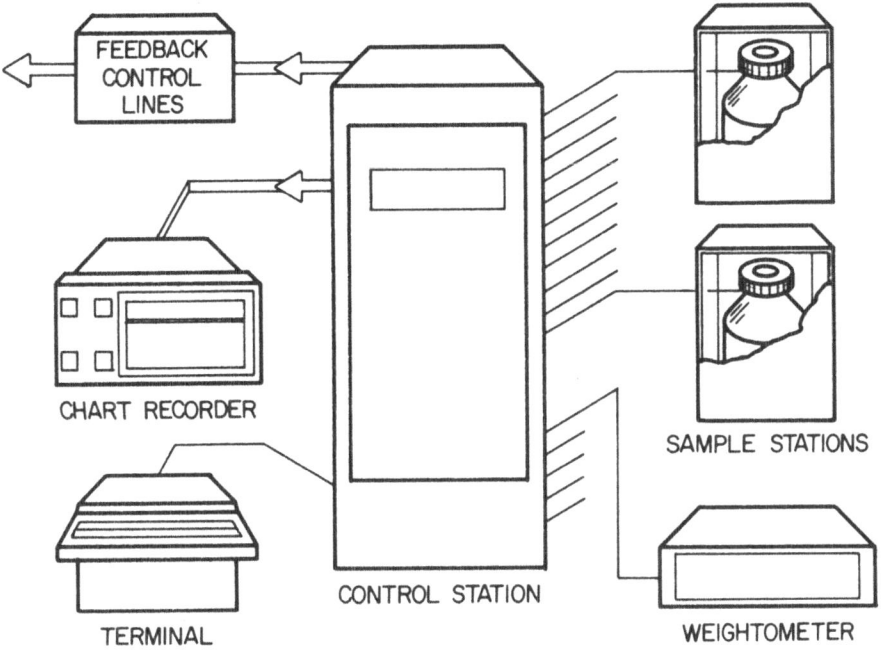

Fig. 1 On Stream Analysis System

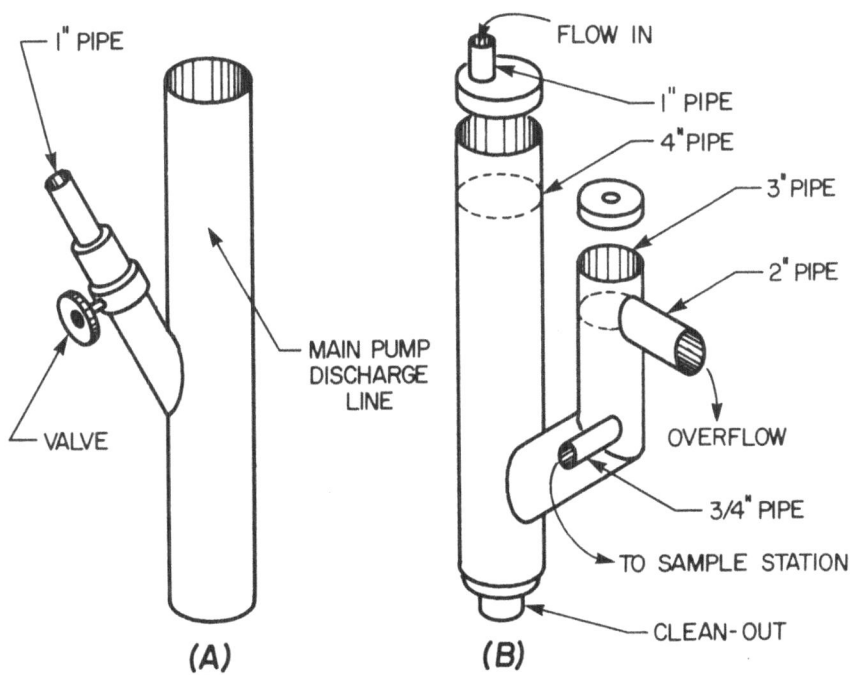

Fig. 2 Thief Sampler

TABLE 1

CALIBRATION EQUATIONS

$$\% \text{ solids} = \frac{A * CR_o e^{- \text{ DAYS}/86.6}}{\text{Scatter/Time}} + B + \sum_i c_i \frac{\text{Element}_i}{\text{Scatter}}$$

$$\% \text{ element} = \frac{\sum_j C_{ij} \frac{\text{Element}_j}{\text{Scatter}} + B}{\% \text{ Solids CALCULATED}}$$

XRF sensitivity is best for elements with Z greater than Manganese. For elements of lower atomic number the problems associated with particle size become important. Marked inter-element effects are usually less significant than say powder samples due to the high dilution of the atoms of the solids by the matrix and by the water carrier.

DISCUSSION

Data for two streams are presented in Figures 3 and 4. These data are derived from actual results accumulated over several days. Deviations from the lab results are due in large part to sampling errors. Other data indicate that for Copper, Lead and Zinc ores in the range 3 to 10%, major constituent concentrations precisions are <5% relative for 15 minute measurements with accuracies of <10% relative. Ore tails precision and accuracies are <0.01% absolute. Silver can be measured to ±10 ppm for silver ore tailings in the range 0-200 ppm.

CONCLUSION

Operating results from the Bondar-Clegg and Co. Ltd. XRF on-stream analyzer indicate that quick and accurate analyses can be obtained in the mill environment thus allowing enhanced float-ation process control to be carried out.

REFERENCES

1.   Watt J.S. "Nuclear Techniques for On-Line Measurement in the Control of Mineral Processing"; IAEA Int. Symp on Nuclear Techniques, Exploration, Extraction and Processing of Mineral Resources; IAEA-SM-216-103; 569-602 (1977).
2.   Cooper H.R. and Vaugh R.L. "Computer Automated X-Ray Fluorescence Assaying"; TRANS SME/AIME; 224:295-301 (1969).

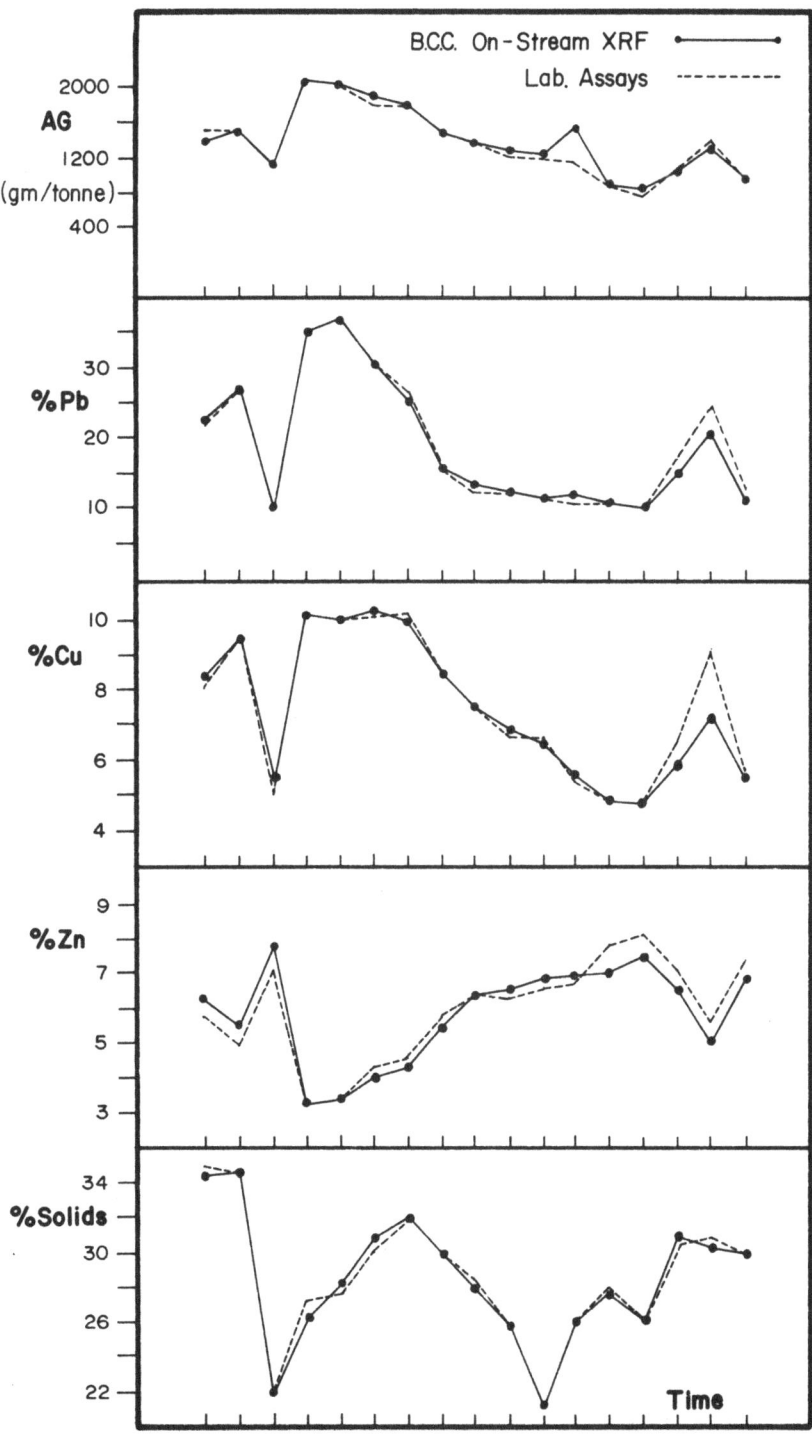

Fig. 3 Lead Concentrate Stream

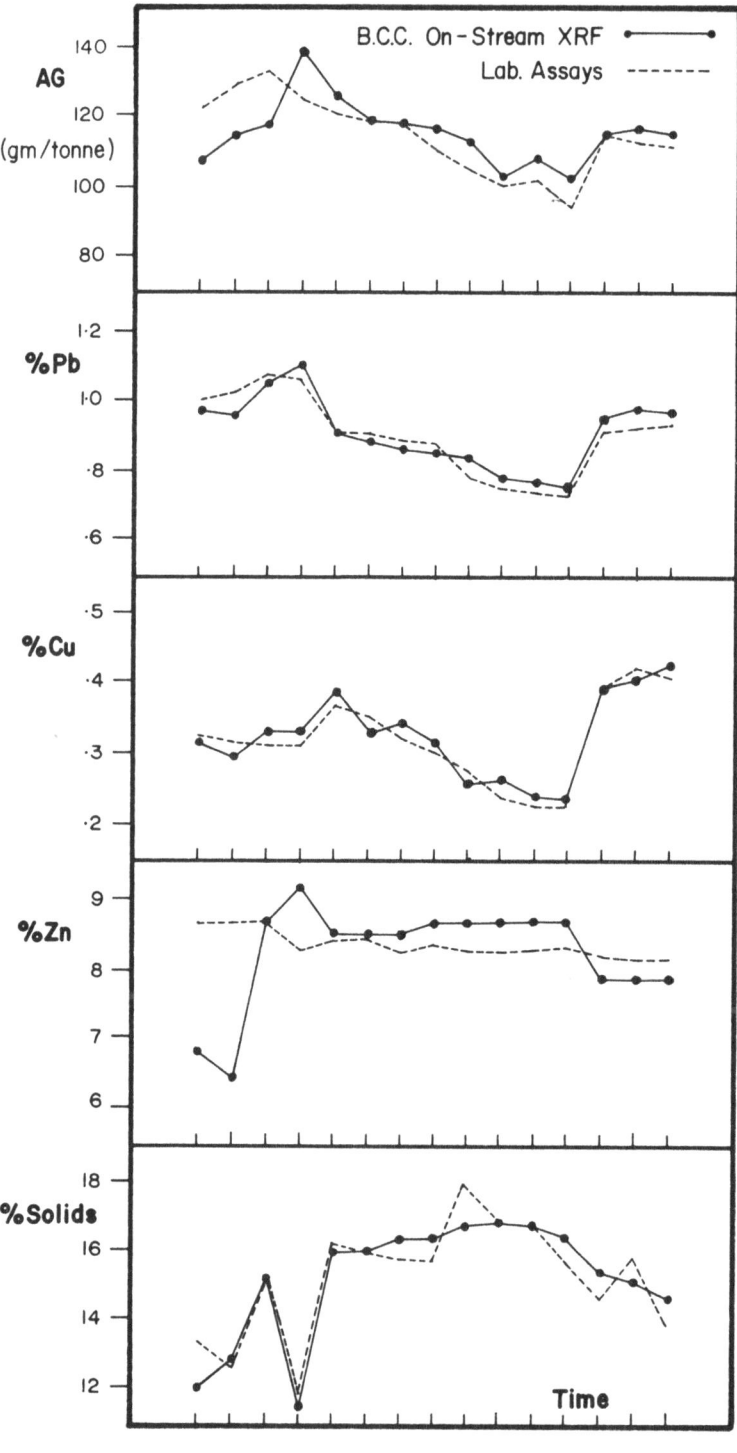

Fig. 4   Copper Tails Stream

X-RAY FLUORESCENCE ANALYSIS OF HIGH Z MATERIALS WITH MERCURIC IODIDE ROOM TEMPERATURE DETECTORS

J. Nissenbaum, A. Holzer, M. Roth and M. Schieber

School of Applied Science and Technology
The Hebrew University of Jerusalem
Jerusalem, Israel

ABSTRACT

Mercuric iodide HgI_2 room temperature solid state radiation spectrometers having 4% energy resolution at 100 KeV detected the x-ray fluorescence (XRF) of the K shell of intermediate and high Z elements. The excitation of the K shells which emit XRF more energetic than 60 KeV was achieved with 7mCi collimated [57]Co and for XRF less energetic than 60 KeV the excitation was done with a 10mCi [241]Am source. The K shell XRF spectra of a 1:1 mixture of U and Th, and also of the single elements of Au, Tb, Ba, Ag, Mo, and Rb are shown. The results prove the feasibility of developing mercuric iodide portable XRF spectrometers which operate at room temperature and which have a wide range of geochemical and industrial applications.

INTRODUCTION

Semiconductor radiation detectors are now routinely used in non-dispersive XRF analysis. High resolution silicon and germanium detectors are available commercially providing the possibility to resolve the K shell fluorescence peaks of all elements with Z>6. However, the necessity of cryogenical cooling is a drawback in many applications, e.g., field studies of geological samples, on-line industrial analysis, scanning electron microscopy, etc. This explains the extensive search for new solid state crystals which have a larger bandgap than Si or Ge so that they can be used at ambient temperatures without producing excessive noise due to high leakage currents. Such a detector is HgI_2 grown from the vapor phase[1]. Mercuric iodide detectors are more successful than room temperature scintillators, since the latter

303

are not capable of separating the Compton scattering peaks from the K shell fluorescence.

GaAs[2,3], CdTe[4,5], and HgI$_2$[6,7] compounds have been comprehensively studied in the past few years, but only mercuric iodide detectors are suitable for high resolution spectrometry in the K and L X-ray range of 1 KeV to 30 KeV[8,9]. A comparison between the spectroscopic resolution of HgI$_2$ and CdTe room temperature semiconductor detectors are discussed elsewhere[7,10].

A portable HgI$_2$ spectrometer has been employed very recently by Dabrowski et al[11] to resolve the characteristic L X-rays of uranium. A different approach to the quantitive analysis of uranium ores uses higher energy UKα and UKβ X-rays. The latter have higher penetration which reduces the homogeneity requirements of the analysed samples. Rather than exciting the L and M shells, which requires high quality detectors and preamplifiers, the present paper describes the X-ray fluorescence of the K shell, using average quality HgI$_2$ radiation detectors and average charge sensitive preamplifiers without any preselected FET.

The present paper reports the typical K shell XRF detected with a 4% resolution HgI$_2$ detector in the energy range of about 10 to 150 KeV, which involves the excitation of uranium, thorium and gold with ^{57}Co and of Tb, Ba, Ag, Mo, and Rb with ^{241}Am.

FABRICATION OF XRF MERCURIC IODIDE SPECTROMETERS

The detectors used for the XRF spectrometers were fabricated from single crystals of HgI$_2$ grown by the temperature oscillation method from the vapour phase[1]. The crystals were cleaved into plates and etched in a 5% KI-water solution until the plates reached the desired thickness of 0.4 mm. The contacts were thin evaporated palladium films and platinum wires of 25μm diameter were attached to the contacts with Aquadag (carbon emulsion). The detector plates were then fully encapsulated in Humiseal and mounted in an aluminum case provided with a berillium radiation window. A charge-sensitive preamplifier with a non-selected FET at the first stage of amplification was used, followed by common nuclear spectroscopy electronics[12]. The detector bias voltages ranged from 1000 to 2000 V according to the transport properties and thickness of each detector.

For the XRF analysis of uranium and thorium, the HgI$_2$ detector had an active contact area of 16 mm^2 and a thickness of 0.4 mm and had an energy resolution of 4% at 100 KeV. For the XRF analysis of Au, Tb, Ba, Ag, Mo and Rb, the HgI$_2$ detector had an active contact area of 18 mm^2 and a thickness of 0.32 mm and an energy resolution of 4% at 100 KeV.

X-RAY FLUORESCENCE ANALYSIS

The excitation was carried out with two sources, with ^{57}Co for the XRF of uranium, thorium and gold and with ^{241}Am for the XRF of Au, Tb, Ba, Ag, Mo and Rb. In the first case, 7 mCi ^{57}Co disk source had been put into a 28 mm diameter lead collimator 1 cm long. The sample powders were mounted in a plastic box at a distance of 3 cm from the source and 1 cm from the collimated HgI_2 detector. The 122 KeV ^{57}Co gamma emission is above the K absorption edge of uranium, thorium and gold (115.6, 109.6 and 80.7 KeV respectively [13]) allowing the excitation of all spectral lines.

A standard 10 mCi ^{241}Am disk source provided with a tungsten alloy rear shielding was used to excite the Kα and Kβ fluorescence of Tb, Ba, Ag, Mo, and Rb with respective K absorption edges of 52.0, 37.4, 25.5, and 15.2 KeV respectively. The use of the 59.6 KeV gamma emission avoided the Kα and Kβ lead fluorescence of the collimator which occurs at about 85 and 75 KeV. However, the high energy α particles emission of ^{241}Am excited some of the WKα and WKβ fluorescence, which occurs at about 60 and 68 KeV respectively and therefore did not interfere with the desired XRF analysis of the elements which occurred between 13.4 KeV for RbKα to 50.2 KeV for TbKβ.

Optimal geometrical conditions were used for both the ^{57}Co and ^{241}Am excitations in order to improve the ratio between the intensity of the excited peaks and background and at the same time to minimize the Compton peak interference with the examined lines. In order to determine these geometrical conditions, different scattering angles ranging from $0°$ to $140°$ in $10°$ steps were obtained.

Using scattering angles of $90°$ minimizes the intensity of the Compton peaks but since it appears at about 95 KeV the Compton peak overlaps with the source of Kα and Kβ peaks of uranium and thorium which are placed in the 89 to 114 KeV region. Only the $130°$ scattering angle shifts the position of the Compton peak to allow the determination of the Kα and Kβ peaks of U and Th. A secondary lead collimator of 4 mm diameter was mounted on the detector's entrance window in order to reduce the spectral dispersion of the primary Compton scattering and thus improve the peak to background ratio.

The XRF and the K peaks of high concentration mixtures of uranium and thorium can easily be analyzed with a HgI_2 detector at room temperature. Fig. 1 shows a 1:1 mixture of $4O_2(NO_3)_2 \cdot 6H_2O$ and $Th(NO_3)_4 \cdot 4H_2O$, taken at a scattering angle of $130°$. The following lines can be seen: UKβ2(114 KeV) as a discernible shoulder UKβ1(111.1 KeV), ThKβ1(106 KeV), UKα1(98 KeV), UKα2 and

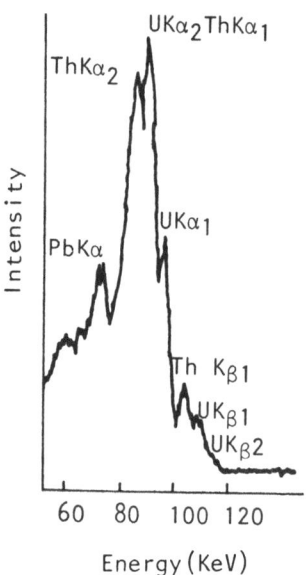

Fig. 1 XRF spectra of a
1:1 mixture of $UO_2(NO_3)_2$
$6H_2O$ and $Th(NO_3)_4 4H_2O$
scattering angle 130°.

ThKα1 lumped together(94 KeV)
ThKα2(90KeV) and PbKα(75KeV).

The distinction between the
Kβ lines of U and Th serve as a
quantitative evaluation of the U
and Th concentration in a given
sample. Preliminary results,
however, show that with a 4%
energy resolution HgI_2 spectrometer at 100 KeV, one can resolve the Kβ peaks of U and Th
mixtures containing low concentrations (down to 0.15%) of U in
the presence of Th incorporated
in low Z matrixes. At concentrations lower than 0.15% U the
Kβ peaks are obscured by the Kα
peaks and Compton scattered radiation. Better energy resolution HgI_2 spectrometer will allow XRF analysis of smaller
concentrations than 0.15% of U
in Th containing ores.

Fig. 2 shows the XRF of gold
excited with [57]Co. The
Kα1 and Kα2 appear as one line
at ∿ 79 KeV and the Kβ1 and Kβ2 appear as another single line at
∿ 69 KeV. No interference of the K shell X-rays of the lead collimator which should be at about 75 and 85 KeV for the Kα and Kβ
respectively. Fig. 2 also shows an iodine Kα escape peak at
about 40 KeV.

The XRF of the K shells which emit at lower energies than 60
KeV were excited with the 59.6 KeV of [241]Am. The K lines of
the pure elements of Tb, Ba, Ag, Mo and Rb were recorded as shown
in Figures 3a to 3e. Again as in the case of gold (Fig. 2) there
is a lumping together of the Kα1 and Kα2 while the Kβ1 are lumped
with the Kβ2 lines. The respective peaks are at about 45.6 and
50.9 KeV for the Kα and Kβ of Tb at 32.0 and 36.6 KeV for the Kα
and Kβ of Ba at 22.0 and 25.2 KeV for the Kα and Kβ for Ag at
17.4 and 19.7 KeV for the Kα and Kβ of Mo and 13.3 and 15.0 KeV
for the Kα and Kβ of Rb.

The lower the XRF energy is for the characteristic K lines,
the smaller the energy separation between the Kα1 and Kα2 or the
Kβ1 from the Kβ2. For the lower energies at 17.4 KeV of Mo Kα
(Fig. 3e) the Kβ lines can be hardly guessed as a tiny shoulder

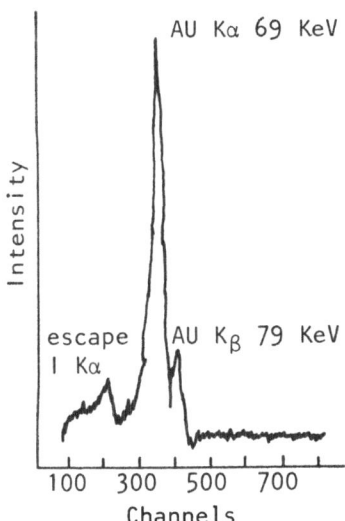

AU Kα 69 KeV

Intensity

escape | AU Kβ 79 KeV
I Kα

100   300   500   700
Channels

Fig. 2   XRF spectra
of gold (scattering
angle 130°).

to the stronger Kα lines. It is
on this low energy X-ray range
where low noise preamplifiers
(8.9) would be helpful for better spectral line separation.

The α emission of the ^{241}Am
source exists also the Kα and
Kβ lines of the tungsten shield
of the ^{241}Am source. The WKα
line is at about 58.6 KeV and is
lumped together with the photopeak of 59.6 KeV of ^{241}Am,
while the WKβ line is at about
68 KeV. The WKβ line can be
seen in most of the XRF spectra
shown in figures 3a to 3e. Iodine escape lines at about 41
KeV for the ^{241}Am photopeak can
also sometimes be seen. There
is no serious interference from
either the escape lines or the
collimator and shield fluorescence with the analytical XRF
of the elements studied in this
paper.

CONCLUSIONS

   The present paper demonstrates that even with medium grade
HgI_2 detectors and charge sensitive preamplifiers, both of which
are commercially available, one can perform XRF analysis of the
K shell of high and medium Z elements at room temperature. A
portable field XRF detector is thus feasible and could be useful
for mining exploration, warehouse metal identification or monitoring of thickness of protecting coatings of metals on substrates.

ACKNOWLEDGEMENTS

   The authors are indebted to Professor S. Ofer of the Racah
Institute of Physics of the Hebrew University of Jerusalem who
provided the ^{57}Co source and to the Yissum Corporation of Hebrew University of Jerusalem for supplying the HgI_2 spectrometers
used in the present work.

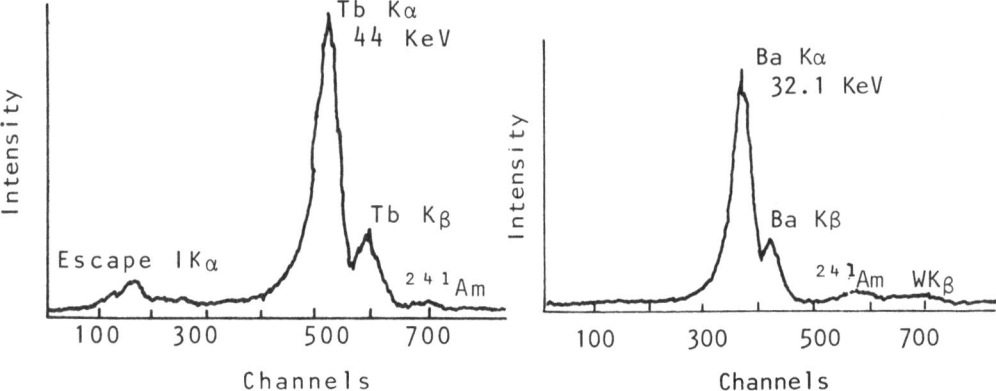

Fig. 3a XRF spectra of terbium.   Fig. 3b   XRF spectra of barium

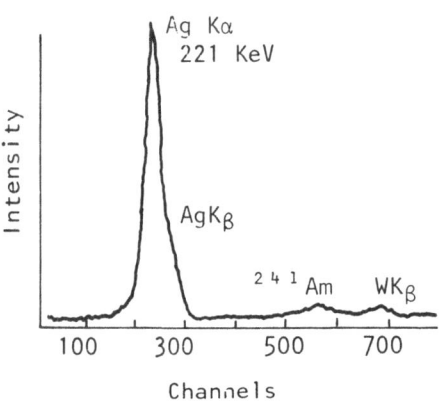

Fig. 3c XRF spectra of silver

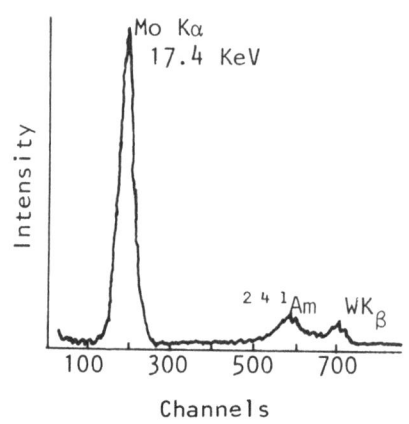

Fig. 3d   XRF spectra of
           molybdenum

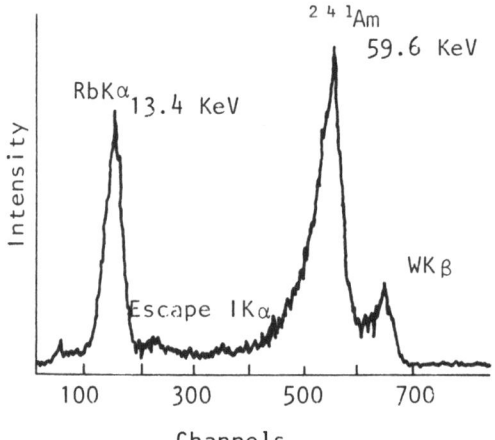

Fig. 3e   XRF spectra of
           rubidium

REFERENCES

1. M. Schieber, I. Beinglass, G. Dishon, A. Holzer, 1977, Journal of Crystal Growth, 42;166.

2. J.E. Eberhardt, R.D. Ryan, and A.J. Tarendale, 1971, Nuclear Instruments and Methods, 94, 463.

3. T. Kobayashi, I. Kuru, A. Hojo, and T. Sugita, 1976, IEEE, Trans. Nuc. Sci., NS-23, No. 7, 97.

4. J. Iwanczk and A.J. Dabrowski, 1976, Nucl. Instruments and Methods, 134, 505.

5. A.J. Dabrowski, J. Iwancyk, W.M. Szymczyk, P. Kokschinegg, J. Stelzhammer and R. Triboulet, 1978, Nucl. Instruments and Methods, 150, 25.

6. M. Slapa, G.C. Huth, W. Seibt, M. Schieber and P.T. Randtke, 1976, IEEE, Trans. Nucl. Sci., NS-23, 101.

7. R.C. Whited and M. Schieber, 1979, Nucl. Instruments and Methods, 162. See a comprehensive review of this subject.

8. A.J. Dabrowski, G.C. Huth, M. Singh, T.E. Economu, and A. L. Turkerich, 1978, Appl. Phys. Letter, 33, 211.

9. G.C. Huth, A.J. Dabrowski, M. Singh, T.E. Economu, and A.L. Turkerich, 1979, Advances in X-Ray Analysis, Edited by C.J. McCarthey et. al., Vol. 22, Page 461, Plenum Press, N.Y.

10. C. Scharager, P. Siffert, A. Holzer, and M. Schieber, 1980, IEEE, Trans. Nuc. Sci., NS-27.

11. A.J. Dabrowski, M. Singh, G.L. Huth and J.S. Iwanczyk, to be published.

12. Such HgI_2 XRF spectrometers and preamplifiers are available commercially from Yissum Corp., Hebrew University of Jerusalem, Israel.

13. The X-ray energies are quoted from the tables published by the Technical Measurement Corporation, North Haven, CT.

APPLICABILITY OF U L X-RAY LINES FOR THE DETERMINATION
OF LOW URANIUM CONCENTRATIONS

J. Parus, T. Żółtowski, J. Kierzek, W. Ratyński

Institute of Nuclear Research

03-195 Warsaw, Dorodna 16, Poland

The determination of uranium in low grade ores, some industrial raw materials (e.g. phosphates) and industrial wastes (e.g.
coal ashes) at concentrations below 200 ppm is very important for
various reasons, and poses a serious analytical problem. In these
materials uranium is often accompanied by elements emitting K lines
which interfere with the most intensive U $L\alpha$ and U $L\beta_1$ lines. The
most serious interferences for U $L\alpha$ lines are from Sr and Rb sometimes present in analysed materials in comparable or excessive quantities. Also Mo, Zr and Y can interfere.

Radioisotopic excitation using 109-Cd (5 mCi activity) has
been applied. The measuring system consisted of Si(Li) detector,
pile up rejector and multichannel analyser. The recorded spectra
have been transmitted for evaluation to the PDP 11/45 computer (a
satellite to Cyber 73 machine). Due to limitations of the methods
for spectra evaluation[1,2] at low concentrations, the library least
squares method (LLSM) has been modified by introducing the inner-
outer iteration approach, minimizing the reduced chi-square value
on amplitudes of library spectra and library spectra channel indexes
(spectra shifts). After the completion of the inner minimization
loop related to the spectra shift, one step of the outer iteration
is carried out. The procedure is halted after attaining the con-
vergency limits (Fig. 1). During the minimization process the following a priori constraints have been employed: (i) on spectra
shift, and (ii) on relative intensities.

Two sets of samples have been analysed: (a) synthetic U+Sr
solutions, and (b) powdered uranium ores from geological prospecting.
First the measurements have been performed with the synthetic solutions containing an excess of strontium (10 mg per milliliter)

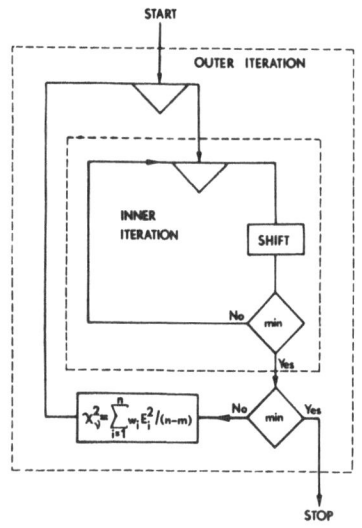

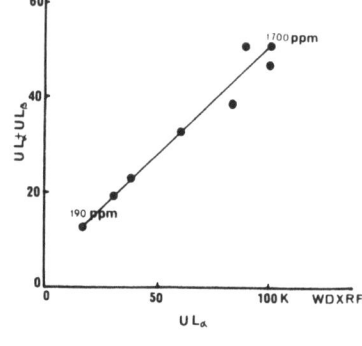

Fig. 1. Doubly Minimized
Library Least Squares Method

Fig. 2. The Relationship
between EDXRF and WDXRF
Results

corresponding to uranium/strontium ratio from 1/1000 to 1/1. Uranium concentration ranged from 10 to 10000 ppm. In the second step a set of samples in powder form from geological prospecting has been measured (U content from 100 to 2000 ppm). Their common feature is the presence of relatively high concentration of molybdenum (between 0.1 and 1%)and in some cases a very high content of the rare-earth elements.

The obtained relationships, using LLSM with inner-outer iteration approach between U Lα lines intensity and U concentration with strontium added, as well as the relationship between a sum of U Lα and U Lβ lines and U concentration in a case of natural samples, have had good linearity in the whole range of concentrations. Fig. 2 shows the relation between a sum of U Lα and U Lβ intensities for geological uranium samples obtained with energy dispersive X-ray fluorescence (EDXRF) and inner-outer iteration LLSM against U Lα intensity from wavelength dispersive X-ray fluorescence (WDXRF).

We have tried to show the advantage of using the radioisotope excited EDXRF to analyse low concentrations uranium samples. It seems to be possible to determine uranium at the tens of ppm level with strong overlaps if the inner-outer iteration approach to the LLSM is used. It is worth mentioning that in the light of current prospecting work for new uranium deposits and resources the EDXRF will be a technique that sometimes should be preferred to WDXRF.

REFERENCES

1. J. C. Russ, X-Ray Spectrom., 5:217 (1976).
2. T. Żółtowski, J. Parus, Z. Banasik, G. Kuc, to be published in Nukleonika, 1980.

APPLICABILITY OF PIXE AND XRF TO FAST DRILL CORE

ANALYSIS IN AIR

Lars-Eric Carlsson and K. Roland Akselsson

Department of Nuclear Physics, Lund Institute of Technology
Sölvegatan 14
S-223 62 LUND, Sweden

Abstract

The properties of particle-induced X-ray emission, PIXE, and
secondary target mode X-ray fluorescence, XRF, applied to the analysis of unprepared drill cores in open air have been evaluated.
Typical detection limits for elements heavier than Mg have been determined for a PIXE-system with an external 2.55 MeV proton beam and for
an XRF-system with Ti, Mo and Tb secondary targets. These two systems
were found to have similar detection limits for most elements in a
typical geological sample. The heterogeneous composition of drill
cores prevents the performance of accurate matrix corrections, though
calculations using fundamental parameters show that in the PIXE
analysis of elements heavier than Ca, these corrections are much less
sensitive to variations in the matrix composition than in the XRF
analysis.

Introduction

A large mineral company in Sweden stores 30 - 40 km of drill
cores in a central storage every year. These cores are visually
inspected for their mineral composition and only small fractions
are prepared for accurate elemental analyses. A technique for fast
analyses of all metals of prospecting interest without time consuming
sample preparation would open new possibilities for successful prospecting efforts.
There are several reasons why energy-dispersive X-ray methods
are suitable for such scanning drill core analysis. They are truly
multi-elemental, giving signals from unexpected elements without any
extra effort, which is an important property in elemental analysis

for prospecting purposes. Further, these methods are fast. With
irradiation times in the range of 1-3 minutes it is possible to
determine both major and trace elements below the 10 ppm level.
X-ray elemental analysis techniques are nondestructive and for some
applications little or no sample preparation is necessary.

Accurate analyses of undiluted pressed pellet samples in vacuum
have earlier demonstrated with both PIXE and XRF using the fundamental parameter technique. Verbeke and Adams (1) used Ti and Mo secondary targets to determine elements heavier than Na in seven rock
standards and stated an accuracy of approximately 10%. We have
earlier used 2.55 MeV protons to determine Li, F and all elements
heavier than neon in six rock standards by simultaneous X-ray and
gamma-ray analysis (2). Accuracies typically better than 5% and 10%
respectively were achieved.

In this work the main purpose has been to establish comparable
detection limits for PIXE and XRF analysis and to evaluate the accuracy of matrix correction for these two modes of X-ray excitation.

Experimental

A homogeneous pellet of geological material was irradiated with
2.55 MeV protons, which emerged out of the accelerator vacuum into
air through a 7 μm thick polyamid foil (Kapton). In fig.1 the principal view of this external beam set-up is showed.

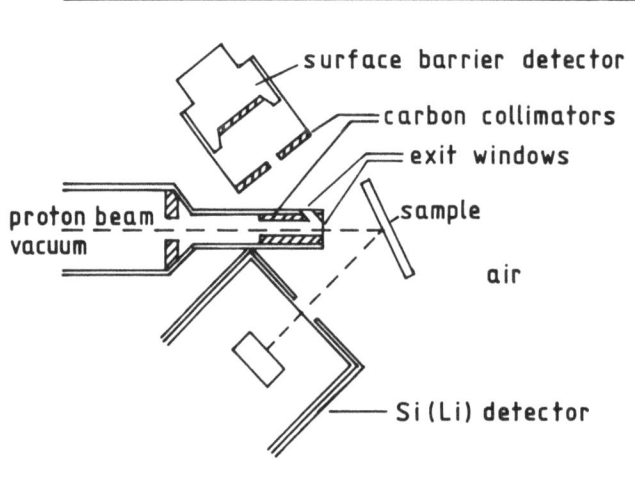

Fig. 1.   A schematic view of the set-up for external beam
           PIXE analysis.

A 25 mm path of air between sample and detector window renders X-ray
analysis of Na and Mg impossible. Beam current on the sample is
monitored by measuring protons backscattered from the exit foil.
In front of the Si(Li) detector a 0.3 mm thick Al absorber with a
small hole was placed to reduce the intensity of X-rays from the
abundant light elements in the sample.

The sample analysed with external beam PIXE was also analysed
by XRF, with a Mo secondary target at the XRF laboratory in
Gothenburg (3). This set-up is designed with a three axis geometry
to make use of the polarization effects of primary bremsstrahlung
in secondary target and sample. By careful collimation of beam paths
and the detector crystal, the background on the low energy side of
the scattered exciting radiation is very low. This analysis was
also performed in air.

Matrix effects in PIXE and XRF analysis

The efficiency calibrations of XRF and PIXE systems are often
performed using thin target standards, thus being independent of
matrix effects. Systems calibrated in this way can be used in
combination with the fundamental parameter technique to do accurate
analysis of thick homogeneous samples. Energy dispersive systems may
easily be designed to measure practically all major elements
simultaneously and no preknowledge of the sample to be analysed is
necessary. In this way the sample composition can be determined by
iterative calculations of matrix correction factors, which convert
the thin target yield (e.g. pulses per ng/cm^2) to a thick target
yield (e.g. pulses per ppm). These correction factors are composed
of one component caused by effects on incoming exciting radiation -
- slowing down of particles (PIXE) and absorption of X-rays (XRF) -
- and of one component caused by absorption of induced characteristic
X-rays. Here two major differences can be found between PIXE and XRF:

1)   The component due to the exciting radiation is for PIXE less
     dependent on matrix composition and generally this component
     works in a compensating manner to the absorption of the induced
     X-rays. This is generally not the case for XRF.

2)   For XRF with Mo and Tb excitation a thicker layer is analysed
     giving stronger corrections for the absorption of the induced
     X-rays.

This advantageous matrix dependence of PIXE is especially important
in the analyses of heterogeneous samples, where only approximate
average concentrations can be calculated from the recorded spectrum.
An element detected could have originated from a matrix quite
different from this average composition. Fig. 2 shows what the error
would be if matrix corrections were made for a pure quartz matrix
when the composition really was half quartz and half iron pyrite or
half quartz and half carbon. For PIXE analysis the X-ray self
absorption and the proton stopping power compensate each other

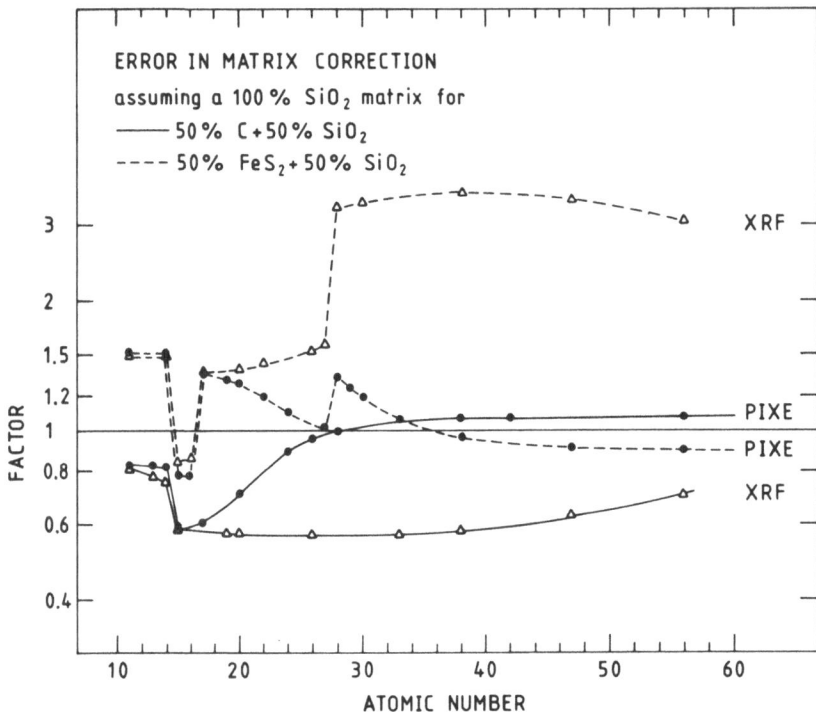

Fig. 2.   Error in matrix correction factors due to the assumption
            of a pure quartz matrix. Filled circles are for PIXE and
            open triangles are for XRF.

around Z = 30 making matrix correction for those elements quite
independent of the matrix composition.

Results

     The spectrum of a homogeneous pellet of a typical geological
material analysed with external beam PIXE is shown in fig. 3. The
elemental composition determined is found in table I. From this
spectrum detection limits corresponding to an irradiation time of
3 minutes and a count rate of 1500 pulses per second have been
calculated. The detection limit was defined as $3\sqrt{B}$, where B is the
background in a 2 FWHM region under the peak of interest. The
characteristic peaks were not included in this background. An XRF
system needs typically three different excitation modes to cover
the same elemental range as PIXE. The detection limits for XRF have
been calculated for three excitations using an irradiation time of
1 minute and a count rate of 2000 pulses per second each. The higher
count rate for the XRF analyses is used because the pulse pile-up
peaks from the scattered exciting radiation do not fall in the

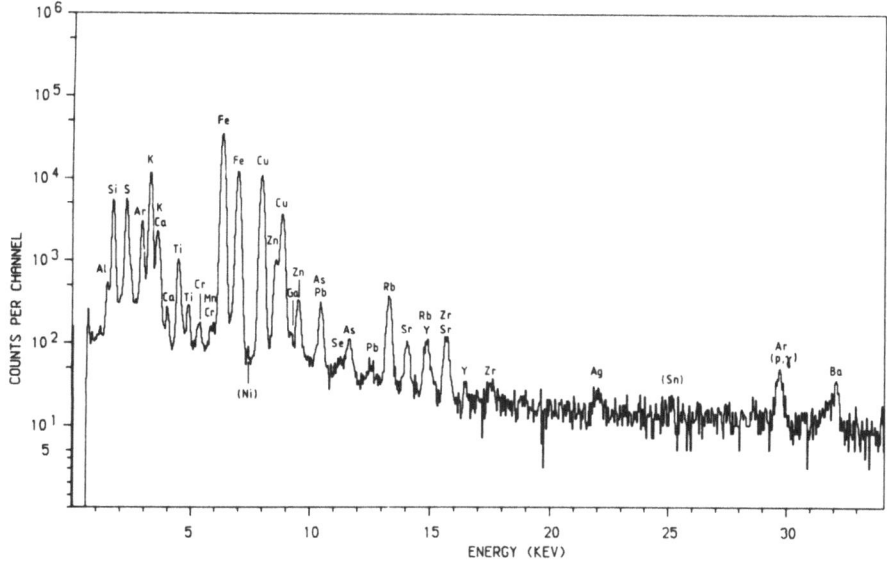

Fig. 3.   A spectrum of a homogeneous pellet of a typical geological
          material analysed in air with PIXE. The proton energy at
          the sample surface was 2.55 MeV. The total number of
          pulses in the spectrum is 500,000.

Table I.   Elemental concentrations and relative uncertainty (%) for
           the typical geological sample used.

Al	5.6 %	(10)		Ga	9 ppm	(17)
Si	22 %	(6)		As	50 ppm	(7)
S	6.4 %	(5)		Se	3 ppm	(30)
K	3.9 %	(5)		Rb	64 ppm	(6)
Ca	0.16 %	(6)		Sr	17 ppm	(8)
Ti	0.17 %	(6)		Y	4 ppm	(25)
Cr	130 ppm	(13)		Zr	41 ppm	(6)
Mn	100 ppm	(40)		Ag	13 ppm	(24)
Fe	5.5 %	(5)		Ba	320 ppm	(10)
Cu	0.44 %	(5)		Pb	11 ppm	(11)
Zn	310 ppm	(5)				

X-ray region of interest. The spectrum from Mo excitation of the
same sample as used in fig. 3 is seen in fig. 4. The detection
limits for Mo excitation are calculated from this spectrum and from
the concentrations given in table I. Detection limits for the Tb
excitation have been evaluated from ref. (4). For both the proton
and the Mo excitation the count rate was limited by the 5.5 % Fe

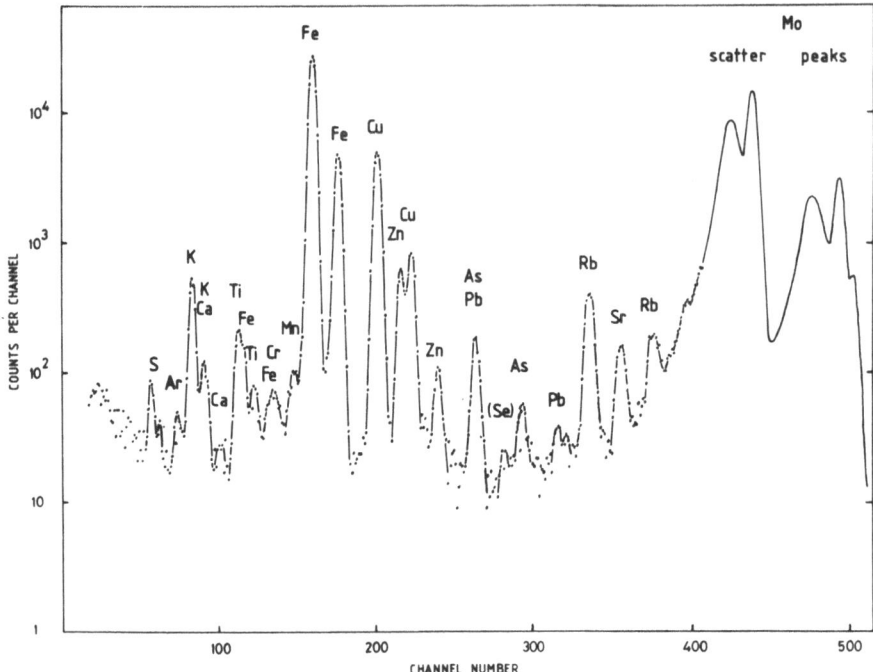

Fig. 4.   A spectrum of the sample used in fig. 3 when analysed with
           Mo secondary target in air. The total number of pulses in
           the spectrum is 500,000.

content. In the case of Tb excitation the count rate practically
always is dominated by the scattered radiation and here the
detection limits will be fairly independent of the matrix composition
        The detection limits for 2.55 MeV proton, Tb and Mo excitations
are given in fig. 5. Also included in fig. 5 is a coarse estimation
of the detection limits using Ti excitation, whick will be very
dependent on Ca, K and S concentrations in the sample. Data from
vacuum analyses in refs. (1) and (5) have been used in the estimation
        Also for the analysis of homogeneous pellets the advantageous
behaviour of matrix correction in PIXE analysis is useful. If a
systematic error were present in the determination of major elements,
the iteratively calculated correction factors would introduce an
additional error. This is illustrated in fig. 6, where the error
in correction factors due to 10% low concentrations of all elements
determined and a corresponding increase of the oxygen content for
the sample listed in table I are plotted. Fig. 2 and fig. 6 show
that PIXE is more suited to be used with the fundamental parameter
technique than XRF.
        In the case of pressed pellet samples XRF analysis of heavier
elements is less sensitive than PIXE to particle size effects,

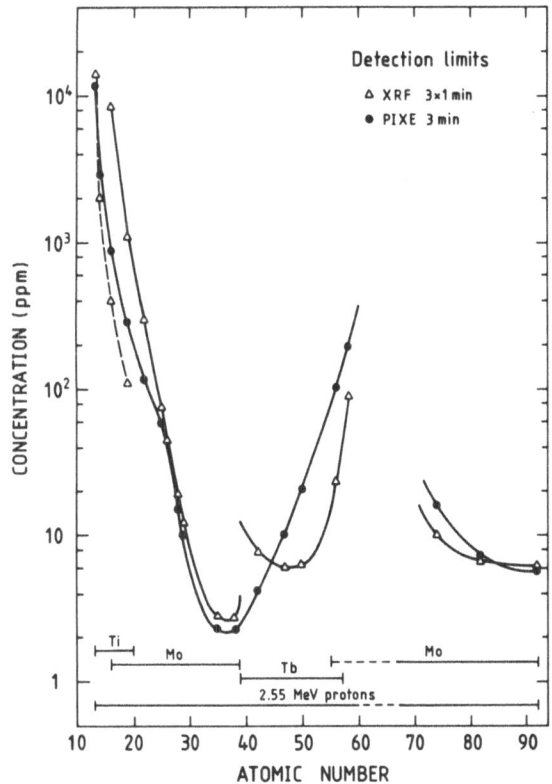

Fig. 5.

Detection limits for PIXE
and XRF. Calculated for
totally 3 minutes
analysis of the sample
used for the spectra in
figures 3 and 4.

where a different composition of small particles at the surface of
the pellet can make this layer non-representative to the bulk sample.
In PIXE more than 90% of the X-rays from the heavier elements come
from a layer of about 30 µm, which means that particle size effects
would influence the analysis of all elements.

Simultaneously with the accumulation of the PIXE spectrum it
is possible to measure the gamma-rays from proton induced nuclear
reactions in the sample. The gamma-ray spectrum in fig. 7 was
recorded simultaneously with the spectrum in fig. 3 using the
external proton beam. This simultaneous use of gamma-ray analysis
makes in possible to determine Li, F, Na, Mg and Al simultaneously
with the X-ray analysis (2). Detection limits for these elements
using the same conditions as in fig. 5 are found in table II.

Conclusions and discussions

Energy dispersive X-ray analysis using protons or monoenergetic
X-rays for excitation are well suited for fast multi-elemental
monitoring of drill cores. The advantageous matrix dependence of

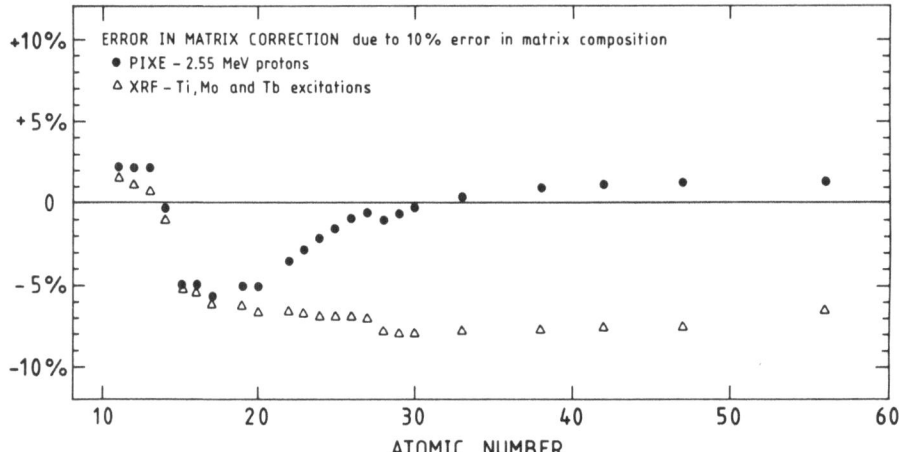

Fig. 6.   Error in matrix correction due to the assumption of 10%
           too low values of all elements determined compensated by
           an increased oxygen content. The sample composition in
           table I was used.

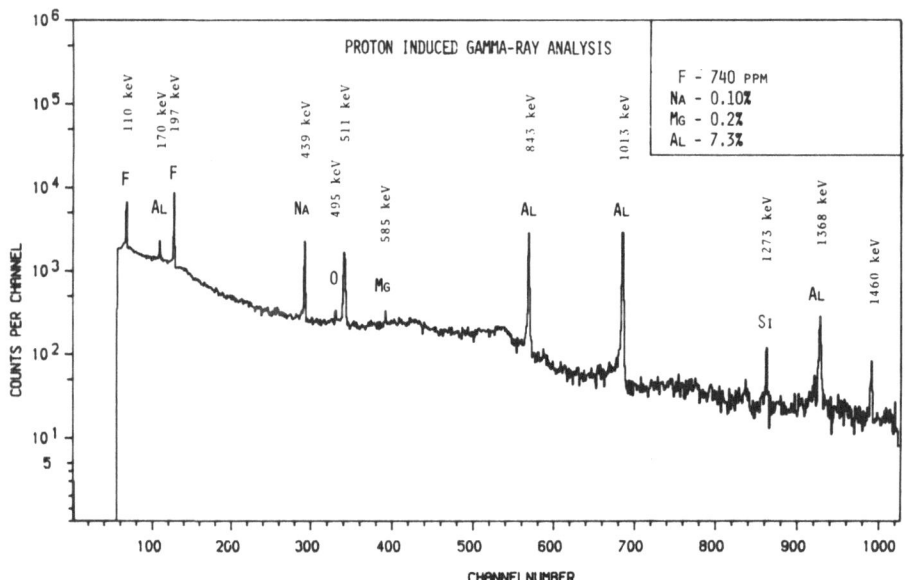

Fig. 7.   Gamma-ray spectrum aquired with a Ge(Li)-detector
           simultaneously with the X-ray spectrum in fig. 3.

Table II.   Detection limits for the gamma-ray analysis.

Element	Detection limit
Li	40 ppm
F	50 ppm
Na	100 ppm
Mg	0.4 %
Al	0.1 %

PIXE makes it possible to do rather accurate determinations of elements heavier than Ti even in heterogeneous samples. With the PIXE technique it is also possible to determine many elements lighter than Si by simultaneous gamma-ray analysis.

The depth of analysis for heavier elements is greater for XRF than for PIXE, which is an advantage when average concentrations in heterogeneous samples are wanted. Another advantage of XRF systems is the commercial availability and relatively low capital cost, though a PIXE system based on one of the small easy-to-handle tandem accelerators that are commercially available today, will have about the same operational cost as an XRF system.

Acknowledgements

The authors are grateful to A. Rindby and M. Öblad at the Department of Physics, Chalmers University of Technology and to L. Malmkvist, G. Sundkvist and N.-E. Marinder at Boliden Mineral AB for their valuable participation in discussions concerning drill core analysis. This work was supported by the Swedish Mining Association and the Swedish Board for Technical Development.

References

1. P. Verbeke and F. Adams, Anal Chim Acta 109 (1979) 85
2. L.-E. Carlsson and K.R. Akselsson, to be published in Nucl Instr and Meth
3. P. Standzenieks and E. Selin, Nucl Instr and Meth 165 (1979) 63
4. R.D. Giauque, R.B. Garret and L.Y. Goda, Anal Chem 49 (1977) 1012
5. M.S. Ahlberg and F.C. Adams, X-ray Spectr 7 (1978) 73

USE OF COMPTON SCATTERING IN X-RAY FLUORESCENCE FOR

DETERMINATION OF ASH IN INDIAN COAL

H. D. Pandey, R. Haque and V. Ramaswamy

Research & Development Centre for Iron & Steel
Steel Authority of India Limited
Ranchi 834002 (India)

ABSTRACT

X-ray fluorescence method using Compton scattering for the determination of ash in coal has been systematically studied for application to Indian Coal. Fluorescent intensities from major constituents such as Al_2O_3, SiO_2, TiO_2, Fe_2O_3, CaO and their various combinations were used in conjunction with the reciprocal of the Cr K_α Compton intensity in a regression equation. Varying degrees of correlations were obtained between the values of ash determined by conventional methods and those calculated from X-ray data. It is found that Ti fluorescent intensity plays a major role in the regression equation and its contribution cannot be ignored if the Cr X-ray tube is used.

INTRODUCTION

Indian coals are usually very rich in ash content and their washability characteristics are rather poor. As a result, coke produced in India contains a higher percentage of ash than that produced in other countries. Yet, the ash in our coke is main-tained within reasonable limits by suitably blending coals from different sources before they are charged into coke ovens. An ef-ficient blending operation requires constant and rapid monitoring of ash in coals to be blended. Since conventional methods of ash analysis are highly time consuming, the faster X-ray fluorescence techniques should be more suitable for this purpose. Of these, the X-ray backscattering method[1-3] for ash analysis is now fairly widely employed in routine applications. However, in a recent study, Renault[4] has outlined the use of Compton scattering for the estimation of ash in coal. A detailed and systematic investigation

was carried out in our laboratory to assess the suitability of this
method for the determination of ash in Indian coals containing ash
over a wide range. The results are presented in this paper.

Using three different X-ray target tubes (viz. Cr, Mo and W)
in wavelength dispersive X-ray spectrometer, Renault carried out
ash analysis on 13 New Mexico coals and obtained varying degrees
of correlations between the reported ash and the ash contents calculated from the regression equations. He obtained a good correlation ($R^2 = 0.94$) using Mo K_α Compton intensity in the regression,
while the R^2-value was only 0.85 in the case of intensities measured using W-tube. In the latter case both WL_α and WL_γ Compton
intensities were incorporated in the regression equation. However,
when Cr-tube was used, the correlation obtained was rather poor
($R^2 = 0.69$). This was improved to an R-square value of 0.85 by selecting a suite of nine samples representing a Ca intensity range
and constructing a general purpose calibration curve.

Renault has considered only the Ca characteristic intensity in
the analysis of his coals using the Cr X-ray tube. However, when
coals contain elements of higher atomic numbers than Ca their
fluorescent intensities need also be included in the regression
analysis. This is especially important when coals contain TiO_2 in
the ash as in the case of Indian coals. This is because Ti has a
higher absorption effect on Cr K_α radiation than Ca. In this
paper, it will be shown that by including fluorescent intensities
of Ca and Ti, a very good correlation can be obtained. R-square
value can be further improved by including fluorescent radiations
from other major elements in the ash, such as Fe, Al and Si. Even
while using the Ca intensity alone, improved R-square values can
be obtained if samples are analyzed in vacuum.

EXPERIMENTAL DETAILS AND REGRESSION ANALYSIS

Eleven coal samples from different sources were obtained from
the Central Fuel Research Institute, Dhanbad (India). The ash contents in these samples ranged from 10 to 44%. The samples were
crushed to -72 mesh size and briquetted under 10T pressure. 0.5 cc
of distilled water was used as binder and the samples were dried
in atmosphere. They were then analyzed under vacuum in the Philips
PW 1450/20 X-ray Fluorescence Spectrometer with a Cr tube at 50 KV
and 45 mA. Each sample was analyzed for Cr K_α Compton, Fe K_α,
Ti K_α, Ca K_α, Al K_α, and Si K_α radiations, the counting time in
each case being 20 secs. While LiF (220) Crystal was used for
measuring the Compton intensity, for Fe, Ti and Ca K_α radiations
LiF (200) was employed. Si and Al K_α lines were measured with PE
and TlAP crystals respectively. A coarse collimator was used in
all the cases. Flow proportional counter was used in all the
measurements except in the case of Cr K_α Compton and Fe K_α

intensities for which both flow and scintillation counters were employed in tandem. No attempt was made to determine the concentrations of individual constituents or to correct for matrix absorption effects.

Table 1. Calculated values of B, C_i, D and R^2 for various
combinations.

i	B $(\times 10^3)$	C_i					D	R^2
		i=Ca	i=Ti	i=Fe	i=Al	i=Si		
Ca	0.717	-0.216					-82.974	0.892
Ti	0.513		0.819				-65.100	0.932
Ca, Ti	0.444	0.134	1.082				-59.490	0.936
Ca, Fe	0.797	-0.348		-0.771			-82.942	0.925
Ti, Fe	0.540		0.850	-0.365			-64.525	0.942
Ca, Ti, Fe	0.553	-0.0216	0.809	-0.392			-65.391	0.942
Ca, Ti, Fe, Al	0.778	-0.378	-0.228	-0.538	15.706		-206.49	0.949
Ca, Ti, Fe, Al, Si	0.789	-0.402	-0.289	-0.570	16.439	-1.454	-107.511	0.950

From the measured values of these intensities the percent ash
in the sample was calculated from the regression equation:

$$\% \ A \ = \ \frac{B}{I_c} + \sum_i C_i I_i + D \qquad (i = Ca, \ Ti, \ Fe, \ Al \ and \ Si) \qquad (1)$$

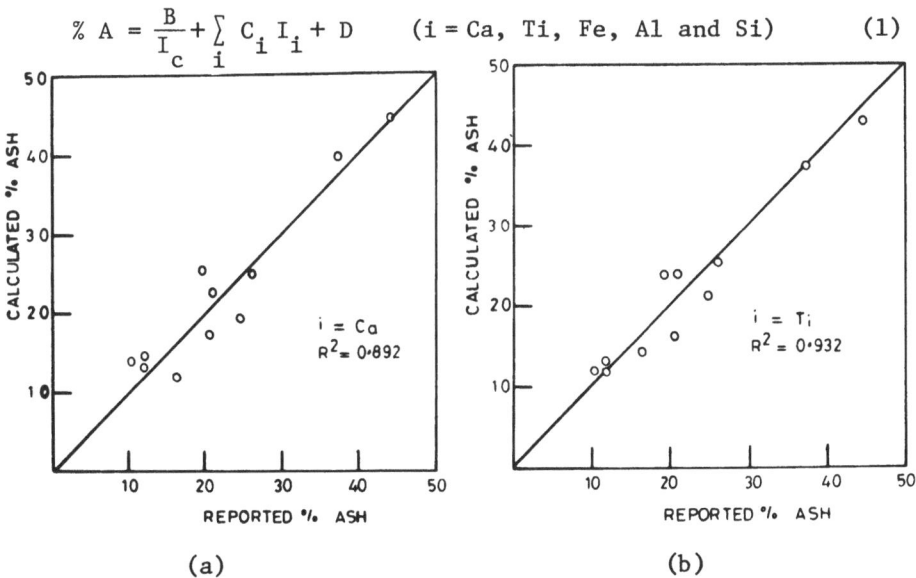

(a)   (b)

Figure 1.   Reported ash versus ash calculated from Eq. (1) with
(a) i = Ca and (b) i = Ti.

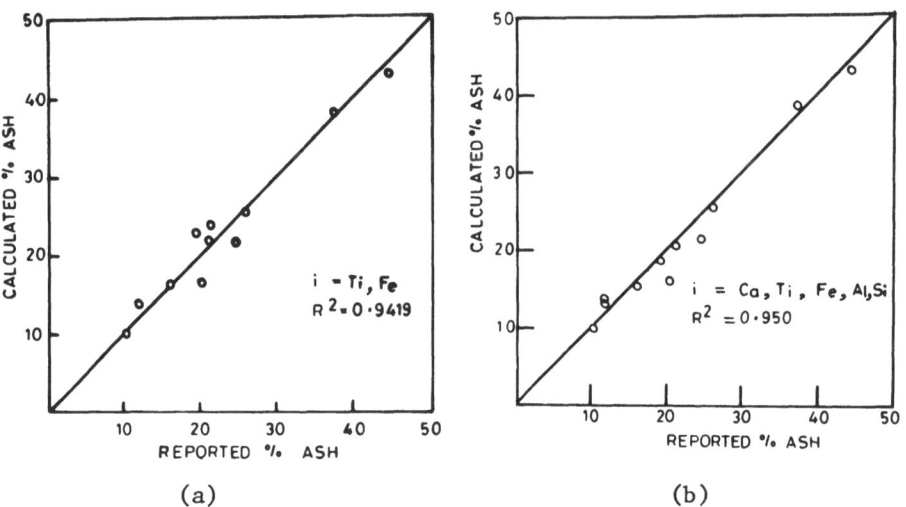

Figure 2.   Reported ash versus ash calculated from Eq. (1) with
             (a) i = Ti, Fe and (b) i = Ca, Ti, Fe, Al, Si.

where D is a constant, B and C_i the coefficients of regression and
I_c and I_i are, respectively, the Cr K_α Compton intensity and K_α
fluorescent intensity of the element i.   Calculations were made
after including in the equation characteristic intensities from
elements taken singly or several together (i.e., i=Ca, i=Ti, i=Ca,
Ti etc.), as shown in Table 1.   The calculated values of ash content were then plotted against the reported values for each of
these combinations, and the corresponding R-square values were
determined.

RESULTS AND DISCUSSION

        The results of the analysis are summarized in Table 1. The
reported values of ash content versus the values calculated for
the various combinations are shown in Figs. 1 and 2. It is seen
from the table that the R-square value calculated with the inclusion of Ca fluorescence alone in Eq. (1) is 0.89. This improvement
in the R^2 value as compared to the value reported by Renault is
probably due to measurements being done in vacuum.   The correlation obtained using Ti intensity (Fig. 1(b)) is superior to the
case where Ca line (Fig. 1(a)) was used.   In fact, it is seen that
when both Ti and Ca are included in the regression equation, R^2
value (0.936) is only marginally better than when Ti alone was
used (0.932). This shows that Ti with its higher mass absorption
coefficient (583 cm²/gm at Cr K_α Compton wavelength) overshadows
the absorption effects of Ca (499 cm²/gm at Cr K_α Compton wavelength). Thus, for a rapid determination of ash in these coals,

only Ti fluorescent intensities need be considered. Further, the examination of R^2-values for all the combinations shows that Ti,Ca, Fe, Al and Si together give the best correlation (Fig. 2(b)). However, the R^2-value in this case (0.950) is not significantly higher than when Ti and Fe (Fig. 2(a)) are considered together (0.942). Thus for an optimum combination of rapidity and accuracy of analysis of coals, Compton intensity together with Ti and Fe fluorescent radiations should be included in the regression equation. From our results it is seen that the accuracy obtained with this combination is well within the 10% variation specified in the standard ASTM method[5] for coals containing more than 12% ash. This variation could be narrowed down to within ±3% as specified in Indian Standards for coals containing more than 10% ash, by incorporating all the five fluorescent intensities in the regression equation (Fig. 2(b)). The results could be further improved by taking into account the effect of bulk density and moisture content in coal. Work on these lines is in progress in our laboratory.

ACKNOWLEDGEMENTS

We are grateful to Mr. Jacques Renault of New Mexico Bureau of Mines and Mineral Resources for useful communications. Our thanks are particularly due to Dr. C. R. Shastry for useful discussions and help in preparing the manuscript of the paper, and to Mr. Dilip Bhargava for his help in carrying out the regression analysis on the IBM 1620 computer.

The authors are grateful to Dr. G. Mukherjee, Director (R&D) for his continued interest and support for the work and permission to publish the paper. We are also thankful to Dr. S. K. Gupta,GM (R&D) for his interest and suggestions. The help rendered by CFRI, Dhanbad is also gratefully acknowledged.

REFERENCES

1.   J.R. Rhodes, J.C. Daglish and C.G. Clayton, "A Coal Ash Monitor with Low Dependence on Ash Composition," Radioisotope Instruments in Industry and Geophysics, 1966, I, 457.

2.   I.S. Boyce, C.G. Clayton and D. Page, "Some Considerations Relating to the Accuracy of Measuring the Ash Content of Coal by X-ray Backscattering," Nucl. Tech. Miner. Resour., Proc. Int. Symp. 1977, IAEA: Vienna, Austria, 135-65 (1978).

3.   F.V. Brown and S.A. Jones, "On Site Determination of Ash in Coal Utilizing a Portable XRF Analyzer," Advances in X-ray Analysis 23, 57 (1980), Plenum Publishing Corp., New York.

4.   Jacques Renault, "Rapid Determination of Ash in Coal by Compton Scattering, Ca and Fe Fluorescence," Advances in X-ray Analysis 23, 45 (1980), Plenum Publishing Corp., New York.

5.   American Society for Testing & Materials, 1977 Annual Book of Standards, pt. 26, 370 (1977).

THE PRESENT STATE OF X-RAY SPECTROMETRY

J. L. de Vries

N. V. Philips

Eindhoven, The Netherlands

If we want to evaluate the present state of X.R.S., we should do this in perspective with other methods, e.g. Optical Emission and Atomic Absorption Spectrometry.

One way to compare would be to estimate the number of papers published or given at conferences on these three methods. Table 1 gives a survey of some meetings. It is evident that X.R.S. is less talked about than the other methods. The rejuvenation of OES, thanks to the Plasma sources, can also clearly be seen.

Table 1. Numbers of Papers

Meeting	Year	OES	AAS	XRS
CSI	1958	33	--	8
CSI	1961	37	5	15
CSI	1965	29	8	8
CSI	1971	70	21	16
CSI	1979	91	61	32
FACSS	1975	31	32	16
PITTSB	1975	38	50	10
FACSS	1978	49	37	11
SPEKTAC	1968	12	3	18

In which application fields is X.R.S. mainly used? We estimate that an average over the last 20 years might be:

	X.R.F.S.	O.E.S.
Minerals	30%	5%
Metals	35% (increasing)	85%
Liquids, etc.	20%	5% (increasing)
Envir. Agri., Food Research	15%	5%

Since 25 years I have collected literature references on X.R.S. Table 2 gives the distribution in these personal files over some application fields in the course of the years. The coverage is by far not complete, only journals, but thought to be rather constant. This picture corresponds roughly with the estimate given, but it is surprising that the total number over two year periods is rather constant.

Table 2. XRS Literature References

TOTAL	YEAR	METAL	MINER.	OXIDE	ORGAN.	ENVIR.
38	1957	11	17	--	8	2
34	1958	11	11	4	6	2
83	1960	28	24	8	15	8
57	1962	25	7	7	10	8
82	1964	28	18	9	16	11
67	1966	32	10	11	12	2
71	1968	21	21	11	11	7
68	1970	11	23	20	9	5
79	1972	17	24	13	7	18
78	1974	22	13	11	5	27
78	1976	18	23	3	7	27
76	1978	27	19	4	6	20
811		251	210	101	112	137

Another way to judge the relative merit of a method is to estimate how many analyses are actually performed. Some recent figures show that the world market for all three techniques is of the order of magnitude of 100 M$ annually; slightly more for AAS, slightly less for OES. Evidently, the AAS instruments are the cheapest, XRS is the most expensive method.

Given these market figures and average prices per instrument, one can make a very rough guess at the number of instruments sold annually. Again, one can hazard a guess, based on practice all over the world, how many samples are actually run every day, on average, and how many elements are determined per sample. Combining those estimates, one ends up with a very rough approximation of the total number of elements determined per day with the instruments sold in one year. This total number comes in the neighborhood of 400 K analyses per day; maybe slightly more for OES, slightly less for AAS. Evidently XRS is a method extensively used, but not talked about much; that means it is well established and consolidated.

An important aspect of any analysis method is its precision and accuracy. Now, XRS has always been known to be a very precise technique, but to need a great number of calibration standards to be very accurate. In the last decade the growing understanding of matrix corrections, their mathematical equations and growing use of computers have enabled XRS to become an accurate method using only a very limited number of standards. A great variety of algorithms have been suggested. In our laboratories we use an alpha correction coefficients method as modified by W. K. de Jongh.[1]

$$C_i = (D_i + E_i R_i) (1 + \Sigma \alpha_{ij} C_j)$$

where i and j refer to the analyte and interfering elements respectively, R_i is the count rate relative to the standard, E_i and D_i are the slope and intercept of the line defining the apparent concentration, C_i is the concentration of the jth interfering element and α_{ij} is the interelement correction coefficient.

The alphas are calculated from theory, but can be refined by regression analysis. In this technique any element can be eliminated, e.g. loss on ignition, an organic matrix, not necessarily only the analyte. A great number of alphas have been calculated in our Application Laboratories over the years.

Table 3 gives an indication of the distribution of α's among fields of application.

Table 3. Calculated Alphas According to de Jongh's Program

	Total	Ferro	Non Ferro	Oxides	Minerals
		Metals			
1973	13	4	4	2	3
1974	36	1	11	16	8
1975	55	10	11	18	16
1976	91	7	31	27	26
1977	58	4	10	21	23
1978	48	8	20	35	15
	379	41	100	134	104

Thanks to their great versatility and large concentration range covered, the α's are mainly applied to oxide materials, minerals and complex alloys.

A few years ago, in our Application Laboratory, a comparison was made between the accuracy achieved for synthetic samples made from pure materials and analysed standards. All samples were homogenized by fusion with Li-tetraborate. The accuracy achieved is often expressed as: s.d. = K $\sqrt{c}$ + 0.01 where c is the concentration. Over the concentration range covered K is minimized by regression. It was found for oxides of 9 elements ranging from Mg to Fe that K values of 0.01 - 0.06 were obtained. However, if the same procedure and analysis is applied to real standards, the results are less good. For 2 British Certified Standards K-values of 0.02 to 0.13 were found. Evidently XRS should not have been used as a reference method for these standards. A study along these lines was undertaken in the U.K. and preliminary results were reported by P. W. Hurley at the 21st CSI, Cambridge, July 1979.

A large variety of oxides with very wide concentration ranges were covered. All samples were fused and only one standard sample was used per element for calibration purposes. The alpha coefficients were calculated and applied to 30 BCS standards, treated as unknowns. The agreement between concentrations calculated from measured intensities and as given by chemical analysis was found to be very convincing, given the large concentration ranges covered. This demonstrates that indeed X-ray spectrometry has progressed to where it now can be considered one of the most accurate methods of chemical analysis.

REFERENCE

1. W. K. de Jongh, X-Ray Spectrometry 2:151 (1973).

IMPROVING THE DETECTION LIMIT IN WAVELENGTH DISPERSIVE XRF

Heikki Sipilä

Outokumpu Oy, Institute of Physics

SF-02200 Espoo 20, Finland

ABSTRACT

The origin of the background under the signal in the wave-length dispersive XRF has been investigated. The methods of mini-mizing the background are presented. The bremsstrahlung is elimi-nated by using a primary filter in front of the X-ray tube. The background generated by the X-ray detector is decreased with a pulse risetime and a pulse interval analyzer. For example, in the fixed uranium spectrometer channel a 500-fold improvement of the peak-to-background ratio was achieved.

Statistically the detection limit C_{DL} is given by $C_{DL}(ppm) = 3\sqrt{B}/N$, where B is the number of background counts and N the number of counts for 1 ppm content. The detection limit can, however, not always be improved adequately only by raising the power of the X-ray tube or by increasing the measurement time since the background count number is not solely due to statistical fluctuations. It is often dependent of the sample. In particular the analysis of ore slurry is difficult because the background is a function of the slurry density. In the fixed channel analyzer it is neither practi-cal and not even possible to measure background accurately for every spectrometer.

In the analyzer of Outokumpu Oy Johansson type curved focusing crystals are used. The advantage of fully focusing crystals is the possibility to use a very narrow secondary slit.

A simple and efficient way to prevent scattering of brems-

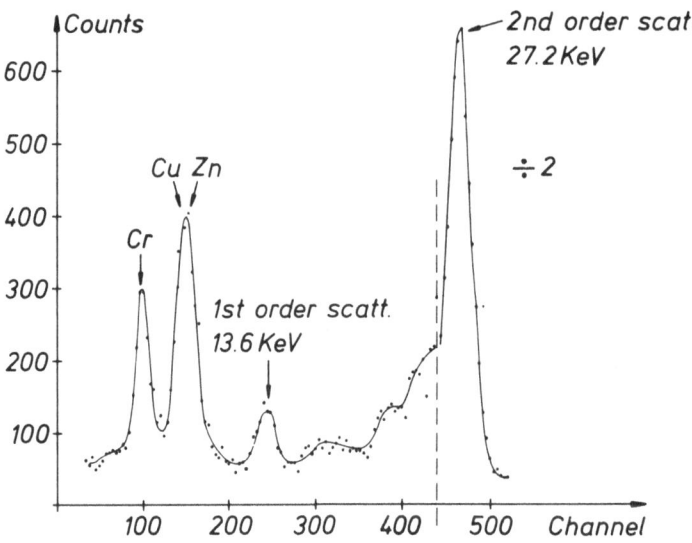

Fig. 1a. Energy spectrum of the uranium spectrometer channel, when
          distilled water was used as a sample.

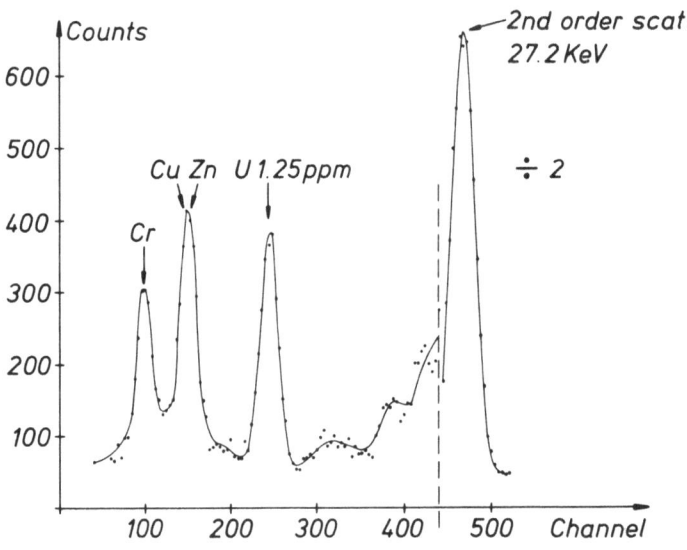

Fig. 1b. Water contains 1.25 ppm U. Measurement times are 3 h.

strahlung from the sample is to filter it before the sample
using a primary filter in front of the X-ray tube. Of course the
intensity to be measured will decrease, but the background will
fall faster. In practice, however, the background will very soon
settle on a level where any increment of the primary filter will
decrease the signal more than the background. There remains an
escape peak problem in the cases of Xe and NaI detectors. Ar-filled
detectors are better in this respect. The wall effect will still
worsen the quality of the spectrum of the proportional counter.
Sipilä and Kiuru /1,2/ have presented a method for the elimination
of this effect.

The principles presented were tested using a uranium $ULα_1$
channel. The ratio of the pulses from metallic uranium and from
a thick nylon sample was used as the measure of the peak-to-background ratio. A Pt-anode X-ray tube was used in the measurements.

Table 1 shows the change of this peak-to-background ratio
when detectors are changed and the filter and the risetime analysis
(RSD) are employed.

Table 1. Peak-to-background ratio in $ULα_1$ channel

Detector	Primary filter	RSD	U(counts)/Nylon(counts)
NaI	–	–	32
Xe - CO_2	0.20 mm Fe	–	2500
Xe - CO_2	0.20 mm Fe	+	2800
Ar - CO_2	0.20 mm Fe	–	5700
Ar - CO_2	0.20 mm Fe	+	16000

Some measurements were made with the improved channel. Figure
1a shows the energy spectrum of the uranium spectrometer channel
when distilled water was used as a sample. Figure 1b shows exactly
the same measurement but the water contains 1.25 ppm of uranium.
The measurement time is 3 h. The detection limit ($3σ$) in those
measurement conditions for uranium was 84 ppb.

The significant improvement of the peak-to-background ratio
indicates that the detection limit of WDXRF analyzer can be improved
essentially.

REFERENCES

1. H. Sipilä and E. Kiuru, U.S. Patent No. 4075486.
2. H. Sipilä and E. Kiuru, Advances in X-Ray Analysis, 20, 555 –
    563, (1977).

BACKGROUND AND SENSITIVITY CONSIDERATIONS OF X-RAY FLUORESCENCE ANALYSIS WITH A ROOM-TEMPERATURE MERCURIC IODIDE SPECTROMETER*

M. Singh, B.C. Clark[a], A.J. Dabrowski**, J.S. Iwanczyk**, D.E. Leyden[b], and A.K. Baird[c], University of Southern California, Medical Imaging Science Group, 4676 Admiralty Way, Marina Del Rey, California.

a. Martin Marietta Aerospace, Denver, Colorado.
b. University of Denver, Denver, Colorado.
c. Pomona College, Claremont, California.

INTRODUCTION

Continued development of mercuric iodide (HgI_2) detectors for x-ray spectroscopy at room-temperature has led to a considerable improvement in energy resolution and a better understanding of the various detector parameters which affect sensitivity. The basic properties of a mercuric iodide detector and some of its characteristics pertinent to x-ray fluorescence analysis have been previously reported (1,2,3). In this paper we present results of studies to determine the shape of peaks and continuum background. Also, the use of HgI_2 in characterizing water pollutants by XRF analysis has been investigated and compared to cryogenically cooled Si(Li) and room-temperature proportional counter systems.

The best value of energy resolution previously reported by us (3) for HgI_2 was 380 eV for the 5.9 keV x-rays of Fe-55. As shown in Fig. 1 a value of 325 eV has now been obtained with a collimated 3mm^2 x 400 μm detector at room temperature and ambient atmosphere.

SENSITIVITY FOR XRF ANALYSIS

In addition to geometrical factors such as detector area, thickness, and solid angle, the sensitivity in XRF analysis is limited by the energy resolution and the background in a spectrum. The ulti-

* Research supported by DOE Contract DE-AS03-76-SF00113 and NASA Contract NAS 1-15943
** On leave from Institute of Nuclear Research, Swierk, Poland

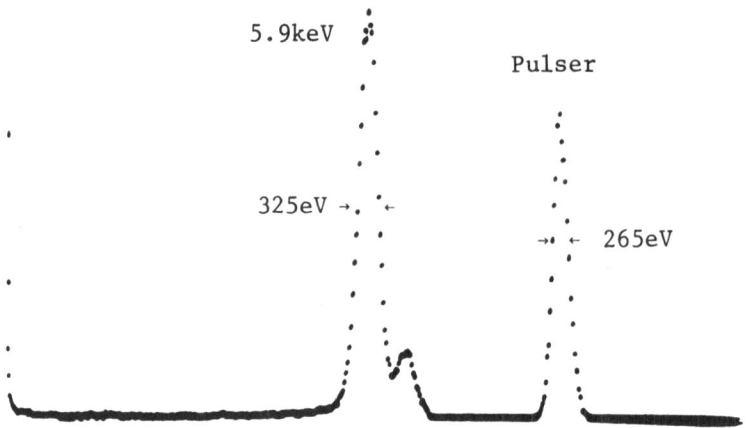

<u>Fig.1</u> Spectrum obtained from Fe-55 source with HgI₂ detector

mate limit in detection accuracy, however, is set by the background
or the peak to background ratio. This limit arises from statistical
fluctuations in the true and background counts recorded within a
characteristic peak.

An XRF spectrum recorded with a Cd-109 source and a bakelite
sample (blank) is shown in Fig. 2. Peaks originating from source
radiation scattered by the sample and "escape" peaks produced following the generation and escape of mercury L_α and L_β x-rays from
the detector are seen. A detailed study of escape peaks in HgI_2
has been reported previously by us (3). The strength and location
of escape peaks can be predicted from their parent peaks, and therefore their contribution in a given spectrum can be subtracted for
quantitative analysis.

CONTINUUM BACKGROUND AND PEAK-SHAPE

In XRF spectra obtained with radioisotopic source excitation,
the continuum background is mainly produced from events within the
HgI_2 detector. These are: a) incomplete charge collection due to
carrier trapping, b) loss of charge in the side surfaces of the
detector due to distortions in the electric field, c) escape of
photoelectrons from the detector, and d) escape of photons after
undergoing Compton scattering within the detector. The result of
these processes is that only a partial amount of incident photon
energy is converted into an electrical signal, thereby giving rise
to peak "tailing" or background in the low-energy region of the
spectrum.

The various interaction probabilities have been computed with
a model in which photons of energy 'E_0' are incident normally onto
the center of the negative electrode of a cylindrical detector of

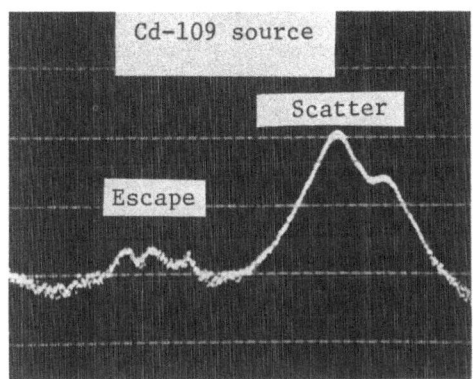

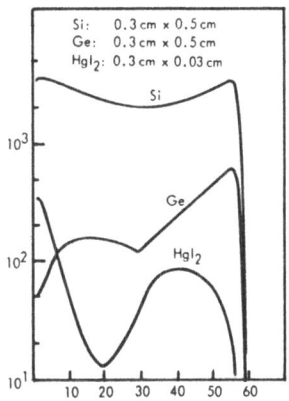

Fig.2 XRF spectrum from bakelite   Fig.3 Background from Comton scat-
sample. Counts are in log scale   ter at 60keV.Horiz. scale:0.2keV/Ch.

thickness 'ℓ' (Fig. 2 of Ref. 3). The probability of photoelectric
or Compton event is calculated as a function of depth 'x' within the
detector. It is asssumed that the photoelectrons are ejected in a
random direction, and that the scattered photons follow the Klein-
Nishina angular distribution. The Bethe-Bloch relationship is used
to compute the energy deposited by the photoelectrons before escaping
from the detector.

It is further assumed that pulses exhibiting a normal amplitude
distribution are produced by the detector, with a mean standard deviation 'σ' defined in terms of the statistical fluctuations in the
charge generation and collection processes and the noise of the
detector-preamplifier electronics, and with a mean amplitude described by a charge collection efficiency function 'η(x)'. The
function η(x) is calculated in a manner similar to Ref. 4 and is a
function of the detector thickness, bias and μτ (mobility x trapping
time) values of the carriers.

With these assumptions, the following expression has been
derived for computing the peak-shape and background due to charge-
trapping and escape of photoelectrons. Loss of carrier charges in
the side surfaces of the detector and effects of detector non-
uniformity have been ignored.

$$\frac{dN(E)}{dE} = \frac{\mu_p}{\sigma} \int_0^\ell \int_0^\pi \exp\left[-\mu_\tau(E_o) \cdot x\right] \cdot \frac{1}{\pi} \cdot \exp\left\{-\frac{\left[E - \eta(x)E_1(x,\theta)\right]^2}{2\sigma^2}\right\} d\theta\ dx$$

Similarly, the continuum background spectrum from Compton scattering
within the detector is given by the following expression.

$$\frac{dN(E)}{dE} = \frac{\mu_c}{\sigma} \cdot \int_0^l \int_0^\pi \exp\left[-\mu_\tau(E_0) \cdot x\right] \cdot P(\theta) \cdot \exp\left\{\frac{-\left[E - \eta(x)E_2(E_0,\theta)\right]^2}{2\sigma^2}\right\}$$

$$\cdot \exp\left[-\mu_\tau(E_3) \cdot S(x,\theta)\right] d\theta dx$$

In these expressions dN(E)/dE denotes the spectral shape of the back-ground indexed in energy E. The symbols which have not been defined earlier are:

μ_p, μ_c, μ_τ = photoelectric, Compton and total absorption coefficients respectively.

$E_1(x,\theta)$ = energy deposited by the photoelectrons after accounting for their escape. $E_1 \leqslant E_0$.

$P(\theta)$ = probability that a photon is scattered between angles θ and $\theta + d\theta$.

$E_2(E_0,\theta)$ = energy deposited by the recoil electron in a Compton scatter event.

E_3 = energy of the scattered photon.

$S(x,\theta)$ = distance traversed by the Compton scattered photon in the detector.

The Compton scatter background spectrum computed from the above equation is shown in Fig. 3 for a typical mercuric iodide, germanium and silicon detector at 60 keV. It is apparent that the lowest background is produced in a HgI_2 detector from this effect.

Typical results of "tailing" and background obtained from the photoelectric effect are shown in Figs. 4-6. Computations can be made for any incident energy, for any detector, as a function of detector thickness, applied bias voltage, or $\mu\tau$ values of electrons or holes. Figs. 4 and 5 illustrate peak shape and continuum background at 60 keV as a function of electron and hole $\mu\tau$ respectively. Fig. 6 shows the peak shapes at 20,60 and 120 keV for a typical HgI_2 detector. One can thus study the effect of each parameter on the background and select an optimum detector.

A comparison of experimental and theoretical peak shapes at 59.5 keV is shown in Fig. 7. A collimated Am-241 source and a 3mm x 173 μm thick HgI_2 detector with measured parameters was used.*
The agreement is quite good. A portion of the tail on the left-most side of the experimental peak is produced by backscattered radiation emanating from within the Am-241 source. Subtraction of this contribution would result in a better fit.

---

* The HgI_2 detector was kindly provided by EG&G Operations Research Group, Santa Barbara, California.

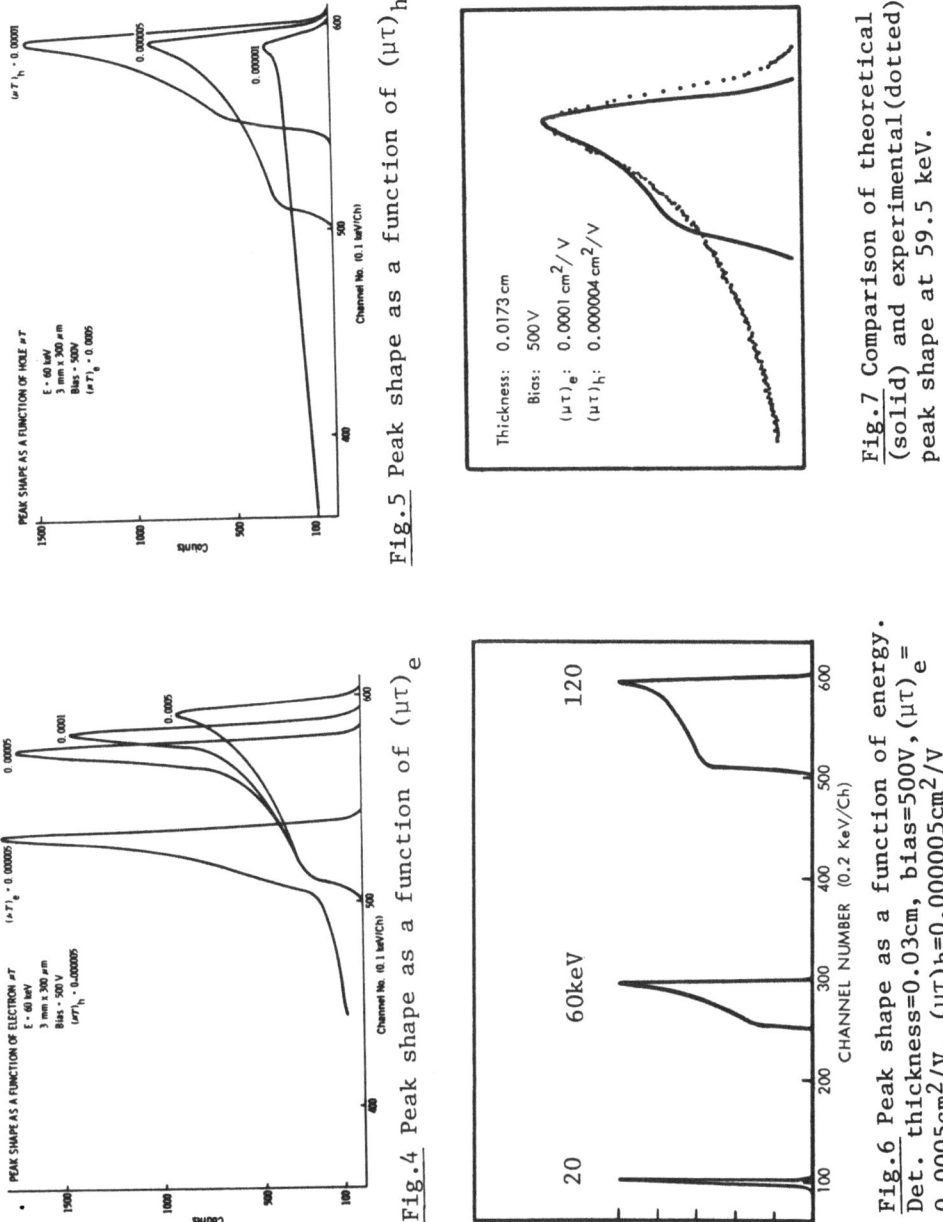

Fig.4 Peak shape as a function of $(\mu\tau)_e$

Fig.5 Peak shape as a function of $(\mu\tau)_h$

Fig.6 Peak shape as a function of energy. Det. thickness=0.03cm, bias=500V, $(\mu\tau)_e$ = 0.0005cm²/V, $(\mu\tau)h$=0.000005cm²/V

Fig.7 Comparison of theoretical (solid) and experimental(dotted) peak shape at 59.5 keV.

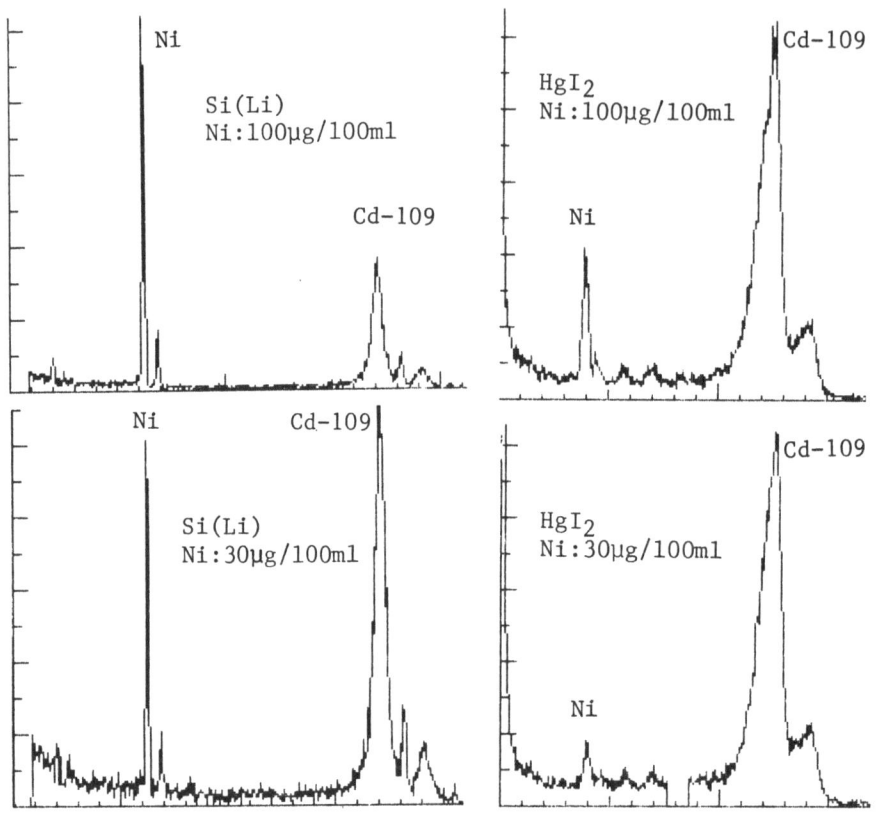

Fig.8 XRF spectra of nickel samples obtained with Si(Li) and HgI$_2$

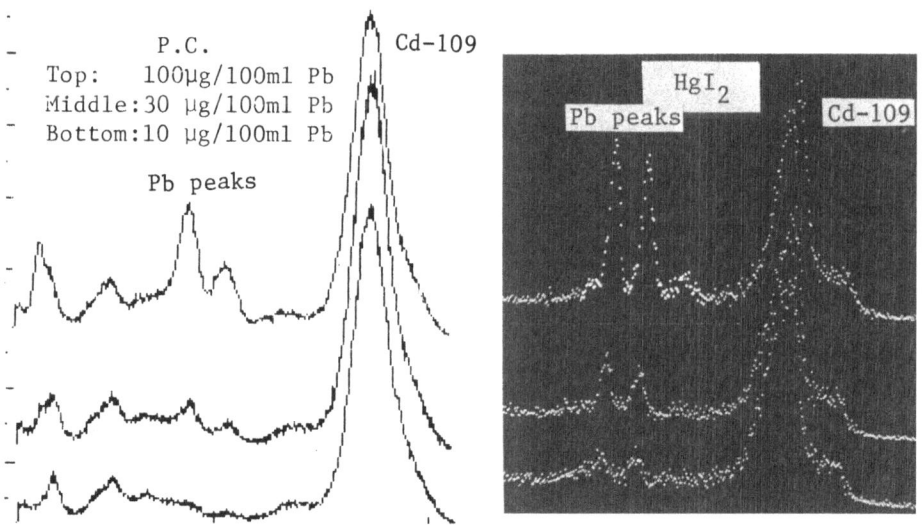

Fig.9 XRF spectra of lead samples with proportional counter and HgI$_2$

APPLICATIONS

The mercuric iodide detector has many applications where portable, miniature, and/or automated systems are needed. We have
studied the utility of the unit in making analysis of thin deposits
of sample, such as would be produced in the preconcentration of dissolved cations from aqueous solution. Even very small element concentrations (0.1ppm) can be determined in this way, making possible
the measurement of environmental levels of many elements. The
numerous potential applications of such a system include a) following
the temporal changes in element levels in streams, lakes and irri-
gation canals; b) routine, periodic monitoring of hazardous element
concentrations in effluent streams from mining and industrial operations, and c) screening of suspect areas for contaminated runoff --
areas such as chemical waste dumps, oil shale workings, coal and
other strip-mines, etc.

Quantitative deposits of various cations were obtained by preparations of dilute standard solutions of the appropriate high-purity
salts, followed by a chemical precipitation reaction and filtration
through a membrane filter (0.45 micron pore size Metricel filter).
The chelating precipitator used was either APDC (ammonium pyrro-
lidinedithiorcarbamate) or Na DDTC (sodium diethyldithiorcarbamate),
depending upon the cation selected. Deposits are isolated nominally
to a 1.7 cm diameter circle on the larger-diameter filters.

As an example, XRF spectra obtained with Cd-109 excitation of a
few nickel and lead samples are shown in Figs. 8 and 9. A compara-
tive study of Si(Li), HgI_2 and proportional counter spectrometers
has been performed for this application. Typical results are shown
in Figs. 8 and 9. It is concluded that room-temperature HgI_2 provides a viable alternative to Si(Li) detector and is superior in
performance to a proportional counter system.

REFERENCES

1. A.J. Dabrowski, G.C. Huth, M. Singh, T.E. Economou, A.L. Turkevich,
   Appl. Phys. Lett. 33(2), 211-213 (1978)
2. G.C. Huth, A.J. Dabrowski, M. Singh, T.E. Economou, A.L. Turkevich,
   Adv. in X-Ray Analysis, 22, 461-472 (1979)
3. M. Singh, A.J. Dabrowski, G.C. Huth, J.S. Iwanczyk, B.C. Clark,
   A.K. Baird, Adv. in X-Ray Analysis, 23, 249-256 (1980)
4. J.S. Iwanczyk, A.J. Dabrowski, Nucl. Instrum. and Methods 134,
   505-512, (1976)

MULTIPLE SCATTERING AND THE POLARIZATION OF X-RAYS

John D. Zahrt

Chemistry Department
Northern Arizona University
Flagstaff, AZ 86011

Richard Ryon

Lawrence Livermore Laboratory
University of California
Livermore, CA 94550

INTRODUCTION

In order to improve the overall performance of energy dispersive x-ray secondary emission spectrometry one can make use of polarized x-rays. We have used polarized x-rays produced by 90° scattering to reduce the background intensity due to the primary x-ray source[1].

An effort has been made to optimize the polarizing scatterer for the elements being analyzed[2]. A discrepancy appears in the theory of such optimization between a simple one point formula and a more elaborate integral formula. Possible reasons for such disparities might lie in 1) collimator geometric effects, 2) multiple scattering and 3) actual primary intensity. The first problem has been dealt with in an unpublished manuscript. This report concerns itself with point 2, multiple scatter, and in particular with double scatter including polarization effects.

Multiple scattering has been treated by many investigators[3] for several different geometries. The consensus of these investigations is that double scatter is very dependent upon the geometry of the system. None of these authors treat the geometry that is used in this laboratory. Most of the recent authors do not treat the prob-

[1] R. Ryon, Adv. in X-Ray Anal., 20, 575-590 (1977).
[2] R. Ryon and J. Zahrt, Adv. in X-Ray Anal., 22, 453-460 (1979).
[3] See for example J. DuMond, Phys. Rev. 36, 1685-1701 (1930) for the classic work and A. Tanner and I. Epstein, Phys. Rev. 13A, 335-348 (1976); 14A, 313-327 (1976); 14A, 328-340 (1976) for current work.

lem of polarization at the detector[4], and of those that do, none treat our geometry.

DERIVATION OF FORMULAE

Consider the geometry of Figure 1. The x-rays are incident on the scattering material of thickness t and are assumed to be perfectly collimated, having passed through a collimator of diameter d, and will exit through a collimator of diameter d. The incident beam can be the primary source of x-rays or the polarized (only partially as we shall see) scattered x-rays. The exit beam is then either the (mostly) polarized source x-rays or the specimen x-rays with a small amount of scattered source x-rays respectively. Single scatter events which are directed into the second collimator must occur in region ABC of Figure 1. With double scatter events the first and second scatter points must be in regions ABDE and ABFG respectively. The scatter thickness t may be smaller than $d/\sqrt{2}$ in which case not all of ABC is seen. This consideration is built into the theory. For circular collimators the cross section ABDEFG shown in Figure 1 varies with x.

Single Scatter

The differential probability of having a single scatter event in ABC, exiting in the $\hat{j}-\hat{k}$ direction is:

$$dP_1 = \frac{dxdy}{2\pi d^2/4} \cdot \sigma_T \ e^{-\sqrt{2}\sigma_T z} dz \cdot f_i \cdot \frac{3}{16\pi} (1 + \cos^2\theta) \ e^{-\sqrt{2}\sigma_T z} \tag{1}$$

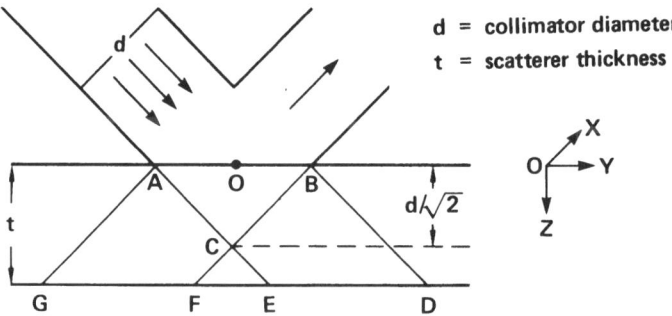

d = collimator diameter

t = scatterer thickness

Figure 1. The experimental collimator plane. The collimators are of cross section d and the scatterer or specimen is of thickness t. This cross section would be the same for circular or square collimators. They would differ only in the x-direction (normal to the plane of the paper).

[4]See however A. Wightman, Phys. Rev. 74, 1813-1817 (1948) or Y. Nishina, Zeits.f.Phys. 52, 869-877 (1929).

which is the probability that the photon enters the scatterer within a neighborhood dx of x and dy of y inside a surface ellipse of
area $\pi \cdot d/2 \cdot \sqrt{2}d/2$, penetrates to a dz neighborhood of z where a
fraction f_i is scattered by the ith mechanism (either Compton,
Rayleigh or both) into an angle θ and then exits in the direction of
the exit collimator. For simplicity we assume no change in absorption cross sections for the change in energy after Compton scattering.
The error introduced by this assumption varies from about 2% at
at 10 Kev to 0.2% at 130 Kev respectively with B_4C as the scatterer.

Double Scatter

The differential probability of double scatter occurring is:

$$dP_2 = \frac{dx_1 dy_1}{\sqrt{2}\pi d^2/4} \cdot \sigma_T e^{-\sqrt{2}\sigma_T z_1} dz_1 \cdot f_i \cdot \frac{3}{16\pi} (1+\cos^2\theta_1) \cdot \sigma_T e^{-\sigma_T r_{12}} dz_2 \cdot$$
$$\frac{dx_2 dy_2}{\sqrt{2}\pi d^2/4} \cdot f_2 \cdot \frac{3}{16\pi} (1+\cos^2\theta_2) \, e^{-\sqrt{2}\sigma_T z_2} \qquad (2)$$

This expression physically follows from two applications of the
single scatter argument. There is however an error in the use of
two independent applications of the Thomson scattering cross section
since an assumption of its derivation is that the incident radiation
be non-polarized. This is true for the first scattering but not the
second. DuMond[3] gives the correct expression which is:

$$2\sigma^2 \left(\frac{3}{16\pi}\right)^2 [(1 + \cos^2\theta_1)(1 + \cos^2\theta_2) - 1] \qquad (3)$$

not

$$\sigma^2 \left(\frac{3}{16\pi}\right)^2 (1 + \cos^2\theta_1)(1 + \cos^2\theta_2)$$

METHOD OF EVALUATION

Since the integrations are 3 fold and 6 fold and dependent upon
other variables the simple methods such as Simpsons rule prove to
be impractival. That is, suppose that each integral was approximated
by 10^2 evaluations then a 6 fold integration would involve $(10^2)^6 = 10^{12}$
evaluations - clearly not feasible. The Monte Carlo method is available, but we believe it to be less efficient than the method of
Gaussian quadratures. We have been able to obtain good convergent
values of integrals (to 1% or better) by evaluating the integrand of
each integral at 4 points or $4^6 = 4096$ total evaluations for a 6 fold
integration.

To evaluate the integral

$$I_1 = \int_a^b f(x) \, dx$$

the integral is written as a sum of n terms as

$$\int_a^b f(x)\ dx = \frac{b-a}{2} \sum_{i=1}^n w_i\ f(x_i)$$

The points to be evaluated are

$$x_i = \frac{b-a}{2}\ z_i + \frac{b+a}{2}$$

where the z_i are the zero's of the nth order Legendre polynomials. A complete table of the z_i and w_i is given by Abramowitz and Stegun[5]. For further discussion of error and of the method in general see A. Cohen or Arden and Astill[6].

RESULTS

Significant, in fact, most, double scattered radiation seen in the exit collimator orignates from the volume from which single scattered does not originate. To reduce multiple scatter all materia not seen by both collimators should be trimmed off.

Figure 2 shows the probabilities of single scatter, double scatter and the ratio of double to single scatter as a function of scatter thickness. Note the rapid rise of P_1 to a constant value at $t=d/\sqrt{2}$. In fact P_1 is nearly a linear function of t. P_2 starts out more slowly for small t and then increases rapidly. P_2 is found to be nearly volume dependent over a wide range of t.

To study the degree of depolarization due to multiple scattering we perform analyses similar to DuMond's[3] and found that radiation with its electric field vector perpendicular to the collimator plane to have an intensity proportional to:

$$E^2 = E_\perp^2 A^4\ [\cos^2 B\ \cos^2\alpha + \sin^2 B\ \sin^2\alpha\ \cos^2\theta_1 + 2\ \sin B\ \cos B\ \sin\alpha\cos\alpha\cos\theta$$
$$+ \sin^2 B\ \cos^2\alpha\cos^2\theta_2 + \cos^2 B\ \sin^2\alpha\cos^2\theta_1\cos^2\theta_2$$
$$- 2\ \sin B\ \cos B\ \sin\alpha\cos\alpha\cos\theta_1\cos^2\theta_2]$$

with $\cos B = \cos\theta_1\cos\theta_2/(\sin\theta_1\sin\theta_2)$ and α is the angle between the first scattering plane and the collimator plane. This is the desired radiation.

Radiation with its electric field vector parallel to the colli-

[5]M. Abramowitz and I. Stegun, "Handbook of Mathematical Functions", U.S. Goverment Printing Office, Dept. of Commerce, 1964.
[6]See for example A. Cohen, "Numerical Analysis", John Wiley, New York 1973 or B. Arden and K. Astill, "Numerical Algorithms", Addison-Wesley, Reading, Mass., 1970.

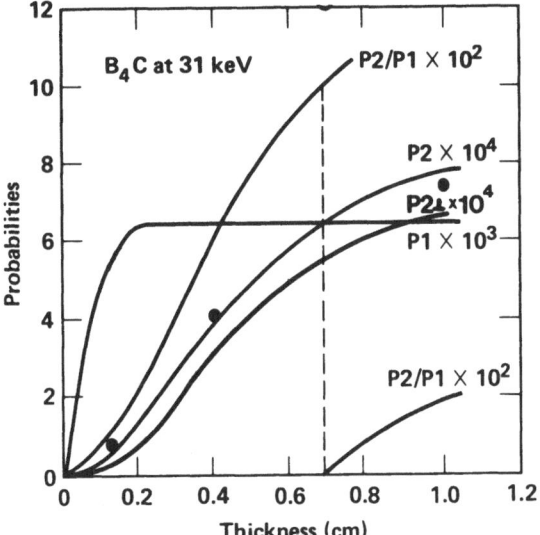

Figure 2.   Probabilities of single scatter (P_1) and double scatter
             (P_2) exiting the exit collimator. P_1 is the probability
             that P_2 be polarized perpendicular to the collimator plane.

mator plane should be totally annihilated by 90° scattering, but because of multiple scatter some of this radiation will emerge to the
exit collimator. The intensity of this undesired radiation is proportional to

$$E^2 = E_{11}^2 A^4 [\cos^2 B \sin^2\alpha + \sin^2 B \cos^2\alpha\cos^2\theta_1 - 2\cos B \sin B \sin\alpha\cos\alpha\cos\theta_1$$

$$+ \sin^2 B \sin^2\alpha \cos^2\theta_2 + \cos^2 B \cos^2\alpha \cos^2\theta_1 \cos^2\theta_2$$

$$+ 2 \sin B \cos B \sin\alpha \cos\alpha \cos\theta_1 \cos^2\theta_2]$$

The total intensity, of course, sums to be proportional to

$$\cos^2\theta_1 + \cos^2\theta_2 + \cos^2\theta_1 \cos^2\theta_2$$

(see the discussion between eqns. (2) and (3).

    Figure 2 also shows the probability that the radiation is polar-
ized perpendicular to the collimator plane (desires radiation). As
can be seen, double scatter can lead to 25% or more undesired radiation (unpolarized) in very thick samples. The three data points indicated in Figure 2 were obtained by observing the intensity of the
double scattered Compton peak of BaK_α scattered from three different
thicknesses of B_4C scattered. These spectra are shown in Figure 3.

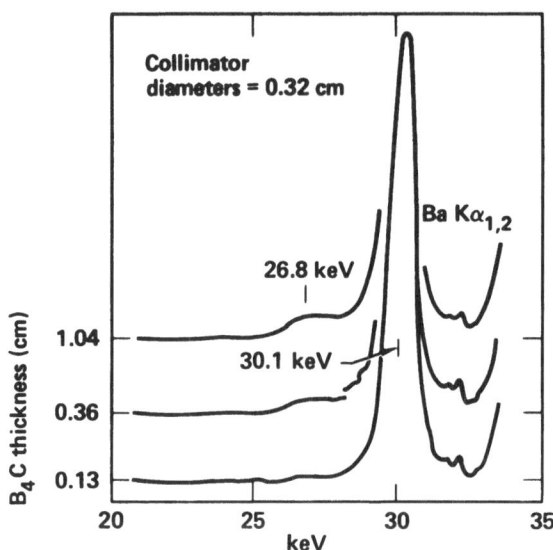

Figure 3. Spectra showing the double scattered Compton peak for
BaK$_\alpha$ scattered from B$_4$C.

CONCLUSION

In order to maintain a high degree of polarization in an experimental design such as Figure 1 it is desired to form the sample and
scatterer into the geometry of the volume seen by both collimators
(we call this an ellipsmid). The ellipsmid can even be truncated a
bit (to perhaps t=d/2 instead of t=d$/\sqrt{2}$). This will reduce P$_1$ by
about 10% but P$_2$ by about 50%. There will be some small amount of
depolarization by multiple scatter for all thicknesses but it is
minimized by going to thinner specimens.

ADVANCES IN LOW-ENERGY ELECTRON-INDUCED

X-RAY SPECTROSCOPY (LEEIXS)

René Bador

Biophysics Laboratory-Pharmaceutical Sciences Unit
Université Claude Bernard - Lyon I
69008 Lyon, France

Maurice Romand, Marlène Charbonnier and Alain Roche

Applied Chemistry Department (C.N.R.S., ERA n° 300)
Université Claude Bernard - Lyon I
69622 Villeurbanne Cedex, France

ABSTRACT

Using low-energy electron induced X-ray spectroscopy (LEEIXS) with a cold cathode tube as excitation source and appropriate calibrations, quantitative surface composition informations are obtained. The results concern anodic film thicknesses and carbon impurities. In addition, qualitative informations on the chemical state of the film components are deduced from the fine structure of X-ray emission bands. The results concern oxygen K emission spectra from CuO and Cu_2O, SnO, SnO_2 and anodized Sn. It is also shown the capabilities of LEEIXS for gaining element-depth profiles. One example pertaining to phosphorus impurities in Al_2O_3 anodic films is given. Finally advantages and disadvantages of LEEIXS over ion-induced X-ray spectroscopy are discussed.

INTRODUCTION

The chemical characterization of surfaces from a few to several hundreds of angströms has received much attention in the recent years. A wide variety of surface analysis techniques including mainly - photoelectron spectroscopy (XPS), Auger electron spectroscopy (AES), ion scattering spectroscopy (ISS and RBS), secondary ion mass spectroscopy (SIMS and IMMA) - have been successfully employed for gaining solution of many research and development

351

problems which are dependent upon surface composition. Among the methods which are available, X-ray emission techniques are not generally accepted as suitable tools for surface characterization studies, the information originating predominantly from the bulk rather than from the surface region. The purpose of this paper is to show how low-energy electron induced X-ray spectroscopy (LEEIXS) can be used to obtain non-destructively, qualitative and quantitative data on surface layers and ultra-thin films.

EXPERIMENTAL CONSIDERATIONS

The surface capabilities of LEEIXS[1-5] are based upon the use of a cold cathode tube producing an electronic excitation of samples under primary vacuum conditions. This device operates as an open window system in which electrons are extracted through an aperture in the anode and impinge directly the specimen[6] without backscattering processes on an intermediary target. Under these conditions only, the electronic excitation can be regarded as monoenergetic and the escape depth from which the X-ray analysis lines are emitted may be obtained by means of significant calculations. In general, 0.5 to 4.5 keV electrons are used and 0.5 to 2 mA electron currents are distributed over areas about 1 cm in diameter. The cold cathode tube is attached to a commercial wavelength dispersive X-ray fluorescence spectrometer equipped with a flow proportional detector and is able to provide performances in the soft and ultra-soft X-ray region. The major components of the LEEIXS system are diagrammed in figure 1. In earlier works[1-5] we have applied the LEEIXS technique to the investigation of thin oxide films, the thicknesses of which are in the range between the monolayer and some hundreds of angströms. This paper presents new results about surface characterization experiments using LEEIXS.

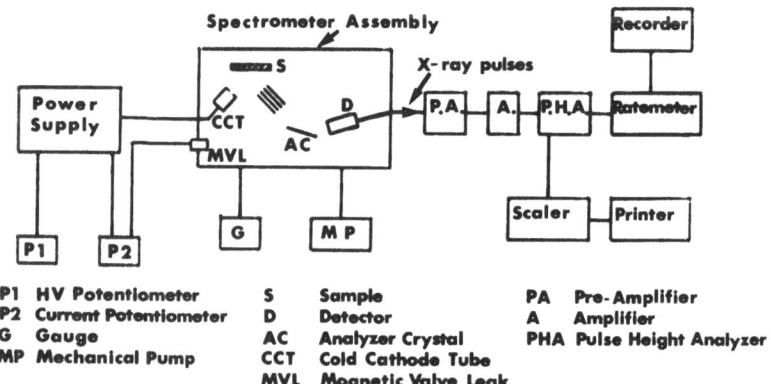

Fig. 1.   Schematic diagram of the LEEIXS instrument.

LEEIXS STUDIES

Long Wavelength Studies

Compared to X-ray fluorescence LEEIXS as well as electron microprobe has relatively high X-ray production cross sections for
long wavelength radiations. Consequently, qualitative and quantitative analysis of light elements such as B, C, N, O and F can be carried out by measuring K emission band intensities.

As a first example, the figure 2 represents the oxygen K band
net intensity as a function of film thicknesses for thin anodic
films of titanium oxide TiO_2. In order to obtain quantitative analysis only the first part of the plot can be easily used. The intensity of the characteristic X-radiation is then nearly propor-
tionnal to the thickness of the oxide layer. Obviously this linear
range may be extended by increasing the electron acceleration potential to an appropriate value which depends upon the analytical
situation. Such measurements are suitable for the study of any

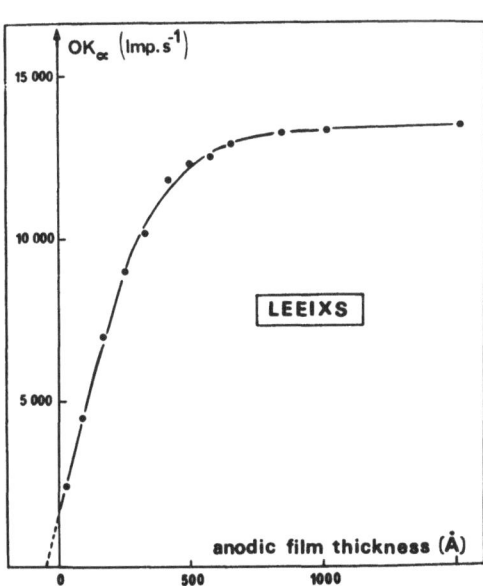

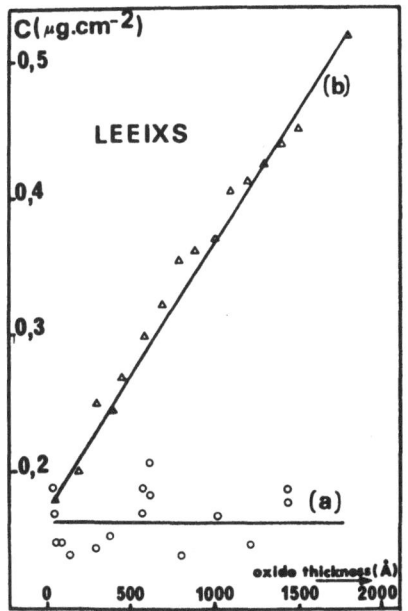

Fig. 2. Oxygen Kα net intensity
for 3 keV electrons from
anodic TiO_2 films of varying
thicknesses.

Fig. 3. Quantitative analysis of
carbon impurities in anodic
TiO_2 films grown in phosphoric
acid (curve (a)) and tartaric
acid (curve (b)) electrolytes.

simple metal oxide / metal system. On non-valve metals, oxygen surface densities as small as 10^{-9} g.cm^{-2} can be detected.

As another example the figure 3 is a plot of the total carbon
content in anodic oxide films grown on titanium substrates versus
film thickness. In the case (a) films are prepared in phosphoric
bathes. The C K signal from these samples is thought to be typical
of an adsorbed carbon contaminant layer. Indeed the C K intensity
is seen roughly independent of the film thickness. Contrarywise in
the case (b) the anodizing processes take place in tartaric baths
and tartaric anions are incorporated in the films. Consequently the
carbon content increases with the film thickness. In these experiments standards are provided by silicon slides which have been carbon implanted in the surface region with determined doses of carbon ions.

In addition to quantitative analyses, qualitative information
on the chemical state of the film components can be deduced from
the fine structure of some X-ray emission bands. Such studies are
made possible owing to the adequate resolution of the wavelength
dispersive spectrometer. By way of example the figure 4 shows the
oxygen K X-ray band spectra for the CuO and Cu_2O copper oxides.
Large changes are seen in these spectra and the corresponding differences would be able to be used to solve practical problems. In
a similar way the oxygen K bands of various tin oxides are presented in figure 5. These bands are obtained respectively from the
SnO and SnO_2 oxides and from a tin substrate which has been anodized in a tartaric acid electrolyte. In every case these spectra
have been unfolded into their constituent components by means of
a DUPONT Model 310 Curve Resolver   assuming that the involved
molecular orbitals are gaussian in shape. In every case the peak E
at about 513 eV is due to a line-like reflectivity structure which
phtalate acid analyzing crystals exhibit[7]. Although no chemical
shift is shown for the A, B, C and D sub-bands the influence of
the chemical state of the tin oxides can be easily established by
a detailed investigation of the A and B sub-bands relative intensities. In SnO, A is nearly as intense as B while in SnO_2 the former
component is much more intense than the latter one. Consequently,
the chemical composition of the anodized film can be identified as
SnO_2. Let us note that an analogous conclusion can be drawn comparing in a simpler way the full width at half maximum of the overall OK bands.

More generally, it is quite apparent from such studies that
soft X-ray band spectroscopy by means of LEEIXS can be a very
powerful tool for characterizing the chemical state of superficial-
ly treated substrates. The line shape can then be regarded as a
"fingerprint" of the surface compounds even though the corresponding fine structures are not analyzed in terms of densities of
state.

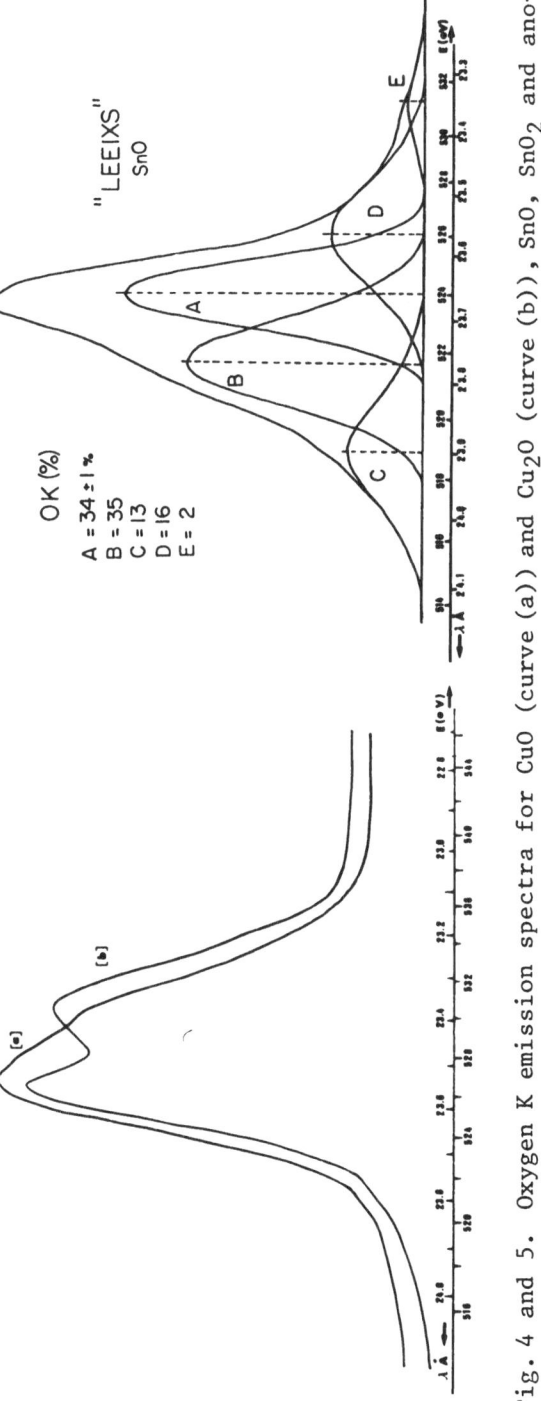

Fig. 4 and 5. Oxygen K emission spectra for CuO (curve (a)) and Cu₂O (curve (b)), SnO, SnO₂ and anodized Sn. All the spectra are obtained at 3 kV operating voltage using a TlAP analyzing crystal.

Sampling Depth

The major factors that affect the sampling (or information) depth are the energy of the electron beam, its incidence angle on the specimen, the matrix composition, and the excitation energy threshold of the analytical line.

By using LEEIXS the sampling depth is linked to the material thickness from which characteristic X-rays are being produced. Then, this parameter depends upon the penetration ranges of the bombarding electrons. Feldman[8] has measured the practical maximum range R of 1 - 10 keV electrons in solids and has formulated the following range - energy relationship

$$R \, (\overset{\circ}{A}) \; = \; 250 \; \frac{A}{\rho} \; \left(\frac{E_o}{Z^{1/2}} \right)^n$$

where $n = 1.2/(1 - \log Z)$
In this relationship E_o is the electron energy in keV, A, Z and ρ respectively the atomic weight, the atomic number and the density of the material under investigation.

Consequently, the corresponding excitation range R_e can be written as follows :

$$R_e \, (\overset{\circ}{A}) \; = \; 250 \; \frac{A}{\rho \, Z^{n/2}} \; (E_o^{\; n} - E_x^{\; n})$$

where E_x is the critical energy for ionization of the atomic inner level that is involved in the concerned emission process.
A similar equation known as Castaing's relationship[9] and commonly used by electron microprobe practitioners can also be taken in order to determine the excitation range R_e

$$R_e \, (\overset{\circ}{A}) \; = \; 330 \; \frac{A}{\rho \, Z} \; (E_o^{\; 1.7} - E_x^{\; 1.7})$$

Table 1 gives a comparison between the oxygen excitation ranges in Al_2O_3 and Nb_2O_5 films using Feldman's or Castaing's equation for 1, 2 and 3 keV electrons. As it can be seen the relative deviations (Δ %) between results are small and nearly constant in the case of Al_2O_3 films but they vary to a large extent versus electron energy in the case of Nb_2O_5 films. For heavy matrix, these changes may be easily explained considering Feldman's relationship and the dependence of the exponent n of the energy with Z. This dependence makes Feldman's equation more universal than the Castaing's one. It is why other developments are undertaken by using the former formula.

Table 1. Oxygen Excitation Range

E_o (keV)	Al_2O_3			Nb_2O_5		
	R_e (Å)			R_e (Å)		
	CASTAING	FELDMAN	Δ %	CASTAING	FELDMAN	Δ %
1	139	149	6.7	109	76	43
2	615	662	7.1	483	421	15
3	1 296	1 400	7.4	1 017	1 030	1.3

Figure 6 shows the change of the excitation depths of carbon
and phosphorus atoms distributed in different materials as function of the primary electron energy. Curve (a) is related to carbon
impurities in TiO_2 films while curves (b) and (c) are concerned
with phosphorus impurities in Al_2O_3 and TiO_2 films respectively.
By way of example the sampling depth of the phosphorus impurities
in alumina layers is less   than 250 Å when 2.5 keV electrons are
used. Let us note that the different starting points of the (a)
and (c) curves correspond with the values of the excitation thre-
sholds of carbon K (0.283 keV) and phosphorus K (2.142 keV) X-ray
emissions. Moreover the different slopes of the (b) and (c) curves
are attributable to the fact that the same impurity (P) is located
in two different matrix, the heavier one (TiO_2) being more absor-
bing for electrons than the lighter one (Al_2O_3).

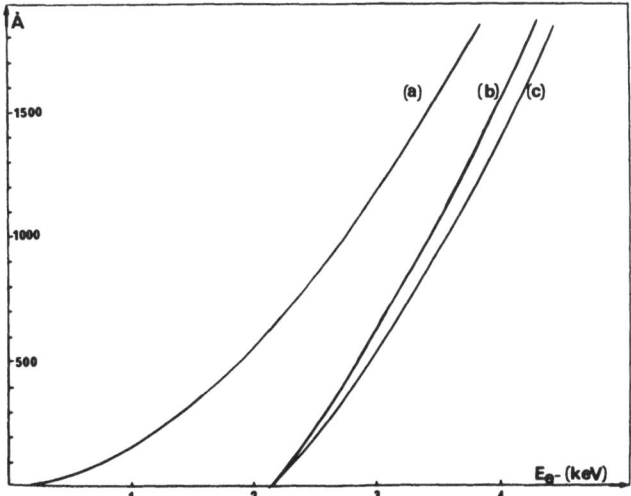

Fig. 6. Excitation
depth vs.electron
energy. (a), (b) and
(c) curves concern C
in TiO_2, P in Al_2O_3
and P in TiO_2 films
respectively.

Impurities Analysis and Elemental Depth Distribution

As has been previously suggested, LEEIXS exhibits a high surface sensitivity for impurity analysis. Thus, phosphorus or sulfur impurities in anodic oxide films can be detected at surface
densities as small as 10^{-9} g.cm^{-2} for fifty seconds analyses.
Using the carbon-containing oxide films studied in the chapter
"Long Wavelength Studies," a three-sigma minimum detection limit
of about 1.10^{-9} g carbon/cm^2 can also be calculated. Actually,
such low carbon surface densities cannot be measured because a
light carbon contamination is unavoidable in analyses performed in
a primary vacuum.

For many practical purposes it is important to carry out elemental depth profiles. Numerous modern surface analytical techniques are available for such studies in conjunction with ion beam
sputtering (AES, XPS, ISS, EMMA...) if not (RBS). In order to test
the potential usefulness of LEEIXS for gaining such information,
it is first necessary to know the depth resolution inherent in this
technique. It may be defined as the difference ΔR_e between two
limit excitation depths R_e associated with a meaningful minimum
change of the electron beam accelerating voltage. From Feldman's
relationship, ΔR_e may be formulated as follows:

$$\Delta R_e = 250 \frac{A \, n}{\rho \, Z^{n/2}} E_o^{n-1} \Delta E_o$$

Table 2 gives some values of ΔR_e. These ones are calculated
for 1, 2 and 3 keV electrons impinging Al_2O_3 and Nb_2O_5 matrix, assuming that $\Delta E_o = 0.05$ keV and taking an incidence angle of the
electron beam equal to $70°$. Clearly, these results and those pertaining to the surface sensitivity of the method prove that LEEIXS
is potentially able to provide information on elemental depth profiles of impurities. Such investigations can be carried out either
destructively by operating well-controlled ion sputter or
chemical etching steps and running experiments at a constant accelerating voltage (unfortunately the former operations can only be
performed outside the X-ray spectrometer and then, the relevant
analysis is of a relatively limited interest) or
non-destructively by increasing step by step the accelerating
voltage. Under these operative conditions thicker and thicker
slices are probed. Obviously, the relevant analyses without specimen erosion or special sample preparation present unequalled advantages but unfortunately the mathematical processing of the physical problem offers too many uncertainties. Moreover the factors
involved such as those used in quantitative electron microprobe
analysis are increasingly inaccurate as the electron energy drops
below 10 keV. Besides, the large energy dependence of the ionization cross section in the concerned energy range must be taken

Table 2.   LEEIXS Depth Resolution

E_o (keV)	ΔR_e (Al_2O_3)	ΔR_e (Nb_2O_5)
1	18 Å	10 Å
2	30 Å	22 Å
3	40 Å	35 Å

into account. Despite these difficulties some in-depth information
relating to the phosphorus distribution in an alumina anodic film
1 100 Å thick have been obtained using a far simpler method requiring no correction. This method needs having standards in which
oxide thicknesses including the phosphorus impurities are similar
to the probe depths in the specimen. An average concentration of
phosphorus is then deduced in every slice (about 150 Å thick) corresponding to two successive limit excitation depths R_e. The
results are illustrated in figure 7. The in-depth profile shows a
phosphorus-free portion near the metal-oxide interface. This
phosphorus depletion in about the one-third inner part of the film
is in agreement with RBS data.[5]

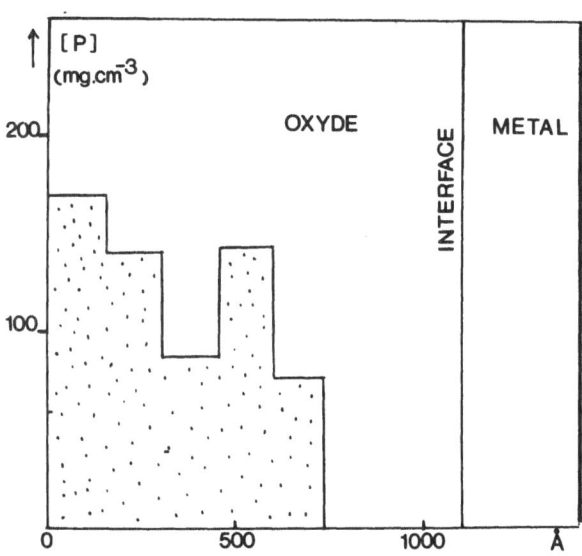

Fig. 7. Phosphorus com-
position profile through
an anodic alumina film
1 100 Å thick.

DISCUSSION AND CONCLUSIONS

      When solids are bombarded with ions, X-rays may also be gene-
rated. The relevant techniques are known as PIX (proton-induced
X-ray) and more generally as IIX (ion-induced X-ray)[10] spectrome-

tries. In these experiments ion beam energies ranged from some
hundreds keV to some MeV.

Both IIX and LEEIXS have relatively high surface sensitivities
for low Z elements. Ion bombardment offers the advantage of providing a negligible continuous background of Bremsstrahlung radiation compared to that for electron excitation, and light elements
analysis can then be made with a "windowless" energy dispersive
[Si(Li) detector] X-ray spectrometer.

In addition, depths over which X-rays are emitted are much
larger in IIX than in LEEIXS and the depth resolution is much
poorer in the former case. Consequently, in order to gain some
elemental depth distribution information, these factors make the
situation in IIX less favourable although ions suffer relatively
small energy losses in getting through thin films, which allows us
to consider the ionization function as being nearly uniform.

Besides, it is worthy of note that a wavelength dispersive
spectrometer used in LEEIXS is able to provide molecular information while solid state detectors used in IIX are quite insensitive
to line shape.

Finally let us point out that IIX can be used in ultra-high
vacuum jointly with some other advanced techniques for surface
analysis while LEEIXS runs only in primary vacuum. This last fact
can appear as a drawback in some studies dealing with characterization of ultra-thin surface layers but it should have to be recognized in the coming years as having serious advantages for study-
ing a wide range of industrial or research problems involving
materials such as metals or alloys, semiconductors, and glasses
prior to or after various treatments.

REFERENCES

1. M. Romand, G. Bouyssoux and R. Bador, Performances of a cold
      cathode windowless tube for thin film studies, in "Abstracts
      of the 19th Colloquium Spectroscopicum Internationale,"
      Philadelphia (1976)
2. M. Romand, R. Bador, A. Roche, G. Bouyssoux, Utilisation de
      sources à cathode froide pour la caractérisation des sur-
      faces, J. Microsc. Spectrosc. Electron. 2:627 (1977).
3. R. Bador, A. Roche, G. Bouyssoux, M. Romand, Analyse de
      films minces et ultra-minces: perspectives nouvelles en
      spectrométrie d'émission des rayons X, Spectrochimica Acta
      33B:437 (1978)
4. A. Roche, R. Bador, F. Buiguez, G. Bouyssoux, M. Charbonnier
      and M. Romand, LEEIXS: a new tool for surface and thin
      films studies, in "Abstracts of the 21st Colloquium
      Spectroscopicum Internationale," Cambridge (1979).

5. A. Roche, A. Cachard, R. Bador, F. Buiguez, M. Charbonnier and M. Romand, Etudes comparatives de couches superficielles à l'aide de techniques spectrométriques d'émission X et de rétrodiffusion d'ions, J. Microsc. Spectrosc. Electron. 4 : 351 (1979)

6. J.J. Sahores, E.P. Larribau and J. Mihura, New results obtained with a regular XRF spectrometer and a portable field spectrometer, equipped with a cold cathode electron tube, in "Advances in X-ray Analysis," C.S. Barett, J.B. Newkirk and C.O. Ruud, Ed., Plenum Press, New York (1973)

7. R.J. Liefeld, S. Hanzely, T.B. Kirby and D. Mott, X-ray spectrometric properties of potassium acid phtalate crystals, in "Advances in X-ray Analysis," B.L. Henke, J.B. Newkirk and G.R. Mallet, Ed., Plenum Press, New York (1970)

8. C.F. Feldman, Range of 1 - 10 keV electrons in solids, Phys. Rev. 117 : 455 (1960)

9. R. Castaing, Electron probe microanalysis, in "Adv. Electronics Electron Phys.," Academic Press, New York (1960)

10. R.G. Musket and W. Bauer, Surface analysis using proton beams, Thin Solid Films 19 : 69 (1973)

ESCAPE PEAK INTENSITIES IN ARGON/METHANE FLOW DETECTORS

M. A. Short and J. Tabock

Ford Motor Company
Research Staff
Dearborn, Michigan 48121

INTRODUCTION

X-ray proportional gas detectors are used in X-ray diffraction and X-ray fluorescence and emission analysis. The amplitude of the electrical pulses produced by these detectors is directly proportional to the energy of the incident X-rays. In most instances, the energy of the incident photons will be utilized completely in producing photo-ionization in the outermost shells of the detector atoms. If the energy of these photons is greater than the absorption edge of the atoms in the detector, a certain fraction of them will produce photo-ionization of an inner shell. This in turn results in the production of a characteristic X-ray. If this characteristic X-ray is absorbed by the detector, then the original X-ray may still be considered to have deposited the whole of its energy within the detector. In this case the amplitude of the detector output pulse will be the same as if the original X-ray photon had been absorbed entirely by photo-ionization of the outer shells. Because of the relatively low absorption of a detector for its own characteristic radiation, there is a significant probability that the detector characteristic X-ray will escape from the detector without producing further photo-ionization. In such cases the energy absorbed by the detector will be reduced by the energy of the emitted characteristic X-ray, and the amplitude of the electrical pulse produced will be reduced by a proportionate amount. Consequently, an energy spectrum of a detector output will show two peaks, one corresponding to the X-ray photons whose energy has been completely absorbed within the detector and a second corresponding to those photons which have resulted in the production of detector characteristic X-rays which have escaped from the detector without producting further photo-ionization. This second peak is the well known "escape peak". The

difference in energy between the main photopeak and the escape peak
is equal to the energy of the characteristic X-ray line of the
detector. This applies equally to argon, xenon and other gas detectors. These and other detector effects have been well summarized by
Bertin.[1]

A continuing question has been to determine what actions should
be taken and what actions can be taken in differing experimental
situations to allow appropriately for the occurrence of escape peaks.
Escape peak pulses are pulses which are "lost" from the main photopeak. Ideally, therefore, they should be counted along with the
main photopeak pulses. The extent to which this may or may not be
accomplished depends on whether the escape peak is partially merged
with the photopeak, is separate from the photopeak, is merged with
the system electronic noise, or is partially or fully merged with
second order escape peaks and/or photopeaks. Jenkins and Hurley[2]
have discussed the problem of interferences between second and higher
order escape peaks and first order photopeaks. The inclusion or
exclusion of the escape peak depends also on the extent to which the
pulse height analyzer window may be controlled. While this is
straightforward in a manually controlled, commercially available,
X-ray system, it is usually more difficult in an automated system.
In many experimental situations the escape peak may be routinely
included or excluded depending on the energy separation of the escape
and photopeaks. If this inclusion or exclusion is made for both
standards and unknowns systematically, then problems seldom arise.

In certain circumstances, however, it is desirable to make a
correction for the lost escape peak pulses which, for some reason,
cannot be included in the pulse height analyzer window. An example
of such an instance has been discussed by us.[3] This involved the
derivation of an X-ray emission intensity for argon by interpolating
between measurements made for chlorine and potassium using an
argon/methane detector. While chlorine and argon do not give escape
peaks with this detector, potassium does. It is clearly necessary to
correct the measured intensity for potassium for escape peak losses
before making the interpolation for argon. This cannot be accomplished unless the fraction of pulses contained in the escape peak is
known.

Some experimental measurements of escape peak intensities of gas
detectors have been made by Jenkins.[4] He concluded that the fraction
of photons resulting in an escape peak was related to the fluorescence
yields of the gases and, for a given gas, was relatively constant over
a wide range of X-ray energies. A basis for the calculation of
escape peak intensities relative to the intensities of the main photopeaks has been presented by Maurice and Ruste.[5] They related the
escape peak intensities not only to fluorescence efficiencies but
also to detector dimensions, gas mass absorption coefficients and gas
densities.

Because of the importance and large scale use of argon/methane in flow proportional detectors used in X-ray fluorescence and X-ray emission analysis, we have made a new series of experimental measurements of escape peak relative intensities associated with argon/methane detectors. These measurements were made for a wide range of incident X-ray wavelengths, for detectors of three different sizes and for two different gas densities.

ESCAPE PEAK MEASUREMENTS

Measurements of the fraction of total counts appearing in the escape peaks (i.e. escape/(main+escape)) were made for wavelengths in the range K Kα to Zn Kα for argon/10% methane, P 10, flow proportional detectors on Philips PW1450 and Siemens SRS-1 X-ray fluorescence analysis units and on a Cameca MBX electron microprobe. The nominal detector dimensions and gas pressures are shown in Table 1.

TABLE 1:   Detector Parameters

	Gas Pressure atmospheres	Detector Length cm	Detector Diameter cm
Philips PW 1450	1.1	8.0	3.0
Siemens SRS-1	1	8.5	2.3
Cameca MBX	3	5.3	2.0

Most of the measurements were made using pure element samples and the excitation conditions were such as to keep the total number of photons entering the detector below about 10^4 cps to remove the necessity of making dead time corrections. On each instrument and for each wavelength, pulse height energy scans were made using narrow energy windows to determine the separation of the escape peak from the main photopeak and from the system electronic noise. In situations where there was good separation of the escape peak, two measurements of integrated intensity were made. First the PHA window was set to include all of the escape peak and a statistically significant number of counts was accumulated. The PHA was then reset to encompass the main photopeak and a measurement made for the same length of time as for the escape peak. These measurements enabled us to make a direct determination of the fraction of total counts in the escape peak. In cases where there was a partial overlap with electronic noise or with the main photopeak, a simple manual deconvolution with an estimated accuracy of 5% was made. Only on the Philips PW1450 was it possible to pull the potassium escape peak out of the system noise, using careful manipulation of the amplifier gain and the detector high voltage power supply. The results of these measurements are shown in Table 2.

TABLE 2:   Percentage of Total Counts in Escape Peak

Radiation	Philips PW1450	Siemens SRS-1	Cameca MBX	Jenkins (ref.4)
Zn Kα		8.8%		
Cu Kα	7.7%	8.8		
Ni Kα	7.9	8.6		
Co Kα	7.8	8.5	6.6%	
Fe Kα	7.7	8.5	6.5	7.7%
Mn Kα	7.7	8.9	6.2	
Cr Kα	7.8	8.4	6.7	
V Kα	7.7	8.3	6.7	8.2
Ti Kα	7.8	8.5	6.8	
Sc Kα	7.7			7.8
Ca Kα	7.9	8.3		7.4
K Kα	7.6			7.3
average:	7.8%	8.6%	6.6%	7.7%

It may be seen that, within the range of wavelengths examined, a constant percentage of counts in the escape peak was obtained for the entire range of wavelengths for each detector. It is clear from these results, that, in the case of the chlorine-argon-potassium emission intensity analysis discussed above, an appropriate correction to the potassium intensity may be made by measuring the fraction of counts in the escape peak for a conveniently accessible element, such as chromium, rather than for the element potassium itself.

CALCULATED ESCAPE PEAK INTENSITIES

Escape peak relative intensities were calculated using the expression given by Maurice and Ruste.[5] This says that the fraction of total counts appearing in the escape peak is given by:

$$(\omega_K \exp(-\mu\rho L))/(1 + 1/r),$$

where ω_K is the fluorescence efficiency of the gas, μ is the mass self-absorption coefficient of the gas, ρ is the density of the gas, r is the K edge jump ratio, and L is related to the effective path length of the gas characteristic X-rays. The expression clearly conveys the notion that the fraction of counts in the escape peak is directly related to the escape of the gas characteristic X-rays from the detector: the higher the gas pressure and the greater the detector size the less will be the escape of characteristic X-rays and the smaller will be the intensity of the escape peak.

Table 3 shows escape peak relative intensities calculated using the above expression together with the average experimental results given in Table 2. L, the X-ray path length, has been taken as detector diameter/4, this giving good agreement of calculated and measured escape intensities for the three detectors.

TABLE 3:   Calculated vs. Experimental Escape Peak Intensities

	L atmos. cm	calculated	experimental
Philips PW1450	0.825	7.8%	7.8%
Siemens SRS-1	0.575	8.4	8.6
Cameca MBX	1.5	6.4	6.6

CONCLUSIONS

It is concluded that each argon/methane gas proportional detector gives a constant fraction of escape peak pulses independent of incident X-ray wavelength. This fraction is different for different detectors and depends on the gas pressure and detector dimensions. The percentage of total pulses appearing in the escape peak was found to vary from 6.6% to 8.6% for the three detectors examined.

Good agreement was obtained between the experimental measurements and the calculations suggested by Maurice and Ruste.

REFERENCES

1. E. P. Bertin, "Principles and Practice of X-Ray Spectrometric Analysis", pp. 219-284, Plenum Press, New York(1978).
2. R. Jenkins and P. W. Hurley, "Escape Peak Interference in X-Ray Spectrochemical Analysis", Canadian Spectroscopy, 13:35(1968).
3. M. A. Short, J. Tabock and D. W. Hoffman, "Electron Microprobe Analysis of Rare Gases in Thin Metal Foils with the Use of Surrogate Standards", Microbeam Analysis - 1980, San Francisco Press, San Francisco(1980).
4. R. Jenkins, "Escape Peak Ratio as a Function of Fluorescent Yield", Proc. 5th Conf. on X-Ray Anal. Methods, p. 54, N.V. Philips Gloeilampenfabrieken, Eindhoven(1966).
5. F. Maurice and J. Ruste, "Ensembles de Comptage en Spectrometrie de Rayons X a Dispersion de Longeur d'Onde et a Selection d'Energie: Principe, Controle, Reglage", pp.13-14, Commissariat a l'Energie Atomique - France, CEA-R-4909(1978).

LEVELING DEVICE FOR FORMING X-RAY SPECIMEN*

Vann Y. Won

Physical Science Laboratories

McClellan AFB CA 95652

A device was made for preparing accurate definition of surface, depth and volume of liquid x-ray fluorescence specimens.

The apparatus used in conjunction with a specimen holder and plastic film window material accurately and consistently forms a flat bobble-free analysis window on the open face of the specimen holder. The specimen holder in the form of a shallow cylinderical cup is slightly over filled and covered by the plastic film. Placement of the mating leveling apparatus over the film squeezes out trapped air bubbles, levels the exposed face of the specimen, draws the plastic film tightly over the exposed face of the specimen and allows easy installation of a film retaining O-ring to maintain the specimen material in a level state within the holder.

The apparatus was used for several years in the x-ray laboratory to prepare oils, hydraulic fluids and other liquid (non-volatile) specimens to determine elements such as chlorine and sulfur under vacuum. Powder samples also may be used in the apparatus. The control of surface flatness, consistent depth, and repeatable volume are the functions required to obtain high accuracy of x-ray fluorescence analysis from the apparatus.

*U.S. Patent 4,115,689

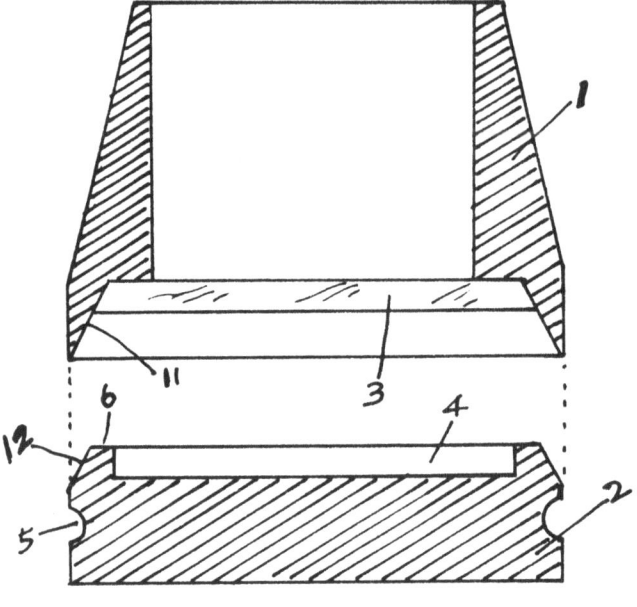

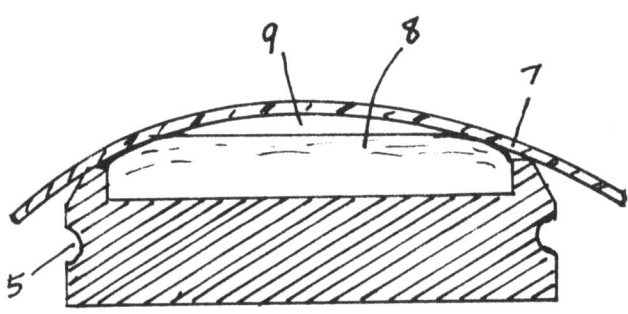

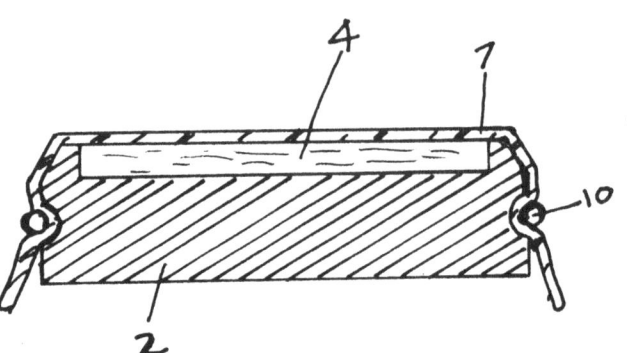

(1) Specimen Loader, Aluminum

(2) Specimen Holder, Aluminum

(3) Transparent Plate, plastic

(4) Placement of Specimen

(5) Groove of Holder (2) is designed to accept and O-ring type elastic band

(6) Top Surface of holder, which is the surface for the plastic plate (3) of loader (1) to rest upon

(7) Plastic Film, mylar film

(8) Over filled specimen material

(9) Entrapped Air Bubbles

(10) O-ring Type Retainer

(11) (12) The surface in a chamfer fabricated to meet the mating surface (11) of loader (1)

MONTE CARLO DETERMINATIONS OF OPTIMAL PHOTON ENERGIES FOR XRF

ANALYSIS OF IODINE IN VIVO

T. Grönberg and S. Mattsson

Radiation Physics Department
University of Lund
S-221 85   LUND, Sweden

INTRODUCTION

After the pioneering work of HOFFER et al.[1] on fluorescent scanning
of natural iodine in the thyroid, several authors have suggested
the use of X-ray fluorescence analysis to follow in vivo the distribution and elimination of stable tracers injected prior to the
investigation. The normal method of performing such studies is to
inject iodine containing contrast media and then follow the variation in iodine concentration by in vivo measurements over the organ or tissue of interest[2,3].
A theoretical treatment of the application of X-ray fluorescence
in in vivo studies has been made by TINNEY[4] by means of analytical
calculations.
Sources of ^{241}Am, emitting mainly 59.5 keV photons, are often used
for the excitation of K_α - X-rays from iodine. The aim of the present work is to study if this photon energy is optimal for the analysis of iodine in various in vivo situations. For this purpose,
we have used Monte Carlo calculations to determine the amount of
K_α-photons generated per primary photon as well as the contribution from interfering scattered primary photons, which together
give an estimate of the minimum detectable concentration.
The results of the calculations of scatter spectra are compared
with experimental data at a primary photon energy of 59.5 keV
(^{241}Am).

METHOD

Monte Carlo simulation: The system to be modelled is a disc-shaped
well collimated radionuclide source, a water phantom with an
iodine-containing volume inside and a collimated detector (Fig. 1).

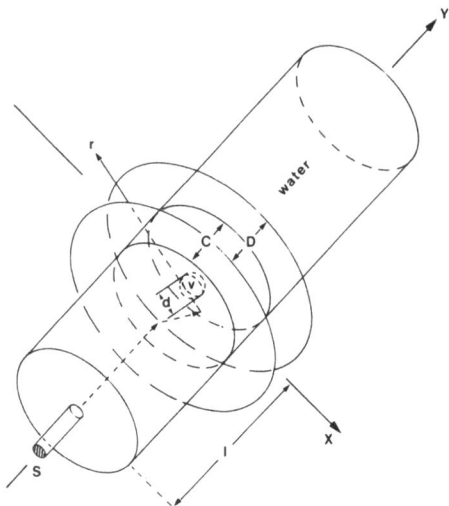

Fig. 1.   Simulated experimental set up

To reduce the number of photons rejected in the calculations, the
collimator and detector were allowed to rotate around the water
phantom (around the y-axis). The MC calculations were carried out on
a micro-computer with an eight-bit processor and a data memory consisting of 16 k RAM. The MC program is described briefly below:
a. Photons are emitted from the source (S) parallel to the y-axis.
b. Each pathlength of a photon in the water phantom is sampled
according to $s=(1/\mu_T(E))\cdot\ln\xi$ where $\mu_T(E)$ is the attenuation coefficient in water and ξ is a uniformly distributed random number[5].
c. In the energy interval of interest (20-60 keV), photons interact
by photoelectric absorption and scattering. Only compton scattering
is taken into account in our calculations. Photon histories are not
terminated subsequent to photoelectric absorption. Instead, photons
gain a weighting factor $W_n=W_{n-1}\cdot(\mu_c(E_{n-1})/\mu_T(E_{n-1}))$, n denotes the
number of interactions.
This is a well known method of variance reduction[6].
d. The compton scattering angle θ is sampled according to the re-
jection method given by KAHN[7]. The angle θ and the rotational angle $\phi=\pi(2\xi-1)$ give the direction towards next point of interaction.
e. When a photon crosses the iodine-containing volume V, the program recordes the pathlength d in V, the photo absorption coefficient of I $(\tau^I(E))$ at energy E of that photon and the weighting
factor W. There might be more than one cross per photon history.
f. If the scattered photon (r) escapes from the phantom through the
collimator opening (C) and the detector (D), its energy E and weighting factor W are registered, creating a scatter spectrum. Photon
histories are terminated if the photons escape the phantom anywhere
except through the collimator opening and the detector or if the
energy E falls below 20 keV.

The cross section data used in the program is taken from HUBBELL[8].

Generation of Kα-photons from I: Since iodine in the in vivo applications appears in trace element concentrations, it would be very
impractical to calculate the absolute number of Kα-photons generated
by the MC-program. Instead, we have derived a quantity P which is
proportional to this number:

$$P = \{\Sigma \ \tau_i^I \ (E) \cdot (d_i/d_{max}) \cdot W_i\}/N_o \qquad \text{Eq 1}$$

Here No is the number of primary photons generated in the source.
The energy E in equation 1 lies between Eo and the K-edge of I.

Calculated scatter spectrum and experimental comparison:
The calculated scatter spectra are converted to detector signal and
compared with measured distributions. The measurements were made by
a Ge-detector and a well collimated Am-241 source.
The energy interval around 28.5 keV is that of greatest importance
since this is the region containing the Kα-photons. The contribution from scattered primary photons in this region is called b.

Detection limit of I: The minimum detectable concentration (MDC) is
proportional to $\sqrt{b}/a$ where a is the net countrate in the Kα-peak
and b the background countrate in this energy interval.
The quantity P discussed above is proportional to a, which means
that a study of $\sqrt{b}/P$ yields the relative variation in MDC.

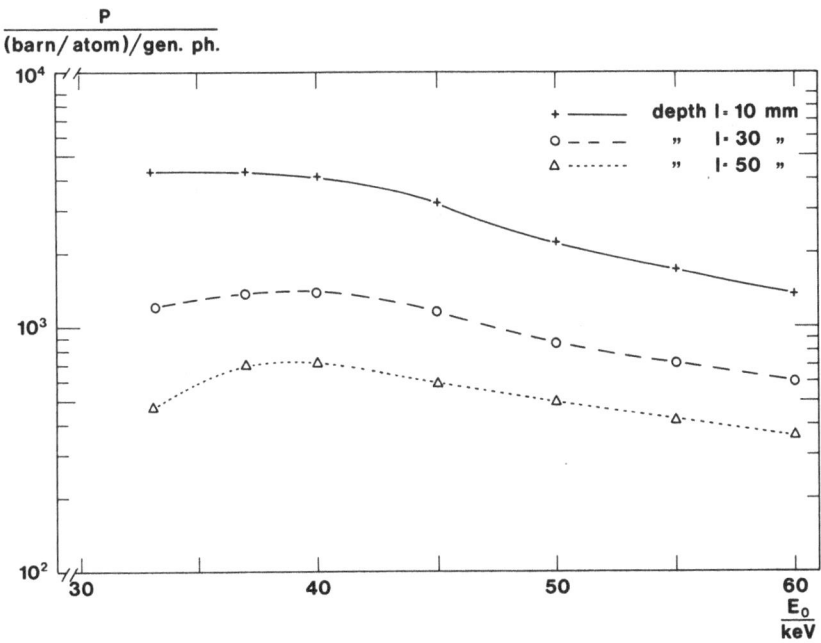

Fig. 2. The variation in P with primary photon energy Eo and depth

RESULTS AND DISCUSSION

Generation of Kα-photons from I: The relative variation in P is given
in figure 2. It is obvious that the increase in τ^I with lower primary
photon energies compensates for the loss of primary photons due to
increasing attenuation in the phantom. Close to the K-edge the attenuation is too high to be compensated for by the increased τ^I which
results in a lower value of P for greater depth. The very pronounced
depth dependence in P is clearly seen in figure 2. The statistical
uncertainty in P is less than 2%.

Calculated scatter spectra and experimental comparison: Figure 3
compares a calculated scatter spectrum (converted to detector signal) and a measured distribution for a depth of 30 mm. As already
mentioned coherent scattering is not included in the calculations.
The calculated contribution b in the energy interval 28-29 keV is
given in figure 4. As expected b increases with decreasing primary
photon energy. The relative value of b is experimentally verified
at Eo=59.5 keV. The statistical uncertainty in b is about 10%.

Detection limit of I: The relative variation in $(\sqrt{b}/P)$~MDC is given
in figure 5. The optimal photon energy for in vivo detection of I
in geometries similar to the one used in this work is about 45 keV.
However, the difference in $\sqrt{b}/P$ between 45 and 60 keV is only about
40%. The statistical uncertainty in $\sqrt{b}/P$ is about 4%.

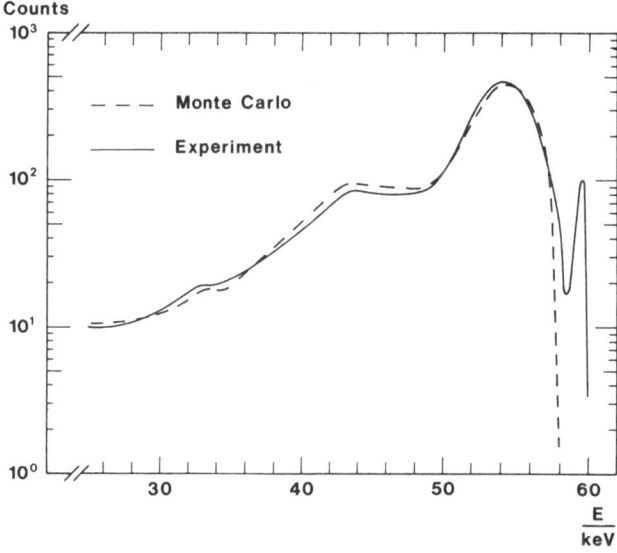

Fig. 3   Comparison of calculated and measured scatter distribution
         at a primary energy of 60 keV.

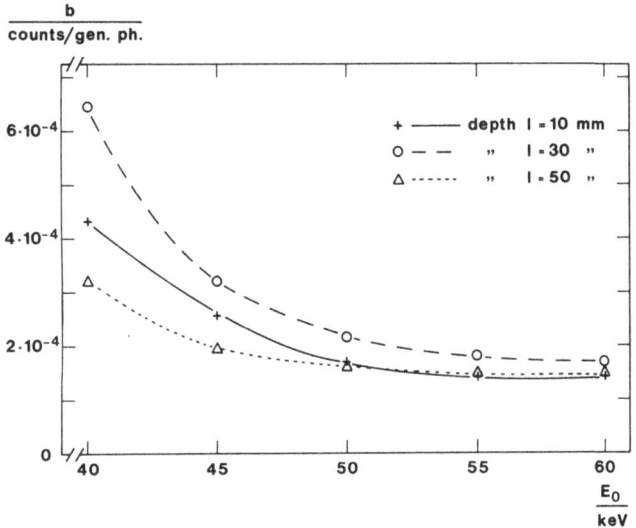

Fig. 4   Scatter contribution in the energy interval 28-29 keV.

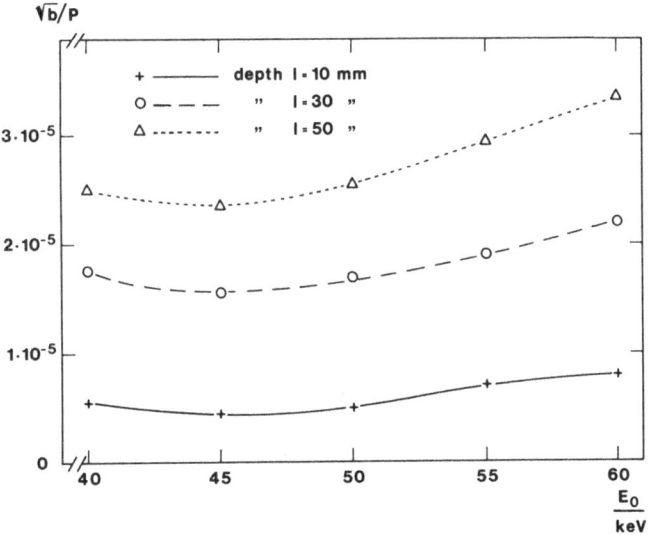

Fig. 5   The relative variation in ($\sqrt{b}$/P)~MDC by primary energy
          Eo and different depth 1.

CONCLUSION

The MC calculations show that for organ depth down to at least 50 mm
the number of Kα-photons from I per primary photon has a maximum for
primary energies very close to the K-edge of I. If, however, one
considers the inevitable scatter contribution the primary energy
which gives the best conditions of measurement is about 45 keV. If
the primary energy is increased to 60 keV (Am-241), the detection
limit is impaired by 40%.
Considering the simplicity and mobility of Am-241 sources, it is
doubtful whether the improvement of about 40% theoretically possible
by changing to the use of an X-ray tube with secondary target is
justifiable.

REFERENCES

1. P.B. Hoffer, W.B. Jones, R.B. Crawford, R. Beck and A. Gottschalk,
   Fluorescent thyroid scanning: A new method of imaging the thyroid, Radiology 90, 342-344 (1968).

2. L. Ahlgren, T. Grönberg and S. Mattsson, In vivo X-ray fluorescence analysis for medical diagnosis, Advances in X-ray Analysis
   23, 185-191 (1979).

3. T. Grönberg, T. Almén, K. Golman, K. Lidén, S. Mattsson and
   S. Sjöberg, Noninvasive estimation of kidney function by X-ray
   fluorescence analysis, (submitted to Phys. Med. Biol.).

4. J.F. Tinney, The application of X-ray fluorescence to in vivo
   biological studies, Ph.D thesis, The University of Oklahoma,
   (1968), Univ. Microfilms, Inc. Ann Arbor, Michigan, USA.

5. D.E. Raeside, Monte Carlo principles and applications, Phys.
   Med. Biol: 21, 181-197 (1976).

6. W.S. Snyder, M.R. Ford, G.G. Warner and H.L. Fisher, Jr., Estimates of absorbed fractions for monoenergetic photon sources
   uniformly distributed in various organs of a heterogeneous phantom, MIRD pamphlet No 5, J. Nucl. Med. 10, Suppl. 3 (1969).

7. H. Kahn, Applications of Monte Carlo, RAND, Santa Monica,
   Calif. (1956).

8. J.H. Hubbell, Photon cross sections, attenuation coefficients,
   and energy absorption coefficients from 10 keV to 100 GeV,
   National Bureau of Standards, NSRDS-NBS 29 (1969).

LEAD AND BARIUM IN ARCHAEOLOGICAL ROMAN SKELETONS MEASURED

BY NONDESTRUCTIVE X-RAY FLUORESCENCE ANALYSIS

L. Ahlgren, J.-O. Christoffersson and S. Mattsson

Radiation Physics Department
University of Lund
S-221 85 Lund, Sweden

INTRODUCTION

Lead is a non-essential toxic metal of considerable topicality. It is accumulated in the skeleton, which contains 90% of the total body burden[1].

It is known that during the Roman period the intake of lead was high. The Romans used kitchen utensils made of lead and lead pipes for drinking water. The most important source of lead for people of the upper social classes was sapa, a compound added to wine to sweeten and perserve it[2].
Using non-destructive X-ray fluorescence analysis, lead and barium concentrations of archaeological Roman bones have been measured. We have earlier used a similar technique to determine lead concentration in the human skeleton <u>in vivo</u> [3].

MATERIAL

Ten Roman skeletons from the Roman-British cemetery at Cirencester in England have been studied. This cemetery dates back to the third and fourth centuries A.D. Samples of rib, skull, fibula, phalanx and metacarpal or metatarsal bone were examined. These bones were selected as representing the different types of bone in the human skeleton, and thus permitting an estimation of the total body burden of lead to be made. Bones from two different periods from Sweden were also analysed (3000 B.C. and 1200 A.D.) to provide a comparison.

METHODS

The quotient between the countrate of characteristic X-rays

of the element under investigation and that of primary photons incoherently scattered in the sample towards the detector has been
used to calculate the concentration of the elements in the bone
sample. The cross-section for incoherent scattering is proportional
to the electron density. Since the number of electrons per unit
mass is almost constant for most of the elements constituting biological matrices, the countrate of incoherently scattered photons
is proportional to the mass of the irradiated sample seen by the
detector. When incoherent scattering is the dominant interaction
mode for the incident photons as well as for characteristic X-rays,
the quotient between the characteristic X-rays and the incoherently scattered photons will be approximately independent of the
density and of the atomic composition of the matrix and hence proportional to the concentration of the trace element.

Since the cross-section per atom for coherent scattering is
proportional to $Z^{2.4}$,[4] the countrate in the coherent peak is related to the bone mineral concentration in the volume considered.
Hence by taking the ratio coherently – incoherently scattered
photons the bone-mineral concentration can be estimated.
Water, saw-dust and bone ash containing known amounts of lead
$(100-10000 \ \mu g \ g^{-1})$ were used to simulate various biological matrices. Measurements on various amounts of the matrices (in thin
walled plastic containers) were carried out to study if the quotient is independent of the mass investigated and the atomic number.
The quotient of the photons coherently and incoherently scattered
from the bone measurements was compared with that of measurements
of bone ash to see how much of the bone samples consisted of inorganic material. The bone samples were dry (room temperature) and
all soil had been removed from them. No other sample preparation
was carried out.

To generate of the characteristic X-rays of lead ($K\alpha_1$ 75.0,
$K\alpha_2$ 72.8) and barium ($\overline{K}\alpha$ 32.1 keV) two ^{57}Co radiation sources
($E\gamma$=122.0 and 136.5 keV) of a total activity with approximately
700 MBq were used. The collimators of the source were made of high
purity tin. The various bones were placed one by one in a perspex
holder and irradiated for 40 minutes. The angle between the primary
and secondary radiation was 90°. The secondary radiation was studied
using a 1 cm^3 Ge-detector connected via a charge-sensitive preamplifier and a pulse-shaping linear amplifier to a 80 MHz analog-
to-digital converter. The collimator in front of the detector was
made of high purity copper covered with tantalum foils of high
purity.

RESULT AND DISCUSSION

Calibration technique

For samples of various mass and atomic composition, the quotient between registrations from characteristic Pb X-rays and incoherently scattered photons is approximately constant for equal

Pb-concentrations. The lead concentration in the bone samples could therefore be calculated by comparing this quotient with that from a measurement of a lead solution of known concentration.

Due to the lower energy of the Ba Kα X-rays this calibration pattern is not generally valid. However, in the range of sample masses used, this quotient was found to be constant to within 5%. The detection limit for lead in the bone samples is $10 \, \mu g \, g^{-1}$ compared with the detection limit of $20 \, \mu g \, g^{-1}$ _in vivo_. The detection limit for Ba in the bone samples is $15 \, \mu g \, g^{-1}$.

Lead distribution in Roman skeletons

Table 1 shows the lead concentration in the different bones studied. In all the bones from the ten graves studied lead concentrations between $60-1000 \, \mu g \, g^{-1}$ were found. Figure 1 shows the distribution of lead in fibula, metacarpal, phalanx and in samples of skull and rib from grave nr 14.

To be able to compare the lead concentration measured in the archaeological samples with those from our _in vivo_ measurements, it is important to know the ratio between the weight of fresh bone and the weight at the time of measurement. By comapring the quotient of coherently and incoherently scattered photons from the archaeological bones with that from a bone ash sample, it is estimated that the Roman bones consists of approximately 80% inorganic material. Using the ratio between ash weight and fresh weight for compact and spongy bone given by Strehlow and Kneip[5] and by the ICRP[6] the ratios between the actual weight of untreated bone and fresh weight of these bones were taken as 0.50 for compact bones and 0.25 for spongy bones.

Table 1.   Lead and barium concentration (μg per g archaeological weight) in bones from different Roman graves.

Grave	Phalanx shaft		Metacarpal end		Metacarpal shaft		Fibula end		Fibula shaft		Rib		Skull		Tooth		Estimated average skeletal concentration (fresh weight)
	Pb	Ba	Pb	Ba	Pb	Ba	Pb	Ba	Pb	Ba	Pb	Ba	Pb	Ba	Pb	Ba	Pb
13	501		1004		440		528		453		288		271				146
9	392	64	515	96	197	30	432	72	217	56	433	88	327	43	116	<15	108
56	284	61	602	95	258	63	452	94	244	68	378	84	251	55			101
35	258		295		288		295		180		256		364				86
65			586	79	243	53	235	53	172	49	354	55	164	26			81
46	196	106	395	100	281	97	447	112	142	56	265	52					78
10	198		268		149		230		174		423		195				76
54	282		265	68	139	46	123	56	113	37	278	73	68	24			51
60	70	59	463	91	152	33	173	68	76	44	173	63					47
14	232		243		90		226		81		113		55				35

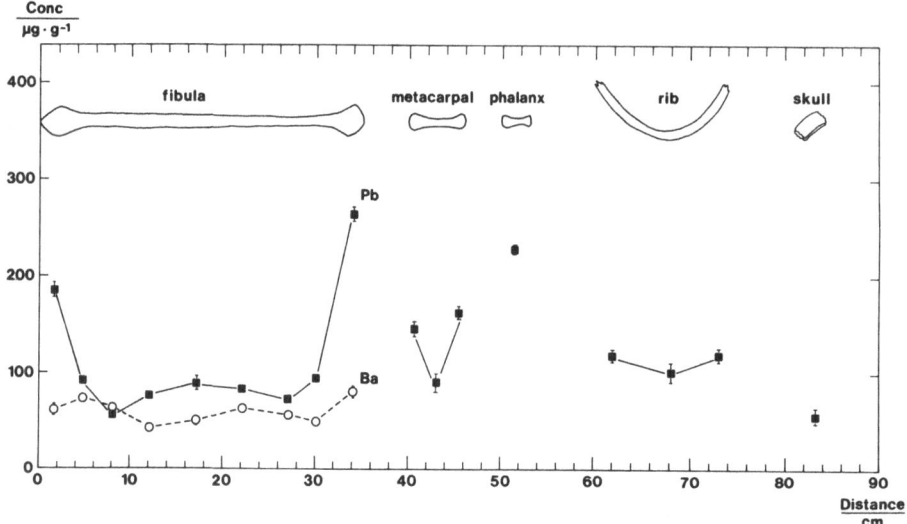

Fig. 1.   Pb and Ba distribution in bones from grave nr 14.

As seen in table 1 and figure 1, the measured concentrations of lead in the end bones are up to three times as high as in the shaft. When the lead concentration in the bone samples is estimated in terms of the fresh weight, the lead concentration at the ends will be even lower than in the shaft.

Archaeological bones from Sweden have also been studied to provide a comparison to the above findings from the Roman-British skeletons. From a pre-historic site at Ihre (≈3000 B.C.) in Öland, southern Sweden, rib samples from four different skeletons were examined. In no instance was any evidence for lead detectable ($<10 \mu g \ g^{-1}$), implying that these people were not exposed to lead to the same extent at the Romans.

The measurements of samples of ribs from seven different skeletons from a Middle Age grave-site (12:th century A.D.) in Stockholm gave the following result. In three samples no lead was detectable ($<10 \mu g \ g^{-1}$). In the other four the concentration varied from 10 to 40 $\mu g \ g^{-1}$ (archaeological weight). This indicates that lead had a certain importance in the Middle Age Swedish daily life.

Barium distribution in Roman skeletons

Barium is in the alkaline earth series of elements along with calcium, strontium and radium. Its metabolic behaviour is thought to be generally similar to that of calcium although some differences are reported. It is accumulated in the leaves of plants[6,7]. The concentration in the plants is dependent on the barium content of the soil upon which they grow [8]. Thus diet determines the intake of barium. The gastrointestinal tract absorbs about 6% of the ingested barium which is deposited in the bone.

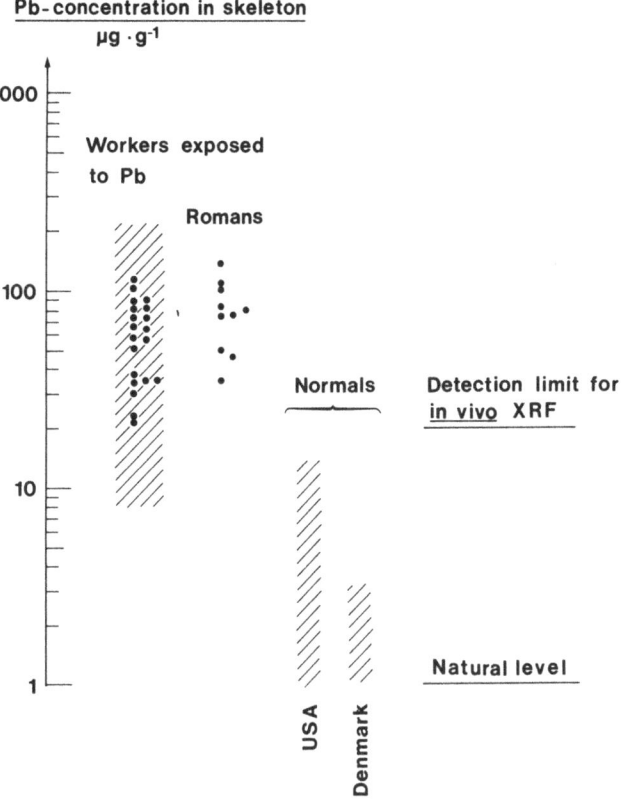

Fig. 2.   Pb-concentration in the Roman skeletons as well as in occupationally exposed "workers"[9] and normals

The distribution of barium in the Roman skeletons is shown in table 1 and figure 1. It lies between 20–100 $\mu g\ g^{-1}$. In comparison to lead it is more uniformly distributed in different bones. The Ihre-findings showed higher concentration of barium 120–440 $\mu g\ g^{-1}$. In the Middle Age ribs, three samples showed detectable amounts of barium 16–20 $\mu g\ g^{-1}$. In the three groups of archaeological skeletons, the different amounts of barium may reflect differences in their diets. The diet of pre-historic people is supposed to contain mostly vegetables with higher barium content than meat. The differences may also be explained by local differences in the barium concentration in the soil used for their food production.

SUMMARY AND CONCLUSION

X-ray fluorescence analysis has shown to be a valuable method for non-destructive analysis of archaeological samples. Lead and barium could be detected down to concentrations of 10 and 15 $\mu g\ g^{-1}$, respectively. Bone samples from Romans (200–300 A.D.) showed concentrations corresponding to a mean value of 80 μg g^{-1} (fresh weight), which is in the same range as occupationally exposed workers today. Figure 2. In comparison, skeletons from pre-historic south Sweden (3000 B.C.) were found to contain detectable amounts of lead (<3 μg g^{-1} fresh weight) while in four skeletons from Middle Age Sweden which showed detectable amounts, a mean concentration of 8 μg g^{-1} fresh weight was found.

The barium concentration (archaeological weight) also shows considerable variations between the three groups of skeletons.

$$
\begin{array}{lll}
\text{Swedes (3000 B.C.)} & : & 340 \ \mu\text{g g}^{-1} \\
\text{Romans (200-300 A.D.)} & : & 70 \ " \quad " \\
\text{Swedes (1200 A.D.)} & : & 17 \ " \quad "
\end{array}
$$

The differences in barium concentration might reflect varying amounts of vegetables (which are rich in barium) in the diet. Information on the lead and barium content of archaeological skeletons can give additional information on the diet and daily life of our ancestor.

ACKNOWLEDGEMENTS

Thanks are due to Dr H.A. Wadldron, London and Dr T. Wadsten, Stockholm for putting the archaeological bone samples at our disposal

REFERENCES

1. H. A. Schroeder and I. H. Tipton, The Human Body burden of Lead, Archives of Environmental Health 17, 965-978 (1968)
2. S. C. Gilfillan, Lead Poisoning and the Fall of Rome, Journal of Occupational Medicine 7, 53-60 (1965).
3. L. Ahlgren, T. Grönberg and S. Mattsson, In vivo X-ray Fluorescence Analysis for Medical Diagnosis, Advances in X-ray Analysis 23, 185-191 (1980).
4. J. Weber and D. J. van den Berge, British Journal of Radiology 42, 378 (1969).
5. C. D. Strehlow and T.J. Kneip, The Distribution of Lead and Zinc in the Human Skeleton, Journal of the American Industrial Hygiene Association 30, 372-378 (1969).
6. ICRP, Report No 23, Report of the Task Group on Reference Man, Pergamon Press, Oxford (1975).
7. N. R. French, Review and discussion of Barium, Radioecology, 557-560 (1963).
8. W. B. Healy and T. G. Ludwig, Barium Content of teeth, bone and kidney of twin sheep raised on pastures of differing barium content, Archives of oral Biology 13, 559-563 (1968).
9. L. Ahlgren, B. Haeger-Aronsen, S. Mattsson and A. Schütz, In vivo determination of lead in the skeleton after occupational exposure to lead, British Journal of Industrial Medicine 37, 109-113 (1980).

ANALYSIS OF REFRACTORY METALS AND WC-BASED HARD METALS BY ENERGY DISPERSIVE X-RAY FLUORESCENCE

Wolfhard Wegscheider and Kurt Müller

Institute for Analytical Chemistry, Micro- and
Radiochemistry, Technical University Graz,
A-8010 Graz, Austria

Hugo M. Ortner

Metallwerk Plansee AG, A-6600 Reutte, Austria

ABSTRACT

The potential of energy-dispersive X-ray fluorescence spectrometry for analysis of refractory metals and WC-based hard metals is investigated. Both, photon excitation by filtered tube radiation and by the characteristic lines of a secondary target are employed. Both excitation systems give good results. If the counting times are adjusted to account for the lower sensitivity of energy dispersive as opposed to wavelength dispersive X-ray spectrometry the detection limit and precision data are comparable. The multielement analyses of interest in these applications that comprise an energy range of 5 keV or more are better handled by direct excitation with filtered tube radiations than either by secondary target excitation or by wavelength dispersive X-ray spectrometry.

INTRODUCTION

The production of high quality refractory metals and tungsten-carbide based hard metals can only be achieved by maintaining a rigorous analytical monitoring at all stages of the production process. Many qualities of sinter-metallurgically manufactured high melting alloys critically depend upon the concentration of trace impurities. These impurities stem partly from the raw materials, but they can also be introduced during the production process.

While topochemical analyses of trace constituents in refractory material are extremely useful for the understanding of their influence, one has to resort to bulk trace analysis for routine work, as topochemical methods, like secondary ion mass spectroscopy, Auger electron spectroscopy and scanning electron microscopy, are not suited to handle large numbers of samples and these methods themselves frequently rely upon bulk trace analyses for quantitation.[1] The two analytical methods employed at present are atomic absorption and X-ray fluorescence spectrometry with crystal dispersion.

This paper reports the results obtained by energy dispersive X-ray fluorescence spectrometry (EDXRF) using the same sample preparation procedures as are in use for wavelength dispersive X-ray fluorescence spectrometry (WDXRF).

EXPERIMENTAL

The EDXRF-analyses were performed on a Philips/Edax EXAM SIX system with a Si(Li) detector having a resolution of 165 eV at 5.9 keV. This system is equipped with a 32 kByte NOVA III minicomputer with dual floppy disk. A Rh-target tube (Watkins-Johnson) was used in pulsed mode in connection with one of two Rh primary source filters: these had a thickness of 25 and 125 µm. When the system was operated in secondary target mode a 3 kW Au-tube was powered by a PW 1140 X-ray generator: The targets were Zr, Mo and Ag foils with matched filters.

The WDXRF-analyses were made on a Philips PW 1220/C or on a PW 1400. The types of matrices studied varied: metal oxides, metals and thin samples of relatively low mean atomic number were analysed. The sample preparation procedure was adapted to ensure adequate sensitivity in the technologically relevant concentration range.

For the determinations of W in MoO_3, of Nb in Ta_2O_5 and of Ta in Nb_2O_5 pressing the oxides to pellets was an adequate way to achieve high precision data. All these analytes were present as oxides and all calculations were done for the oxides.

The analysis of the rhenium/tungsten alloy in the concentration range of 1-30 % Re was accomplished by dissolving 1 g alloy in a mixture of 5 ml HNO_3 (65 %) and 5 ml HF (40 %). This solution is made up to 100 ml with distilled water.

Traces of first row transition metals in molybdenum and tungsten raw materials, intermediate and end products down to the low ppm-range are determined after preconcentration. The preconcentration is a co-crystallization of Mn, Fe, Co, Ni, Cu and Zn as PAN-complexes (1-(2-pyridylazo)-2-naphthol) together with an excess of PAN. The original method[2] simplified by Ortner, Lassner and Hertroys[3] consists of a dissolution step of the oxides, adjustment to pH 10 with 5 ml of NH_4Cl/NH_3-buffer. The PAN is added as 0.1 % solution in ethanol. The solution is boiled for 3 min, the precipitate collected on a cellulose filter, dried and mounted for analysis. While in WDXRF a Au-lined dye is used to hold the filtered precipitates flatly, sandwiching the filter between two sheets of Mylar on a Somar cup is preferred in EDXRF. In this paper, the traces are preconcentrated from a WO_3-matrix as well as from distilled water to get a valid assessment of the deconvolution scheme required.

The sample preparation procedure for the determination of Ta and Nb in Mo and MoO_3 was adapted from Wurzinger and Müller[4] : 1 g MoO_3 is suspended in 50 ml distilled water, 1 mg Co is added in solution, as well as 2.5 ml of a cellulose fiber slurry. The dissolution of the molybdenum oxide and the coprecipitation of the traces together with the cobalt-hydroxide is accomplished by adding 0.7 g solid sodium hydroxide. After heating the solution and keeping it at boiling temperature for 5-10 min the filter paper fibers with the adsorbed hydroxides are filtered off. The filter paper is dried and then ready for mounting on a Somar cup as described for the PAN-precipitate. A summary of the analytes, spectral lines, concentration ranges and matrices is found in Table I.

Table I

ANALYTICAL PROBLEMS STUDIED IN THIS PAPER

Analyte	Spectral line	Range	Matrix
W	$L_{\alpha 1}$, L_β	80-4000 ppm	MoO_3
Nb	K_α	14-350 ppm	Ta_2O_5
Ta	L_β	410-4100 ppm	Nb_2O_5
Re	$L_{\alpha 1}$	1-30 %	W/Re
W	$L_{\alpha 1}$	70-100 %	W/Re
Mn, Fe, Co, Ni, Cu, Zn	K_α	2-100 ppm	W, WO_3
Ta	$L_{\alpha 1}$, L_β	5-150 ppm	Mo, MoO_3
Nb	K_α	5-150 ppm	Mo, MoO_3

RESULTS AND DISCUSSION

The determination of W as WO_3 in MoO_3 pellets is currently performed by WDXRF using a LiF-crystal (200) and a Au-tube (50 kV, 40 mA). The EDXRF-analysis was attempted by direct tube excitation (Rh-tube, 21 kV, 990 µA, 25 µm Rh-filter) and by secondary target excitation with a Zr target (Au-tube, 60 kV, 30 mA). The latter mode gave significantly better results because the Rh-tube voltage had to be kept below the absorption edge of Mo in order to avoid the over-loading of the detector with K-radiation of the matrix (Table II). The secondary target mode employing the Zr-target gave by far the better sensitivity; the $L_{\beta 1}$, $L_{\beta 2}$ multiplet at 9.7 keV was the opti-mal analytical line because of the closeness of the exciting radi-ation to the relevant absorption edge as well as because the $L_{\alpha 1}$-line was overlapped by a Cu-blank from the instrument. A compari-son with the data obtained on the same set of samples using WDXRF shows that the better sensitivity is not sufficient to reach the same low detection limits and good precision. Tantalum oxide (Ta_2O_5) in niobium oxide (Nb_2O_5) and Nb_2O_5 in Ta_2O_5 can readily be deter-mined by EDXRF (Table III). Again, the determination of the lower energetic spectral lines ($L_{\beta 1}$, $L_{\beta 2}$ of Ta in Nb_2O_5) is favorably accomplished by using Zr as secondary target as almost the same detection limit can be achieved as by using the WDXRF. The detection limits and precision data are given for all analyses for the data acquisition times indicated in Table III. The low power excitation conditions for the analysis of Ta in Nb_2O_5 were 20 kV, 990 µA, 25µ m Rh-filter ($L_{\alpha 1}$ line), while 49 kV, 990 µA and a 125 µm Rh-filter were employed for the analysis of Nb in Ta_2O_5. In the secon-dary target mode an Ag target was taken for the Nb determinations.

Table II

PERFORMANCE CHARACTERISTICS OF THE DETERMINATION OF WO_3 IN MoO_3

| | EDXRF | | WDXRF |
	direct tube	sec.target	
sensitivity cps/%	24	117^a/158^b	5820
analysis time s	100	100	3 x 10
instrumental detection limit ppm	60	45^a/40^b	63
standard deviation ppm	128	20^a/13^b	56

[a] L α1   [b] L β1 ; Lβ2

Table III

PERFORMANCE CHARACTERISTICS OF THE DETERMINATION OF Ta_2O_5 IN Nb_2O_5
AND Nb_2O_5 IN Ta_2O_5

Ta_2O_5 in Nb_2O_5	EDXRF		WDXRF
	direct tube	secondary tube	
sensitivity cps/%	12.6	107	1044
analysis time s	400	200	2 x 20
RSD at 0.1 % %	2	0.9	4.4
instrumental detection limit ppm	40	55	103

Nb_2O_5 in Ta_2O_5			
sensitivity cps/%	185	85	6130
analysis time s	100	400	2 x 40
RSD at 300 ppm %	1.0	1.2	6.2
instrumental detection limit ppm	40	13	24

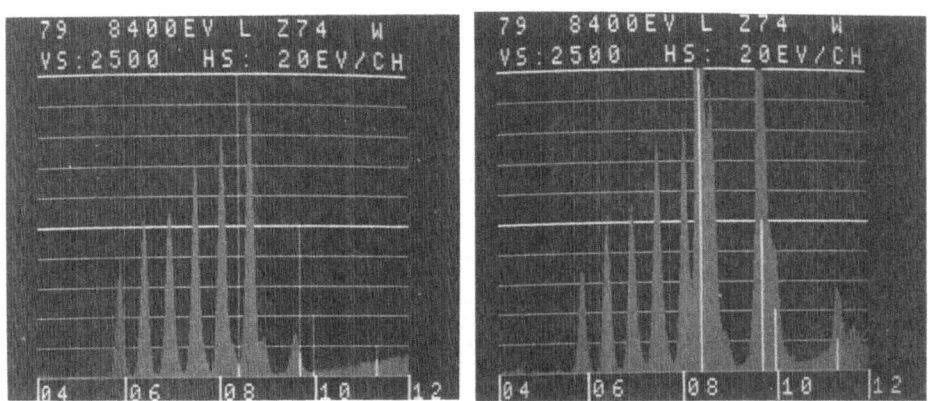

Figure 1:   Spectra of PAN-precipitates containing 50 μg of Mn, Fe,
Co, Ni, Cu, Zn

A aqueous matrix
B aqueous matrix containing W

The life times of analysis with EDXRF were chosen for the testing situation that existed in the course of this work. As precision and detection limits appear to be superior to those from WDXRF the acquisition times can be reduced accordingly. To permit a better assessment of these figures it should be pointed out that more up-to-date WDXRF-equipment may give a better performance in terms of stability and sensitivity. Very recent data run on a PW 1400 support this statement.

Similar to the WDXRF-analyses of the W-Re-alloy the dissolved samples were measured and calibrated against standards made up by dissolving appropriate amounts of pure W-and Re-powder. The direct tube excitation system was employed with the primary radiation from the Rh-tube (25 kV, 990 μA) filtered with a 25 μm Rh-filter. For an analysis time of 100 s the detection limit was 40 ppm Re in the solution. Given a 20 s counting time the detection limits for wavelength dispersive X-ray fluorescence are 23 ppm (PW 1220/C) or 14 ppm (PW 1400). The standard deviation was 0.2 % Re; this corresponds to a relative standard deviation of 20 % for the lowest Re content to be measured (1 %). The relative standard deviation for the determination of W was 0.3 % at 70 % W content. A careful inspection of the residuals of regression analysis suggests an instrument drift contributing to the random error in this particular run.

From the point of view of energy dispersive detection multi-element samples are even more rewarding applications. Chemically these are of interest because the different chemical behaviour of matrix and trace constituents can effectively be utilized to separate the traces from the matrix and - in the same operation - to get close to the optimum of an infinitely thin layer of the analyte. It is also worth noting that two disadvantages of EDXRF as compared to WDXRF are much less severe for such samples:
a) count rate limitations hardly gain any significance, and
b) the better instrumental detection limits of WDXRF are overridden by a detection limit that is largly governed by the precision of sample preparation rather than Poisson variability of the background[5].

The determination of traces of first row transition elements in W and WO_3, for instance, requires a preconcentration procedure. As shown in Figure 1, however, not all the tungsten can quantitatively t removed. The remaining amount is not more than one order of magnitude higher than the trace elements but it is highly variable. Particularly the elements Ni (W L_1) and Zn (W L_α) need to be corrected for the overlapping W-signal. This is done according to a model proposed by Shen, Russ and Stroeve[6]

$$C_i = K_i \left(I_i + \Sigma_j \alpha_{ij} C_j \right) + \Sigma_j B_{ij} I_j + B_i$$

that permits the simultaneous correction for background (B_i), overlapping peaks (B_{ij}) and absorption and enhancement effects (α_{ij})
from measuring the corresponding gross intensities over 1.2 FWHM (I_i).
Although the model appears to be a rough one, it works satisfactori-
ly in the present situation. <u>Figures 2 and 3</u> give the effect of
overlap correction for Ni and Zn for W respectively. The direct tube system was used for exciting the characteristic radiation: the
Rh-tube was operated at 40 kV, 990 μA with the 25 μm Rh filter.

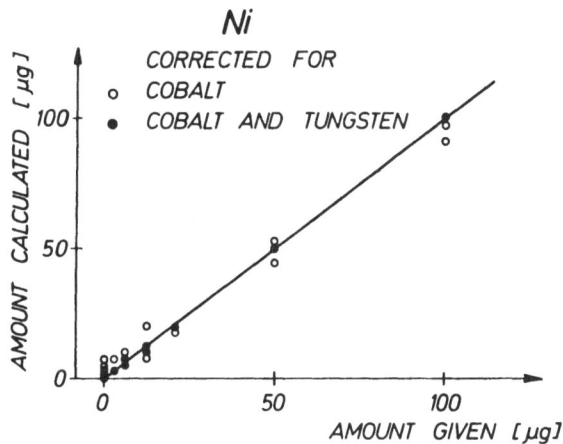

Figure 2:   Effect of overlap correction for the determination of Ni

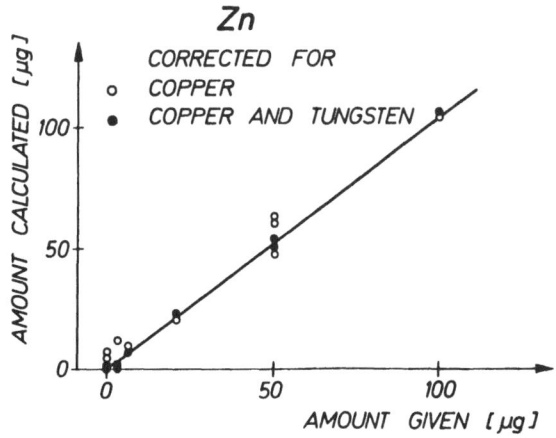

Figure 3:   Effect of overlap correction for the determination of Zn

Table IV

SENSITIVITIES AND DETECTION LIMITS FOR ELEMENTS PRECONCENTRATED BY P,

ELEMENT	EDXRF		WDXRF	
	sensitivity	det. limit	sensitivity	det. limi
	cps/μg	μg	cps/μg	μg
Mn	0.70	0.30	97	0.45
Fe	0.93	0.40	88	0.79
Co	1.1	0.25	110	0.55
Ni	1.5	0.17(0.40)[a]	126	0.53
Cu	1.7	0.29(0.35)[a]	112	0.67
Zn	2.0	0.21(0.38)[a]	124	0.65

[a]
  detection limit in presence of
  residual traces of W

Table V

PERFORMANCE CHARACTERISTICS OF THE DETERMINATION OF Ta AND Nb IN Mo(

TANTALUM	EDXRF		WDXRF
	direct tube	second.target	
sensitivity cps/μg	0.493 $(L_{\alpha 1})$	0.453 $(L_{\beta 1}, L_{\beta 2})$	6.7 $(L_{\alpha 1})$
instrumental detection limit μg	2.4	1.3	1.2
standard deviation μg	7.8	5.5	3.8
analysis time s	100	100	20
NIOBIUM			
sensitivity cps/μg	1.67 (Kα)	0.588 (Kα)	26.7 (Kα)
instrumental detection limit μg	1.1	0.65	0.76
standard deviation μg	4.6	4.2	4.3
analysis time s	100	100	20

No absorption or enhancement effects were encountered for the PAN-precipitates. The sensitivities and detection limits for EDXRF as compared to WDXRF are given in Table IV. The much greater sensitivity of WDXRF does not reflect on the detection limits because of the reason stated above. In fact, even elements overlapping with W have slightly lower detection limits in EDXRF. A comparison of the counting time shows that 200 s is sufficient for EDXRF, and 20 s have to be spent for each emission line and possibly also for background measurements on each line, so that EDXRF has a true advantage in this respect, as well. The instrumental detection limits are corroborated by sample preparation error, but no more than a factor of 1.5 (for Fe a factor of 3, i.e. a detection limit of 1.2 µg can be expected). The relative standard deviation is 5-7 % for all elements except Fe, for which 10-15 % were measured.

Another problem requiring preconcentration is the determination of Ta and Nb in MoO_3. With the direct tube excitation the simultaneous determination of Ta and Nb is possible using a tube voltage of 35 kV, a current of 990 µA and the 25 µm Rh filter. The performance characteristics are given together with those from secondary target excitation and from WDXRF in Table V. The secondary targets are Zr for the determination of Ta and Ag for the determination of Nb. The standard deviations as given are assumed to be constant in the range of 10-150 µg. The instrumental detection limits are given for the analysis times indicated and can be lowered to about 300 ng by increasing the counting time; this value is estimated to be the limit obtainable on grounds of variations in sample preparation. The same low detection limits in EDXRF can only be measured if the PW 1400 with its better sensitivity as compared to the PW 1220/C is employed.

In conclusion, it can be stated that in a routine laboratory of the refractory metals industry EDXRF has the capability of adequately measuring all analytical problems so far run by WDXRF. Generally, with secondary target excitation the detection limits are even somewhat lower if the counting times are adjusted to account for the lower sensitivities. This, however, in multielement determinations will not lead to an overall decrease in sample throughput. Precision and accuracy data give an indication that the data reduction scheme used is also adequate for these problems. As it lends itself very simply to routine analysis situations where the problems are changing several times a day an user - friendly set-up is ensured, as well.

ACKNOWLEDGEMENT

This research was supported in part by instrument grant No. 3543 from the Fonds zur Förderung der wissenschaftlichen Forschung, Vienna.

REFERENCES

1. H.M. Ortner and E. Lassner, Einsatz moderner instrumenteller
   Methoden zur Spurenanalyse in hochschmelzenden Metallen,
   Mikrochim. Acta Suppl. 7:41 (1977).

2. R. Püschel und E. Lassner, Anwendung der RFA in der Spurenanalyse, III. Bestimmung von Fremdmetallspuren in hochrei-
   nem Molybdän und Wolfram mit Hilfe der RFA nach Anreicherung
   durch Fällung mit PAN, J. Less-Common Metals 17:313 (1969).

3. H.M. Ortner, E. Lassner and P. Hertroys, Experiences with
   Automated X-Ray Fluorescence Spectrometry in the Analysis of
   Refractory Metals, X-Ray Spectrom. 4:2 (1975).

4. H. Wurzinger and K. Müller, Bestimmung von Spuren Titan und
   Zirkonium in Molybdän und Wolfram bzw. deren Sauerstoffver-
   bindungen durch Röntgenfluoreszenz-Spektralanalyse (RFA),
   Fresenius Z. Anal. Chem., 284:101 (1977).

5. L.A. Currie, Detection and Quantitation in X-Ray Fluorescence
   Spectrometry, in "X-Ray Fluorescence Analysis of Environmental Samples", T.G. Dzubay, ed., Ann Arbor Publ., Ann Arbor (1977).

6. R.B. Shen, J.C. Russ and W. Stroeve, Modelling Intensity
   and Concentration in Energy Dispersive X-Ray Fluorescence,
   in "Adv. in X-ray Anal.", Vol. 22, ed., Plenum, New York
   (1979).

AN IMPROVED SAMPLE PREPARATION AND ANALYSIS TECHNIQUE FOR THE DETERMINATION OF MINOR ELEMENTS IN CATALYTIC MATERIALS BY RADIO-ISOTOPE INDUCED X-RAY FLUORESCENCE

John J. LaBrecque

Instituto Venezolano de Investigaciones
Científicas - IVIC, Apartado 1827
Caracas 1010-A, Venezuela

SUMMARY

An improved method for sample preparation and determination of minor elements in catalytic materials by
radioisotope induced X-ray fluorescence is presented.
The sample preparation is simple and rapid, it utilizes
both and internal standard procedure as well as a thin
film technique. Synthetic standards are prepared simply
by mixing thoroughly the appropriate amounts of the elements of interest as their respective oxides with the
matrix and the internal standard. The above standards
are used to calculate the relative-fluorescence ratio
factors (F_{JL}) for each element of interest relative to
the internal standard.

Some of the different types of deviations (sample
preparation, standard counting, instrumental, etc) will
be shown for typical determinations applying experimental
methods. Finally the application of this technique for
the determination of Cobalt, Nickel, Molybdenum, Platinum
and Palladium in catalytic materials is presented.

INTRODUCTION

The present X-ray Spectrometers, both wavelength
and energy dispersive, have a very high inherent precision (reproducibility). But it is not always possible
to convert this reproducibility to analytical accuracy
of the same order. To produce results with good analytical accuracy from the direct intensity measurements from

an X-ray spectrometer, it is necessary to: 1) use a set
of calibration standards with a chemical composition similar to your sample of which the elemental composition
is known to a high degree of accuracy and 2) to be able
to eliminate or compensate for the interelement effects.
It has been noted by Bertin (1) that the principal objective of an analytical method is to eliminate, minimize
circumvent or correct for the effects of the matrix (the
matrix absorption-enhancements effects).

Most X-ray fluorescence methods employ calibration
standards, and the intensity data are transferred to analytical concentrations by either a calibration curve or
mathematical realtionship derived from measurements of
standards. Some attempts have been made of an absolute
X-ray spectrometric analysis employing only a mathematical correction but because the mathematical corrections
are very complex only limited success had been achieved.

The technique described herein employs two known
procedures to control the absorption-enhancements
effects: 1) The thin film method: The samples are prepared so thin that absorption-enhancements effects substantially disappear. For high Z-elements when using
the K-lines for analysis in energy-dispersive X-ray fluorescence the sample thickness can approach 1mm and still
have almost 100% transmission. 2) An internal standard
method: this procedure improves greatly on the more
conventional comparison-standard method by quantitative
addition to the samples and internal standard an element
having excitation, absorption and enhancement characteristics similar to those of the element of interest in
the matrix. The calibration factors involve the intensity ratio of the element of interest and the internal
standard lines. And because this is a ratio only the
original amount of sample and internal standard are neede
for the calculation of the analytical concentration.
This implies that the sample doesn't need to be completel
or quantitatively transferred to the sample holder. Thus
error in weighing small amounts of samples and internal
standards and homogenization are eliminated as well as
the error in transferring the sample completely or quantitatively to the sample holder. Also a slight error in
reproducing the sample-source detector geometry is eliminated too.

EXPERIMENTAL

Preparation of standards and calculations of the relative fluorescence ratio-factor (F_{JL}).

The appropriate amounts of the elements of interest as their oxides are added to the matrix and internal standard to cover the different ranges of concentrations in the sample. All the reagents are first ground to less than 106 µm size (150 mesh) and dried at 102°C before weighing in a minivial (≈7 ml) with 4-5 steel balls for mixing in a SPEX MILL for 30 minutes. Because of their size, 6-10 minivials can be placed in the mill at the same time. The homogeneity is checked by taking three portions of the mixture and placing them in independent sample holders and each portion analyzed three times for the calculation of the relative fluorescence ratio-factor. If the deviation is minimal then they are assumed homogeneous. The sample holder is a 2 x 6 cm piece of IBM computer card with a 1 cm aperture for inserting the sample between two pieces of Scotch Magic Tape.

The relative fluorescence-ratio factors (F_{JL}) are calculated using the following equation:

$$F_{JL} = \frac{\text{intensity of internal standard}}{\text{intensity of peak of interest}} \times$$

$$\frac{\text{wt. of element of interest}}{\text{wt. of internal standard}} \times A$$

that is, F_{JL} is equal to the integral of counts above background for the internal standard peak divided by the integral of counts above background of the peak of interest times the weight of the element of interest (mg) divided by the weight of the internal standard (mg) times A; where A is a conversion factor, e.q. to convert to percentage, the pure element to oxide, etc.

Sample preparation and calculation of concentrations of elements of interest.

Three to five grams of representative sample are ground to completely pass a 150 mesh sieve (106 µm) before a 800 mg portion is mixed with 200 mg of internal standard with steel balls in a minivial using a SPEX MILL. The internal standard is previously ground and dried at 102°C before weighing. Another portion of

sample of about 2-3 grams is used for the moisture determination. Usually three separate portions of the
sample mixture are placed in separate sample holder for
analysis. The concentration of the element of interest
is calculated employing the following equation:

$$\%X = \frac{\dfrac{\text{wt. of internal standard}}{\text{sample weight}} \times F_{JL}}{\dfrac{\text{intensity of element of interest}}{\text{intensity of internal standard}} \times}$$

where F_{JL} is the relative fluorescence-ratio factor defined before.

Photon induced X-ray fluorescence system

The X-ray fluorescence system used in these studies
is based on a PDP-11/05 (16K) processor which is employed
as a multichannel analyzer as well as to control the
complete system and perform the calculations. This
system is described in full elsewhere (2). The detector
is a high resolution Si(Li) semiconductor with a measured
resolution about 150 eV (FWHM) at 5.9 keV. The excitation system consists of the radioisotope: 1) ^{109}Cd (7
mCi) or 2) ^{55}Fe (50 mCi) or 3) ^{241}Am (100 mCi) in the
annular geometry. Finally the whole system has been
automated for analysis by means of a pre-written computer
program in Flextran, a language developed by Tracor
Northern. The commands in these programs have been described before (3).

APPLICATIONS

The simultaneous determination of Co-Ni-Mo in hydrodes-
ulfurization catalysts: A statistical evaluation of the
errors involved in this technique have been partially
separated experimentally and the results are shown in
Table I.

TABLE I.   Different Types of Deviations for a Typical Co-Ni-Mo catalyst.

Type of Deviations	Method of Calculations	$\bar{X} \pm SD$ CoO%	NiO%	MoO$_3$%
Total	Five inde-pendent prepared samples	1.40±0.07	2.70±0.1	11.3±0.5
Sample preparation standard, counting instrumental and operational	Five inde-pendent portions of the same sam-ple prep-aration	1.37±0.05	2.70±0.1	11.1±0.3
Standard counting, instrumental and operational	the same portion of the sample preparation measured 5 times	1.40±0.07	2.70±0.1	11.5±0.1

Finally, the results of this technique are compared with conventional flame atomic absorption, conventional neutron activation and prompt-gamma analysis. The data values are in excellent agreement with the technique described herein and are shown in Table II.

REFERENCES

1. E. Bertin, "Principles and Practices of X-Ray Spec-trometric Analysis, Plenum Press, New York. (1975).

2. J.J. LaBrecque, "The Use of a PDP-11/05 Based Data Acquisition and Analysis System (TN-11) for Automated Isotope Analysis and Non-dispersive X-ray Fluorescence Analysis in the Petroleum Industry", Proceeding of the III International Conference on Computers in Chemical Research, Education and Technology, P. 66-81.

3. J.J. LaBrecque, Journal of Radioanalytical Chemistry 53 (1979) 221.

TABLE II.   The comparison of results for Co-Ni-Mo by different methods.

Sample Code	CoO%*				NiO%*				MoO₃%*			
	This Work	AAS^a	NAA^b	p-γ^b	This Work	AAS^a	NAA^b	p-γ^b	This Work	AAS^a	NAA^b	p-γ^b
A	1.43	1.25	1.54	1.23	2.75	3.09	3.03	2.68	15.17	14.76	15.70	13.92
B	2.01	2.34	1.87	1.74	4.04	4.14	3.82	4.21	9.32	10.80	8.87	9.04
C	3.67	3.97	2.59	2.89					16.08	15.92	15.29	16.14
D	3.71	3.57		3.19					13.07	13.01	14.61	14.49

*The Values are the mean of three or more determinations

a These values are from Applied Spectroscopy 30 (1976) 625

b These values are from Analytical Chemistry 48 (1976) 1969

GLASS AND GLASS RAW MATERIALS ANALYSIS USING A PHILIPS PW1600

WAVELENGTH DISPERSIVE X-RAY SPECTROMETER

D. R. Jones IV and G. D. Bowling

Owens-Corning Fiberglas Corp.

Granville, Ohio   43023

The Inorganic Analytical Laboratory of Owens-Corning Fiberglas is routinely using a Philips PW1600 x-ray spectrometer for the analysis of glass and glass raw materials. The spectrometer is equipped with 11 fixed channels, two scanning channels, and an energy dispersive detector. Operation of the spectrometer and data compilation is accomplished using a Digital Equipment Corp. PDP11/34 computer. The x-ray is controlled by an RSX-11M operating system, and the Philips "Alphas" software package is used to process data using the Lucas-Tooth Pyne (LP), Lachance-Traill (LT), or Rasberry-Heinrich (RH) correction models.

A statistical examination of inter- and intra-pellet variation of various preparation methods for glasses and raw materials has led to the conclusion that acceptably precise results for raw materials can be achieved by grinding the raw material to pass a 200 mesh sieve, glasses to pass a 400 mesh sieve, and pressing a pellet of the material using collodion as a binder. The manufacturing processes of Owens-Corning preclude the use of solid glass disks, even though such disks gave the best precision in this study. For glass, the additional sample preparation time needed to achieve this small particle size was worth the improvement in precision.

For these glasses the variation in replicate analyses show a relative standard deviation for silica as SiO_2 of 0.04, 0.54, and 0.18 for a solid glass disk, and pellets made of a <200 mesh and <400 mesh powdered glass, respectively. For these same types of preparation, the results for calcium as CaO are 0.03, 0.49, and 0.08 respectively. For magnesium as MgO they are 0.67, 2.1, and 0.40 respectively.

It has been found that the LT or RH models give the best result:
in correcting for interelement effects present in glasses and raw
materials. As with all empirical x-ray correction programs, the accuracy of the standards one uses to define the curve are of utmost
importance in the accuracy of the data generated for unknowns. The
number and types of Standard Reference Materials available national1:
or internationally for the generation of standard curves is not sufficient to meet our needs. Twenty to thirty internally generated
standards were used for the development of each correction program
for glasses and raw materials.

Using well characterized wet analyzed standards, correction pro
grams for sand, silica flour, clay, zircon, rutile, limestone, rhyo-
lite, nepheline syenite, ulexite, colemanite, and glass have been
developed and are in routine use. The accuracy of these programs,
defined as how close the XRF results come to the wet chemical result
for an "unknown" standard, meets our needs for rapid chemical analys
The pooled deviation of several analyses are presented in Table I.

Table I

Pooled Deviations, XRF Analysis of
Glasses and Raw Materials

Material	σP	
Ulexite	.15	
Sand	.08	Equations used:
Neph Syenite	.23	deviation: $\sigma = |X_{XRF} - X_{WET}|$
Zircon	.19	
Limestone	.03	pooled deviation $\sigma P = \sqrt{\dfrac{\Sigma \sigma^2}{n}}$
Glass 1	.30	
Glass 2	.28	
Glass 3	.40	

In conclusion, the Philips PW1600 XRF instrument has been shown
to be capable of acceptable accuracy in the analysis of glasses and
glass raw materials. Sample preparation time is the limiting factor
for analysis of a large number of samples. The use of the Energy
Dispersive channel allows qualitative/semi-quantitative data to be
gathered on the same samples prepared for wavelength dispersive anal.
sis.

Acknowledgment

The authors thank R. R. Burrell for his help in this work.

A GENERAL METHOD OF BLANK SUBTRACTION FOR

QUANTITATIVE X-RAY FLUORESCENCE INTENSITY MEASUREMENTS

Alan P. Quinn

Corning Glass Works
Research and Development Laboratories
Corning, New York 14830

ABSTRACT

Residual analyte intensity exists in a spectrometer when, for example, the x-ray source is contaminated with the analyte. A blank specimen, identical to the sample but devoid of the analyte, may be used to determine this residual analyte intensity, and, thereby, allow quantification of that intensity due to analyte in the sample. It is generally considered that a single pure material could serve as a blank for many analytes in many matrices. In practice, at least for low average atomic number matrices such as glasses, the reflectivity of similar samples may differ substantially, rendering inaccurate this simple blank subtraction. In such cases a blank is needed for each sample, where the reflectivity of the blank is matched to that of the sample. In order to eliminate the need for generation of blanks to match each of the samples, a general method to quantify the net residual analyte intensity as a function of specimen reflectivity by measurement of two pure material blanks is proposed herein.

INTRODUCTION

Precise determination of net analyte intensity is important in x-ray spectrometry when employing empirical correction methods,[1] and crucial when fundamental parameters methods[2] are to be invoked. In measuring net analyte line intensities it is often necessary to quantify the residual intensity (or background) at an analyte line wavelength, especially when analyte concentrations are low and analyte intensities approach the lower limit of detection. When net residual analyte line intensity is present, perhaps due to analyte radiations from a contaminated x-ray source being scattered

by the specimen, or due to analyte emission of spectrometer components, estimation of the background by measurements near the
analyte line wavelength is insufficient. In such cases, the analyst
may quantify residual analyte intensity by measurement of a specimen devoid of the analyte (i.e., a blank) at the analyte line wavelength. This approach assumes, however, that the blank and the
sample are equally reflective at the analyte line wavelength. That
is, that the combination of 1) attenuation of the incident primary
x-rays, 2) the efficiency of scattering, and 3) attenuation of the
scattered x-ray photons, is equivalent for the blank and the sample.
For low average atomic number materials, such as glasses, substantial differences in reflectivity may result from apparently
minor compositional differences. In Figure 1, the reference specimen differs from the unknown specimen only by five weight percent
zirconia (ZrO_2), yet the reference is nearly twice as reflective
at wavelengths near the nickel K-alpha. Accurate determination of
net nickel K-alpha intensities by the above method would require
in-type blanks for each specimen, where the reflectivity of each
blank would be matched to that of a specimen. In fact, additional
blanks might be required for analysis of other analytes in other
wavelength regions.

In order to develop a more general method, let us consider the
mechanisms which yield net residual analyte intensity.

MODE S: Specimen Reflectivity Dependent

In certain cases, net residual analyte intensity (NETR) arises
from x-ray source contaminants, whose characteristic radiations are
reflected, i.e., scattered and attenuated, by a specimen. In
general, the reflectivity difference between specimens is equivalent both at, and near, the analyte line wavelength. Figure II
illustrates this specimen reflectivity dependent mode where neither
the graphite specimen nor the silica specimen contain calcium. The
net residual calcium intensity of the graphite specimen is approximately three times that of the silica specimen, and intensities
near the calcium peak for the graphite specimen were also three
times greater than for the silica specimen. One would expect, under
such circumstances, that the net residual analyte intensity (NETR)
of an unknown would be

$$NETR_{UNK} = NETS_{UNK} = NET_{BLANK} \cdot \frac{R_{B_{UNK}}}{R_{B_{BLANK}}} \qquad (1)$$

where NETR is the net residual analyte intensity
        NETS is the specimen reflectivity dependent NETR
    and R_B is the count rate near the analyte line

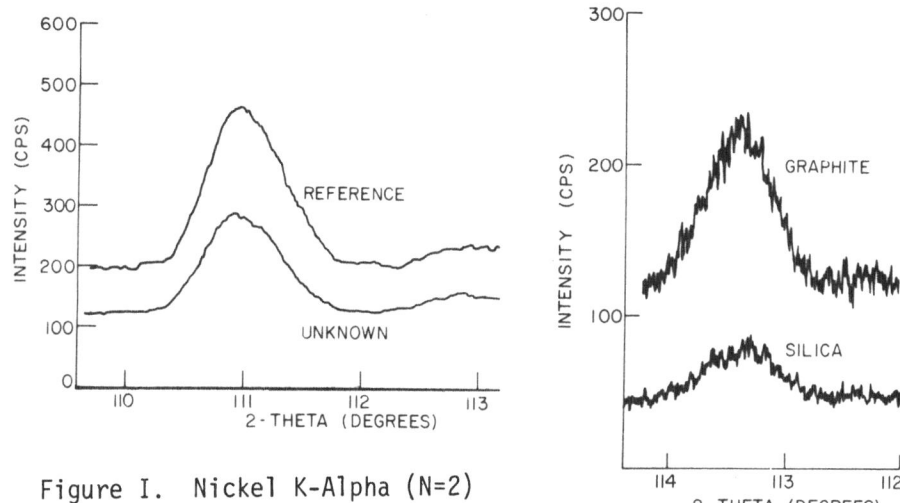

Figure I. Nickel K-Alpha (N=2)

Figure II. Calcium K-Alpha

MODE F: Specimen Independent

Net residual analyte intensity may also arise from spectrometer components, such as specimen masks. Here the analyte may be
present, be excited, and its x-ray fluorescence be detected. There
is virtually no specimen dependence with this mode. The wavelength
scans over iron K-alpha in Figure III show essentially identical
net residual analyte intensity for the two blanks, graphite and
silica, although specimen reflectivity differs by a factor of three.
With this spectrometer one would expect any specimen to have a net
residual analyte intensity equivalent to that of the blank. That
is,

$$NETR_{UNK} = NETF_{UNK} = NET_{BLANK} \qquad (2)$$

General Blank Subtraction Method: Concept

It is reasonable to presume that both modes may coexist. That
is, that

$$NETR_{UNK} = s(NETS_{UNK}) + f(NETF_{UNK}) \qquad (3)$$

where s and f are weighting coefficients representing the fraction
present of each mode. Equations (1) and (2) may be substituted
into equation (3) to yield

$$NETR_{UNK} = s(NET_{BLANK} \cdot \frac{R_{B_{UNK}}}{R_{B_{BLANK}}}) + f(NET_{BLANK}) \qquad (4)$$

The weighting coefficients may be determined for a spectrometer under given operating conditions by employing two general blanks of differing reflectivity. Equation (4) may be rewritten to give

$$NET_{BLANK2} = s(NET_{BLANK1} \cdot \frac{R_{B_{BLANK2}}}{R_{B_{BLANK1}}}) + f(NET_{BLANK1}) \qquad (5)$$

Assuming that net residual analyte intensity results primarily from the two aforementioned modes, i.e., that net residual analyte intensity may be independent of the specimen, specimen reflectivity dependent, or both, then s + f = 1. Substitution of this relation into equation (5) yields

$$NET_{BLANK2} = s(NET_{BLANK1} \cdot \frac{R_{B_{BLANK2}}}{R_{B_{BLANK1}}}) + (1-s)(NET_{BLANK1}) \qquad (6)$$

The two assumptions required by the proposed blank subtraction are, then, as follows. First, that specimen reflectivity near an analyte line wavelength accurately describes the reflectivity of that specimen at the analyte line wavelength. Second, that all mechanisms which produce net residual analyte intensity are either specimen independent (mode F) or dependent upon specimen reflectivity (mode S), therefore, s + f = 1.

General Blank Subtraction Method: Illustration

Given the task of analyzing nickel at the parts per million level in glasses which differ only in zirconia level with a range of zero to five weight percent zirconia (the case given in Figure 1), two general blanks (pure SiO_2 and pure $Ta_2O_5 \cdot SiO_2$) were measured (Figure IV). Four glasses of known nickel content were measured in addition to the two blanks, both at, and near, the nickel K-alpha wavelength (LiF (200), second order reflection). The results are given in Table I. Application of equation (6) gave s = 0.27 and f = 0.73, i.e., twenty-seven percent of nickel net residual intensity was specimen reflectivity dependent, and seventy-three percent was not.

Table II compares various methods of dealing with net residual analyte intensity.[3] Considering the resultant average deviations from the known nickel levels, several conclusions may be drawn. First, ignoring net residual analyte intensity ('no blank') yields poor accuracy. Second, use of a single general blank yields better agreement when that general blank has reflectivity similar to the

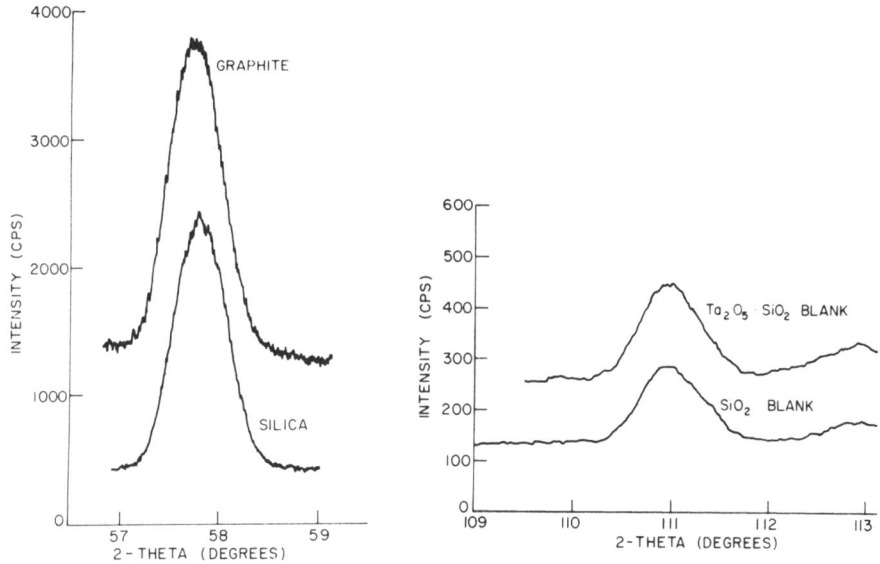

Figure III.   Iron K-Alpha   Figure IV.   Nickel K-Alpha (N=2)

Table I.   Ni K-alpha Intensities (counts per second)

Sample	Nominal ZrO_2Wt.%	Analyzed ppm NiO	Peak Wavelength	Near Peak* Wavelength	Net
'A'	5.	183.	362.9	184.5	178.4
'B'	3.	92.	305.6	164.2	141.4
'C'	0.	40.	222.2	110.3	111.9
'D'	0.	(0.)	206.9	113.8	93.1
SiO_2	0.	(0.)	228.7	126.8	101.9
$Ta_2O_5.SiO_2$	0.	(0.)	363.3	237.6	125.7

*Interpolated from high two-theta and low two-theta measurements
( ) denotes assumed value

analysis samples (as with SiO_2) than when a general blank of substantially different reflectivity is employed (as with $Ta_2O_5.SiO_2$).
Third, the proposed method utilizing two general blanks yields
more accurate values than other general methods.

Additionally, the accuracy advantage of the two general blank
method of about thirty percent over the single silica blank method
is consistent with the values of the s and f weighting coefficients.
The single silica blank method assumes only mode F exists (mode S =
0), whereas mode S was found to constitute about thirty percent.
Thus, the two approaches should differ only by about thirty percent.

Table II.   Comparison of Blank Subtraction Models.
NiO Concentration in Parts Per Million.*

Sample	Known**	No. Blank	SiO$_2$ Blank	Ta$_2$O$_5$·SiO$_2$ Blank	Two Blank Method
'B'	92.	131.	85.	49.	81.
'C'	40.	102.	21.	-43.	35.
'D'	(0.)	85.	-19.	-100.	-15.
Average Deviation		62.	-14.	-75.	-10.

*Sample 'A' used as reference
**Colorimetric, ( ) denotes assumed value

Presumably the accuracy advantage of the proposed two general blank
method over the single general blank method would be greatest for
analytes where mode S predominates.

It should be noted that success with the proposed method relies
upon precise measurements both at, and near, the analyte line wavelength which are free of spectral interferences. For example, the
count rate near an analyte line wavelength may be contaminated for
an unknown specimen, perhaps due to contribution of a higher order
reflection poorly discriminated against by inappropriate pulse
height selection. In such a circumstance, an erroneously high net
residual analyte intensity would be predicted under mode S for the
unknown specimen.

SUMMARY

A general method of blank subtraction is proposed which employs
two pure blanks of differing reflectivity and is applicable to trace
analysis of most analytes in low average atomic number materials.
Resultant analyses are more accurate than with other general blank
subtraction methods.

REFERENCES

1. G.R. Lachance, X-Ray Spectrometry 8, 190 (1979).

2. J.W. Criss, L.S. Birks, and J.V. Gilfrich, Analytical Chemistry
    50, 33 (1978).

3. Data Reduction by CORRAL Computer Program, A.P. Quinn, Advances
    in X-Ray Analysis, 22 293 (1979).

A RAPID AND PRECISE COMPUTER METHOD FOR

QUALITATIVE X-RAY FLUORESCENCE ANALYSIS

T. C. Huang, W. Parrish and G. L. Ayers

IBM Research Laboratory
5600 Cottle Road
San Jose, CA. 95193

ABSTRACT

This paper describes a computer method for wavelength dispersive (WD) qualitative X-ray fluorescence (XRF) analysis. It determines the elements, spectral lines, wavelengths, reflection angles
and peak intensities of the first and second order reflections in
less than half a minute of time using an IBM Series/1 minicomputer.
The resolution and precision are significantly better than the
energy dispersive (ED) method and when combined with high speed
computer recording the speed is comparable.

INTRODUCTION

The recent introduction of computer controlled high-speed data
collection techniques[1] and the development of a precision peak
search method of data reduction for X-ray powder difractometry[2] have
made it possible to develop a similar computer method for WD X-ray
qualitative fluorescence analysis. The use of manual methods may be
difficult and takes considerable time. The method described here
runs on an IBM Series/1 computer in less than half a minute of time
to obtain the desired spectral information. When combined with high
speed data collection, the speed is comparable to the ED method and
has much greater precision because of the higher peak to background
ratios and resolving power. It also has the practical advantages of
much less overlapping reflections and ability to measure weak peaks.

METHOD

The 2θ-reflection angles and intensities of peaks recorded above
a selected background threshold are first determined. The peak posi-

tion is calculated from the first derivative of a five-point cubic convoluting function[3]. A peak is identified in the region where the slope changes from positive to negative as 2θ decreases, and is located at the 2θ angles where the slope is zero.

The second derivative had also been tried, but it was found more difficult to interpret than the first derivative. A five-point convolution was found to be superior to seven- or nine-point convolution because the latter tend to flatten peaks. A cubic polynomial was used to account for possible asymmetry in the observed profile. The peak intensity is obtained by fitting an asymmetric Lorentzian curve to the experimental data near the top of the peak.

The corresponding elements, spectral lines and calculated wavelengths are determined from Bragg's law and the lattice spacing of the analyzing crystal. The wavelengths of the characteristic X-ray lines are stored in the computer. Only the Kα, Kβ, Lα1, Lβ1 and Lγ1 are used because these are the major radiations used in XRF.

EXPERIMENTS AND RESULTS

A "super alloy" consisting of nine elements: Cr, Mn, Fe, Co, Ni, Cu, Zn, Zr and Gd was chosen to illustrate the method. An IBM Series/1 computer controlled spectrometer[4] and a W-target X-ray tube operated at 45 kV were used. The spectrum collected with steps of Δ2θ=0.05° and counting time of 1 second is showed in Figure 1. Either a linear or semi-log plot of the observed intensities can be selected by the user. The semi-log plot has the advantages of enhancing weaker peaks so that peaks with intensities differing by several orders of magnitude can be visible simultaneously in the same plot.

The results of peak search analysis of this alloy are given in Table 1. Peaks are listed in order of decreasing 2θ. To account for random and possible systematic errors, the X-ray lines occuring within ±0.075° of an identified peak are included in the "possible element-radiation" column. Comparing the identifications of the peaks in this column, the "true" elements present can easily be determined. For example, peak #4 can be identified as the second order peak of Zr-Kα1 by noticing the presence of the second order Zr-Kα2 at peak #5. All nine elements present in the specimen were identified in less than a minute of time.

There is a possibility that some peaks will not be listed because they are (1) a different spectral line from the five used, (2) rejected by the threshold limits, (3) higher order of reflection not used in this method. For example, two peaks, one on each side of peak #8 (see Figure 1a), are unlisted because they are the Gd-Lβ2 and Gd-Lβ3, respectively, which are not included in the five spectral lines used in this program.

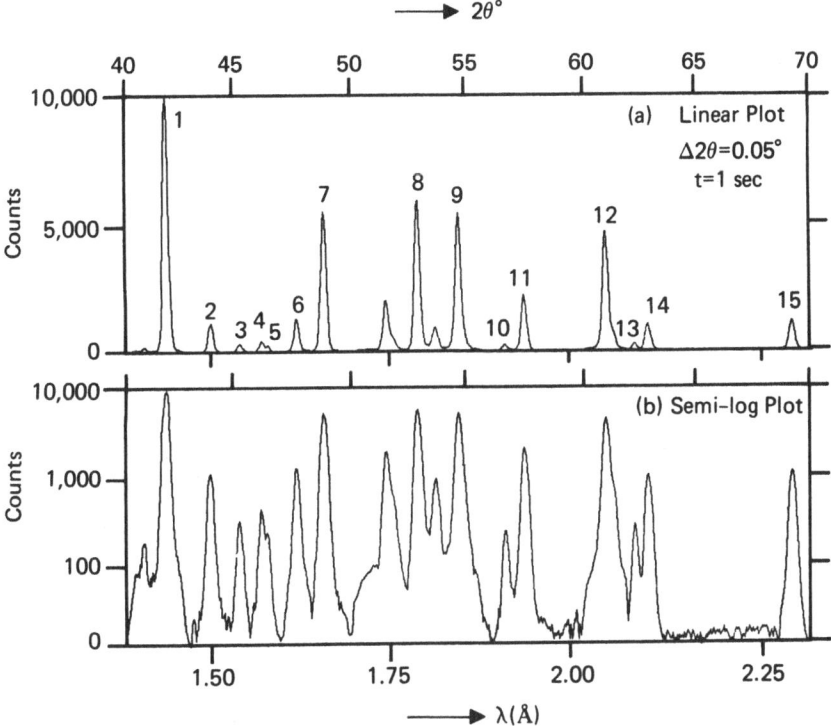

Figure 1. Spectrum of the Cr-Mn-Fe-Co-Ni-Cu-Zn-Zr-Gd "Super Alloy".

Table 1.   Printer List of Peak Search Results

PEAK	ANGLE	I(COUNTS)	LAMDA(A)	POSSIBLE ELEMENT-RADIATION		
1	41.794	10270.	1.4367	ZN-KA		
2	43.743	1162.	1.5005	NI-KB		
3	45.018	347.	1.5420	CU-KA		
4	45.947	452.	1.5721	HF-LA	ZR-KA1(2)	
5	46.192	239.	1.5801	ZR-KA2(2)		
6	47.471	1378.	1.6213	CO-KB	LU-LA	
7	48.653	5605.	1.6593	NI-KA	EU-LG	Y -KA1(2)
8	52.785	6249.	1.7905	CO-KA		
9	54.610	5685.	1.8478	GD-LB		
10	56.674	265.	1.9119	MN-KB		
11	57.518	2322.	1.9380	FE-KA		
12	61.126	5148.	2.0482	GD-LA	CE-LG	
13	62.419	327.	2.0872	CR-KB		
14	62.989	1112.	2.1043	MN-KA		
15	69.361	1210.	2.2919	CR-KA		

2D(A) OF THE ANALYZING CRYSTAL IS 4.0280

A histogram plot of the observed spectrum is also available at
the printer. Figure 2 is a small portion of this type of plot. The
intensities are printed on a logarithmic scale as pairs of dots and
an asterisk for the highest value. The wavelength and peak angle
determined by the program are listed for each identified peak. The
values of I(obs) and corresponding 2θ are printed in the adjoining
columns. The allowed radiation column is more extensive than in
Table 1 because it includes all possible spectral lines for the
entire recorded range, rather than only those which fall within
±0.075° of the identified peaks.

The greater detail of the plotted data is useful for locating
backgrounds free of interference from neighboring peaks and absorption edges, selecting peaks without overlapping and for indentifying
weak peaks with intensities below the threshold limits. If one
wishes to do quantitative analysis these data can be used to select
the angles for accurate intensity measurements for input into the
LAMA program[5,6,7].

Figure 2. Printer Plot of Portion of the "Super Alloy" Spectrum.
           Compare with 44.8° to 46.5° Region of Figure 1.

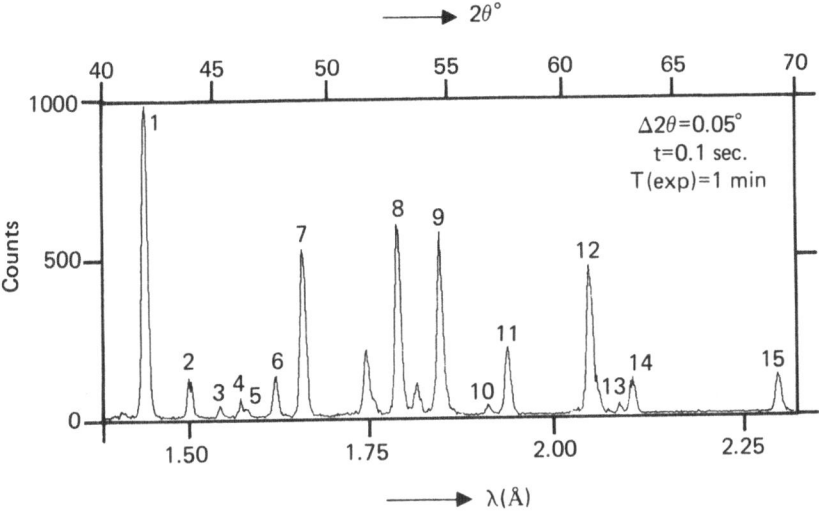

Figure 3. A High Speed Scanning XRF Spectrum of the "Super Alloy".

SPEED AND PRECISION

The observed spectrum of a high speed scan with steps $\Delta 2\theta = 0.05°$ and t=0.1 second is given in Figure 3. The same fifteen peaks, previously obtained, were identified by the program. Notice that the total data collection time T was only about a minute. The data reduction time is less than half a minute for an IBM Series/1 computer and only a fraction of a second for an IBM 370/168 computer.

As in other measurements, the counting statistical accuracy has a major effect on the precision attained as showed in Table 2. It is more difficult to locate weak peaks than those having high intensities. A study of five repeated runs showed that the peak angles determined by this method had a precision of 0.001° for peaks of 10,000 counts and decreased by an order of magnitude when intensities reduced to about 150 counts. The precision in peak intensity determination is given in column 3. For comparison, the precision calculated from one standard deviation $\sigma (=\sqrt{I})$ is listed in column 4, and the peak search results are slightly better.

Table 2. Precision of the Peak Search Method.

I(counts)	$2\theta(°)$	$\Delta I/I(\%)$	$\sigma/I(\%)$
10,000	0.001	0.5	1.0
5,000	2	1.2	1.4
1,000	3	2.0	3.2
150	11	7.7	8.2

CONCLUSION

   A peak search method has been developed for qualitative
X-ray fluorescence analysis. This method is capable of handling
peaks of a wide range of intensities and it is rapid and precise.
It showed that the WD method can be used for high speed data
collection and the peak search method for rapid data reduction of
X-ray fluorescence data. The combination of these two methods
offers a new technique for qualitative X-ray fluorescence analysis
with resolution and precision superior to, and speed comparable
with the conventional ED method.

REFERENCES

1. G. L. Ayers, T. C. Huang and W. Parrish, High Speed X-Ray
    Analysis, J. Appl. Cryst. 11:229 (1978).

2. W. Parrish, T. C. Huang and G. L. Ayers, Comparison of Peak
    Search and Profile Fitting Methods in X-Ray Powder Diffrac-
    tometry, American Crystallographic Association Meeting,
    Eufaula, Alabama (1980).

3. A. Savitzky and M. J. E. Golay, Smoothing and Differentiation
    Data by Simplified Least Squares Procedures, Anal. Chem.
    36:1627 (1964).

4. W. Parrish, G. L. Ayers and T. C. Huang, A Minicomputer and
    Methodology for X-Ray Analysis, Adv. X-Ray Anal. 23:313
    (1980).

5. D. Laguitton and M. Mantler, LAMA I - A General Fortran
    Program for Quantitative X-Ray Fluorescence Analysis,
    Adv. X-Ray Anal. 20:515 (1977).

6. D. Laguitton and W. Parrish, Simultaneous Determination of
    Composition and Mass-Thickness of Thin Films by Quantitative
    X-Ray Fluorescence Analysis, Anal. Chem. 49:1152 (1977).

7. T. C. Huang, Quantitative X-Ray Fluorescence Analysis of Thin
    Films Using LAMA-2, to be published in X-Ray Spectrometry.

MONTE CARLO STUDIES ON THE DESIGN OF RADIOISOTOPE SOURCE HOLDERS

AND SHIELDS FOR EDXRF ANALYZERS

J. M. Doster and R. P. Gardner

Center for Engineering Applications of Radioisotopes
Nuclear Engineering Department
North Carolina State University
Raleigh, N.C. 27650

ABSTRACT

Inherent in the use of radioisotope sources with secondary fluorescers is the background produced by scattering of the source photons from the exciter system. A Monte Carlo program has been developed that is capable of simulating the backscattered photon spectrum as a function of the system geometry, including shielding and collimation variations. This computer program generates the scattered photon spectrum incident on both the sample and detector. The program is applied to a commercially available exciter system to study the effect of specific geometric design changes on the scattered spectrum.

INTRODUCTION

The exciter system examined is the New England Nuclear Model NER-492 H with a 1 Ci Am-241 (60 kev) radioisotope source. The manner in which the exciter system components are arranged leave little room for modification, therefore the proposed shield change was to locate an annular scattering "trap" beneath the fluorescer target (see Figure 1) in hopes of increasing the path length of scattered photons through the tungsten shield and thereby reducing the number of them reaching the sample. In the limit, if the trap could be made infinitely large, one should obtain the limiting ratio of characteristic target X rays to backscattered source photons that is dictated by the target thickness.

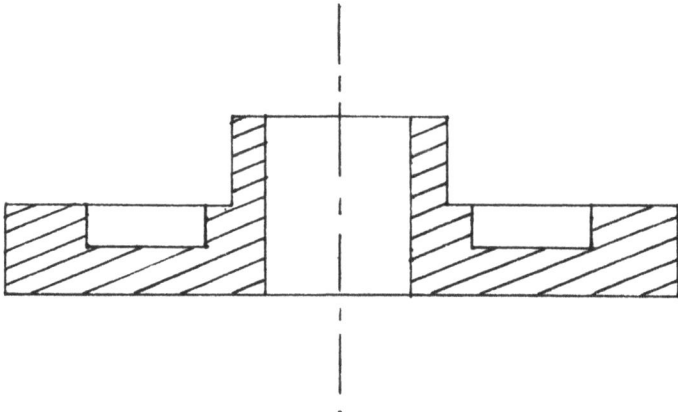

Fig. 1.   New shield design

The purpose of this work was to develop a Monte Carlo simulation model to examine this type of design change and then use it to
study the effect of trap size and location on the intensity of the
backscattered radiation. The model was designed to predict the
photon spectrum reaching the sample from both fluorescence and
scattered radiation along with the scattered directly to the
detector.

MODEL DESCRIPTION

The model generates source photons that are directed at both
the fluorescer target and the central collimator. Those arriving at
the collimator are allowed to scatter freely through the collimator
body until they leave through one of the outer surfaces. At each
interaction point, both Rayleigh and Compton events are forced to
occur and the resulting photons to reach the sample and detector.
The history is terminated when the source photon scatters free of
the collimator.

Source photons arriving at the fluorescer target are forced to
interact in four different modes: (1) production of fluorescence
X rays, (2) coherent scatter to the sample, (3) incoherent scatter
to the sample, and (4) penetration through the fluorescer to the
tungsten shield. Arriving at the shiled, the photons are again
forced to scatter to the sample and produce fluorescence X rays in
the secondary target. It is while passing through the shield that
the trap may be encountered.

X-ray energies are usually low enough such that the effect of electron binding energies cannot be neglected in the calculation of scattering angles and changes in wavelength. These effects are also included in this model.

MODEL PREDICTIONS

The simulation was carried out for a molybdenum target with various trap sizes and locations. The effectiveness of the traps was evaluated in terms of a signal-to-noise ratio, where signal is defined to be the intensity of the characteristic X rays from the fluorescer target at the sample surface and noise is defined as the intensity of the backscattered radiation integrated over energy at the sample surface. These values can be compared to the maximum attainable which is that obtained for a thin molybdenum target with no backing. The signal-to-noise ratios for the standard system (no trap) and for various trap configurations including the limiting case of an infinite trap are given in Table 1.

The ability of traps of practical finite size to reduce scatter is marginal at best. From these results, changing existing designs is unwarranted. More effective means of reducing scatter are possibly available through altering the composition of the shielding material. This type of change could be easily investigated with this model.

Table 1. Signal-to-noise ratio for various trap
          configurations

	S/N	Geometry
(1)	6.83	Standard
(2)	6.79	Trap Center = 1.4605 cm Trap Radius = 0.1000 cm Trap Depth = 0.2380 cm
(3)	6.89	Trap Center* = 1.3605 cm Trap Radius = 0.0500 cm Trap Depth = 0.3175 cm
(4)	7.11	Trap Center = 1.3605 cm Trap Radius = 0.1000 cm Trap Depth = 0.3175
(5)	6.74	Trap Center = 1.3605 cm Trap Radius = 0.1500 cm Trap Depth = 0.3175
(6)	7.66	Infinite trap

AUTHOR INDEX

A

Abell, M. T., 37
Ahlgren, L., 377
Akselsson, K. R., 313
Anderson, C. A. F., 265
Ayers, G. L., 407

B

Bador, R., 351
Baird, A. K., 337
Barrett, C. S., 231
Borgonovi, G., 197
Bowling, G. D., 399
Bras, S., 139
Brown, A., 111
Burleson, J. R., 271
Burns, R., 161

C

Carlsson, L.-E., 313
Castex, L., 139
Charbonnier, M., 351
Chris, M. D., 239
Christoffersson, J.-O., 377
Clark, B. C., 337
Crable, J. V., 37

D

Dabrowski, A. J., 337
Dalton, J. L., 289
D'Antonio, P., 63
Desper, C. R., 161
de Vries, 73, 329
Dollberg, D. D., 37
Doster, J. M., 413

E

Edmonds, J. W., 111
Epperson, D., 197
Ereiser, J., 297

F

Foris, C. M., 111
Frank, G., 297
Freeborn, W. P., 265

G

Gardner, R. P., 413
Gazzara, C. P., 277
Gehringer, R. C., 253
Göbel, H. E., 123, 187
Grönberg, T., 371
Guy, J. W., 297

H

Haque, R., 323
Hasegawa, K., 149, 155, 167, 173
Hashizume, H., 173
Holzer, A., 303
Houghton, G., 197
Huang, T. C., 209, 407
Hubbard, C. R., 99

I

Ikeda, S., 177
Injaian, V. M., 253
Iwanczyk, J. S., 337

417

SUBJECT INDEX

A

Absorption and enhancement, 394

Absorption effects, 42

Adhesive joints, stress in, 231

Al_2O_3, 358

Algorithms for XRPD structure analysis, 8

Alpha coefficients, 331

Aluminosilicate, 49

Aluminum oxide mixtures, 271

Am-241 source, 371

Amorphous
  alloys, 245
  carbon, 70
  crystalline peaks, 239
  state, from ion implantation, 212
  structure determination, 63

Apatite, 12

Appleman's program for cell dimensions, 132

Applications of XRD, numbers of, 78

Archaeological samples, XRF of, 377

Area imaging proportional counter, 161

As_2O_3, 116

Asbestos, 37

Ash in coal, 323

Automated crystal orienter, 283

Automated XRPD search/match, 91, 130

Automation,
  centrally in XRD lab, 121
  in dust analysis, 39
  (see also Position sensitive detector)

B

B_4C, scattering from, 349

Ba in skeletons, 377

Background,
  fitting, XRPD, 6
  in quantitative XRF, 401
  scattering, XRD, 65
  with HgI_2 detector, 338
  XRD, 28

Ball bearings, 215

Blank subtraction in XRF, 401

421